Microorganisms for Sustainability

Volume 41

Series Editor

Naveen Kumar Arora, Environmental Microbiology, School for Environmental Science, Babasaheb Bhimrao Ambedkar University, Lucknow, Uttar Pradesh, India

Microorganisms perform diverse roles on our planet most of which are important to make earth a habitable and sustainable ecosystem. Many properties of microorganisms are being utilized as low input biotechnology to solve various problems related to the environment, food security, nutrition, biodegradation, bioremediation, sustainable agriculture, bioenergy and biofuel, bio-based industries including microbial enzymes/ extremozymes, probiotics etc. The book series covers all the wider aspects and unravels the role of microbes towards achieving a sustainable world. It focuses on various microbial technologies related to sustenance of ecosystems and achieving targets of Sustainable Development Goals. Series brings together content on microbe based technologies for replacing harmful chemicals in agriculture, green alternatives to fossil fuels, use of microorganisms for reclamation of wastelands/ stress affected regions, bioremediation of contaminated habitats, biodegradation purposes. Volumes in the series also focus on the use of microbes for various industrial purposes including enzymes, extremophilic microbes and enzymes, effluent treatment, food products.

The book series is a peer reviewed compendium focused on bringing up contemporary themes related to microbial technology from all parts of the world, at one place for its readers, thereby ascertaining the crucial role of microbes in sustaining the ecosystems.

Pradnya Pralhad Kanekar •
Sagar Pralhad Kanekar

Diversity and Biotechnology of Extremophilic Microorganisms from India

Springer

Pradnya Pralhad Kanekar
Hydrotech Laboratory
Pune, India

Sagar Pralhad Kanekar
Hydrotech Laboratory
Pune, India

ISSN 2512-1901 ISSN 2512-1898 (electronic)
Microorganisms for Sustainability
ISBN 978-981-19-1572-7 ISBN 978-981-19-1573-4 (eBook)
https://doi.org/10.1007/978-981-19-1573-4

This Springer imprint is published by the registered company Springer Nature Singapore Pte Ltd.
The registered company address is: 152 Beach Road, #21-01/04 Gateway East, Singapore 189721, Singapore

DEDICATED TO OUR BELOVED NATION
BHARAT

Preface

Extremophilic microorganisms are a unique group of microorganisms capable of growing at extreme conditions like high or low temperature, highly alkaline or acidic pH, high salt concentration or atmospheric pressure, low water activity, etc. They possess a property of adaptation to the surrounding environment. They evolve different mechanisms to cope up with the harsh conditions they live in. Their habitats are far remote from the natural usual environments, such as soda lakes, glaciers, thermal springs, salterns, and acid mine waters. Based on the type of harsh environments they live in and the extreme conditions at which they grow, they are termed as halophiles, thermophiles, alkaliphiles, psychrophiles, acidophiles, barophiles, and xerophiles. Their physiological properties make them sources of novel biomolecules, biomaterials, and metabolites. These wonderful microorganisms have attracted attention of researchers all over the globe. The microbiologists had to create new taxa to accommodate them in the microbial world. New records of the organisms and their habitats have been described.

India is endowed with a rich biodiversity including microorganisms. Geological and geochemical activities have created different habitats like soda lakes, namely Lonar Lake in North Maharashtra, Sambhar lake in Rajasthan, thermal springs in Ratnagiri and Thane Districts of Western Maharashtra and Himachal Pradesh, acid mine water areas in Madhya Pradesh, solar salterns in west coast of India, and many more. India is blessed with a large coastal area providing saline environment for halophilic microorganisms. Microbiologists in India have extensively studied these versatile microorganisms from different extreme environments. They have published their work mostly as for specific groups like halophiles or thermophiles or acidophiles. To the best of our knowledge, there is no book or a monograph describing all such extremophilic microorganisms from India.

It was thought worthwhile to compile the information available on these microorganisms in the form of a book. The proposed book covers different groups of extremophilic microorganisms through chapters as Halophiles, Alkaliphiles, Thermophiles, Acidophiles, Psychrophiles, etc. The fundamental studies on these organisms like mechanism of halophily, alkaliphily, thermophily, etc., their habitats, taxonomy, diversity, and their biotechnological potential carried out by researchers all over the world have been reviewed in the first part of each chapter. Emphasis is then given to occurrence of such microorganisms in various parts of India, their diversity, taxonomy, and biotechnology. Indian researchers also have extensively studied these novel organisms from extreme environments and published. All this research work from India is described mentioning the authors. References are appended at the end of each chapter. Wherever possible, photographs of the habitats (extreme environments) and tables enlisting isolated microorganisms along with their habitats and corresponding reference are provided.

The purpose of the present book is to make young researchers aware of the attempts made so far to isolate these different microbes and take further the work to explore their biotechnological and industrial potential. It is likely that some changes might have occurred in conditions of the habitats like increase or decrease in salinity or alkalinity of soda lakes, increase or decrease in the temperature of thermal springs, and so on. Such sites could be revisited and new isolations be done. In most of the earlier research, culture-dependent techniques were used. Earlier findings could be expanded by employing molecular techniques. This would open up new fields to work on.

The book will reveal India as the treasure of extremophilic microorganisms and a further path to young researchers.

Pune, India

Pradnya Pralhad Kanekar
Sagar Pralhad Kanekar

Acknowledgments

We could write this book only because we had an opportunity to work on this amazing group of microorganisms at Maharashtra Association for the Cultivation of Science (MACS)—Agharkar Research Institute (ARI), Pune, Maharashtra State, India. We are grateful to our parent Institute for providing all the necessary facilities to study alkaliphilic, thermophilic, and halophilic microorganisms from different extreme environments.

Pradnya Kanekar would like to express gratitude to the late Dr. S. H. Godbole, her teacher, research guide, and Ex-Director, MACS-Research Institute, the late Dr. A. D. Agate, Ex-Director, MACS-ARI, for inculcating the concept of microbial culture collection and preservation, and the late Dr. Kalyan Banerji, an eminent virologist and ex-president, MACS, Pune, for his overall guidance.

Pradnya Kanekar is thankful to Senior Scientists Dr. B. N. Gangili, Ex- Prof. Agharkar Chair, MACS and Dr. P. K. Ranjekar, NCL, Pune for their suggestions from time to time.

Pradnya Kanekar sincerely thanks the funding agencies, Department of Science and Technology (DST), Ministry of Science and Technology, Ministry of Environment and Forests, Department of Biotechnology (DBT), Council of Scientific and Industrial Research (CSIR), Ministry of Science and Technology, Govt. of India, for supporting the work on these wonderful microorganisms.

Pradnya Kanekar would like to thank her colleagues Dr. D. R. Ranade and Dr. K. M. Paknikar for the co-operation in the long-term project (2000–2010), All India Co-ordinated Project on Taxonomy (AICOPTAX)–Centre for Research on Bacteria and Archaea (CRBA) funded by Ministry of Environment and Forests, Govt. of India, Dr. Sabita Dey who worked on thermophilic actinomycetes, Dr. Ratnakar Chitte (Thermophiles), Dr. P. K. Dhakephalkar (present Director of MACS-ARI, Pune) for help in molecular taxonomy of microorganisms, the late Dr. Mrs. S. Sarnaik for her intimate association with all the work, Dr. S. B. Bhosale (work on haloarchaea and psychrophilic methanogens), Dr. T. Y. Yeole (work on psychrophilic methanogens), Dr. Smita S. Nilegaonkar (biochemical and industrial applications of alkaliphiles), Mr. Pranav Kshirsagar (scale up production of polyhydroxyalkanoate, PHA), Mrs. Anita S. Kelkar for methodical documentation of records of these organisms, Dr. D. G. Naik (chemical characterization of antimicrobial compounds of alkaliphiles), her students Dr. Amarja Joshi (alkaliphiles), Dr. Devipriya Mujumdar (thermophiles), Dr. M. Ponraj (alkaliphiles), Dr. Suchitra

Borgave (alkaliphiles and thermoalkaliphiles), Dr. Snehal Kulkarni (haloalkaliphiles and haloarchaea), Dr. Rebecca Thombre (alkaliphiles and haloarchaea), and Dr. Siddharth Deshmukh (halophiles) for their contributions. She would also like to mention the staff of MACS-ARI, staff and students of the then Microbial Sciences Division for their help in one or the other way.

Sagar Kanekar would like to thank Dr. P. K. Dhakephalkar, his research guide, and Council of Scientific and Industrial Research (CSIR), Govt. of India, for support through CSIR-Senior Research Fellowship to work on biotechnological aspects of halophiles.

Pradnya Kanekar sincerely thanks the Principal and other authorities of Modern College of Arts, Science and Commerce, Shivajiganar, Pune, Maharashtra State, for providing facilities to carry out work on halophilic archaea under the CSIR Emeritus Scientist scheme of CSIR, Gov. of India.

Pradnya Kanekar takes this opportunity to thank her colleagues in the field of Microbiology. Dr. Yogesh Shouche from NCCS Pune for help in molecular identification of these microorganisms, the late Dr. B. N. Johri, Dr. G. B. Pant University Pantnagar, the first coordinator of AICOPTAX project, Dr. T. Satyanarayana from Delhi University, Delhi, the second coordinator of the AICOPTAX project, Dr. Anil Tripathi from Banaras Hindu University, Banaras, and Dr. Anjana Sharma from Rani Durgawati University, Jabalpur, Members of the AICOPTAX project, Dr. Tapan Chakraborty from IMTECH, Chandigarh, Dr. S. R. Dave from Gujarat University Ahmedabad, and Dr. S. K. Apte from Bhabha Atomic Research Center (BARC) Mumbai, Maharashtra, for their interactions with our group.

The book could be written because of the extensive contributions made by a number of researchers from our country on these unique groups of microorganisms. We greatly acknowledge them for their contributions. It is likely that we have missed work of some researchers from our country, mainly due to no access to their data on security grounds. We acknowledge all such contributions to the knowledge of extremophiles.

We thank Dr. Navin Kumar Arora for accepting our book proposal under the book series "Microorganisms for sustainability." We also thank Springer Nature for publishing this book.

We sincerely thank Mr. Suhas Pralhad Kanekar, Hydrotech laboratory, Pune, for his moral support.

We thank our relatives, family friends, and friends for their good wishes.

Successful completion of any project is not possible without the support of family members. We happily make a mention of our family members, Suhas, Shilpa, Prachi, and specially Aditya who took interest in the subject at the age of 10 years.

Pradnya Pralhad Kanekar
Sagar Pralhad Kanekar

Contents

About the Authors

Pradnya Pralhad Kanekar is at present Technical Advisor to Hydrotech Laboratory, Pune. She received her Ph.D. in Microbiology from the University of Pune in 1980. She joined MACS-Agharkar Research Institute (MACS-ARI) in October 1974 after a short service as a Lecturer in Microbiology at Govt. College, Aurangabad, and Ahmednagar College, Ahmednagar. Dr. Kanekar headed Microbiology Department of MACS-ARI from 1988 followed by Microbial Sciences Division of MACS-ARI till 2011. She also headed MACS-ARI as Acting Director during 2009–2010. She has successfully guided 16 Ph.D. students in Microbiology at the University of Pune. After retirement in June 2011, she was Emeritus Scientist of MACS-ARI from July 2011 till June 2012. She was awarded CSIR Emeritus Scientist scheme from June 2013 till March 2017 at Modern College of Arts, Science and Commerce, Shivajinagar, Pune.

The research areas of her interest include extremophilic microorganisms, microbial diversity, microbial biotechnology, bioremediation, and biodegradation of toxic organic chemopollutants.

Dr. Kanekar is a Fellow of the Academy of Environmental Biology (FAEB) since 1990 and received Life Fellow Award, Archana Gold Medal and Citation in 1995 for her contribution to development of microbial technology for industrial wastewater treatment. She is also a Fellow of Maharashtra Academy of Sciences (FMAS) since 2009 and Honorary Fellow of Society for Applied Biotechnology (FSAB) since 2011. She was honored with Women Bioscientist Award by SAB in 2011. Dr. Kanekar served as President, Pune Chapter, Association of Food Scientists and Technologists of India (AFSTI) during 1988–1990, President, Pune Unit, Association of Microbiologists of India (AMI) in 2009, and Convener, Pune Branch, Indian Women Scientists Association (IWSA) during 1999–2005. She was Convener of National Conference of Women in Science in 2001 and Silver Jubilee Meet of IWSA, Pune Branch, in 2005.

Dr. Kanekar has 85 research papers published in peer-reviewed journals, two books as one of the editors, 5 Indian patents granted, 17 book chapters published, 25 papers published in proceedings of national and international conferences, one proceedings of the national conference edited, and over 125 papers presented in international and national conferences to her credit. She has successfully implemented 2 international (Indo-Swiss and Indo-US) projects, 12 projects sponsored by CSIR, DBT, DST, MoEF, DRDO, Govt. of India, and 4 projects by industries. She has visited Russia, England, Switzerland, France, Czechoslovakia, Austria, and US for project meetings and international conferences.

Sagar Pralhad Kanekar is at present working as a Microbiologist at Hydrotech Laboratory, Pune, a Water and Food Testing Laboratory. He is involved in microbiological analysis of water and food. After passing B.Sc. and M.Sc. in Microbiology at Savitribai Phule Pune University (SPPU), Pune, he joined MACS-Agharkar Research Institute on CSIR-NMITLI project as a Research Student in 2008 followed by an Institutional project on MACS Culture Collection. He was awarded CSIR Senior Research Fellowship by CSIR, Govt. of India, from April 2011 till March 2014. He received his Ph.D. degree in Microbiology from SPPU in 2015. He served as Asst. Prof. at Biotechnology Dept. of Modern College, Ganeshkhind, Pune, for six months and then joined Hydrotech Laboratory in 2016.

He has two research papers published in peer-reviewed journals, six book chapters, and eight research papers presented at international and national conferences to his credit. He participated in International Conference on Halophiles at the University of Connecticut, USA, in 2013 and presented his work on halophiles from the Andaman Islands, India. He also participated in national conferences in India and presented his work on halophiles.

Introduction

1

Abstract

Extremophilic microorganisms are the microbes that thrive under extreme and harsh conditions like high temperature (thermophilic), low temperature (psychrophilic), alkaline pH (alkaliphilic), acidic pH (acidophilic), high concentration of salt (halophilic), low availability of water (xerophiles), and high hydrostatic pressure (piezophilic or barophilic). Some of them may exhibit resistance/tolerance to more than one extreme condition and are termed as polyextremophiles, e.g., thermoacidophilic, haloalkaliphilic, and cryopiezophilic. They occur in extreme environments like soda lakes, thermal springs, glaciers, acid mine waters, salt pans, hydrothermal vents, deserts, and rocks. These organisms develop different strategies at molecular, cellular, physiological, and genetic level to cope up with extreme environment. By virtue of their physiological properties, they become source of novel biomolecules (extremozymes), e.g., Taq polymerase from *Thermus aquaticus*, and biomaterials (carotenoid, polyhydroxyalkanoate, or biodegradable plastic).

India is endowed with a variety of extreme environments like alkaline soda lake of Lonar, salt lake of Sambhar, thermal spring of Manikaran, and desert of Rajasthan. Almost all groups of extremophilic microorganisms have been reported from India and explored for their biotechnological potential. A number of novel genera and novel species of extremophilic microorganisms are described that contribute to the knowledge, new to Science, from India.

Keywords

Extreme environments · Soda lakes · Extremophilic microorganisms · Archaea · Heat shock proteins · Cryoprotectants · Compatible solutes · Extremozymes · Taq polymerase · Lonar Lake · Hot spring of Manikaran

P. P. Kanekar, S. P. Kanekar, *Diversity and Biotechnology of Extremophilic Microorganisms from India*, Microorganisms for Sustainability 41,
https://doi.org/10.1007/978-981-19-1573-4_1

The beauty of Mother Earth lies in her treasure of biodiversity. While animal and plant biodiversity add to the visible beauty of Earth, microscopic organisms add to the mystery of the origin of life on Earth. Various geological events and geochemical reactions have created amazing environments such as thermal springs, soda lakes, glaciers, islands, deserts, mines, hydrothermal vents, etc. Most of these environments are not "normal" and present harsh or extreme conditions like high or low temperature, high concentration of salt, acidic or alkaline pH, high hydrostatic pressure, etc. Extreme is a relative term, which is viewed compared to what is normal for human beings (Satyanarayana et al. 2005). All such extreme environments are populated by microorganisms which are able to thrive in such harsh conditions and are termed as extremophilic (from Latin "extremus" and Greek "phila" meaning loving extreme environments) microorganisms. Microbes were the first inhabitants of the Earth and its only inhabitants for about half of the history of the biosphere. Microbes transformed that atmosphere to create the environment we now consider normal. They continue to be major players in most ecosystems. Through symbioses with other organisms, microbes opened many major evolutionary pathways. The microbes of extreme environments inform us about how we might find life and even what to look for. Extremophiles may also provide a glimpse of what the earliest forms of life looked like.

Environmental niches, which had hitherto been considered to be devoid of life, have been found in recent times, to be replete with life forms. These life forms have shattered the age-old notion of the limits of life (Satyanarayana and Noorvez 1996).

The discovery of extremophiles began in the 1970s with the discovery of bacteria growing in environments such as the hot springs of Yellowstone National Park and around the hydrothermal vents located on the ocean floor, where the bacteria are in fact the fundamental basis of the specialized ecosystem that is fueled by the vents. These discoveries greatly increased the awareness of the possibilities for bacterial life on Earth and elsewhere. Indeed, the growth of some extremophiles occurs in environments similar to environments which exist on planets such as Mars. Thus, extremophilic bacteria might conceivably not be confined to Earth (Encylopedia.com).

1.1 Extremophiles and Extreme Environments

Extremophiles thrive in habitats which for other terrestrial life forms are intolerably hostile or even lethal, for example, hot niches, ice and salt solutions, acid mine wastes, saline and alkaline lakes or several other habitats that were previously considered inhospitable for life. They have been found at depths of 6.7 km inside the Earth's crust, more than 10 km deep inside ocean at pressures of up to 110 MPa, from extreme acid (pH 0) to extreme basic conditions (pH 12.8) and from hydrothermal vents at 122 °C to frozen sea water at −20 °C. For extreme environmental conditions investigated, a variety of organisms have shown that they not only can tolerate these conditions, but they often require those conditions for survival (Rampelotto 2013). These organisms inhabit nearly every conceivable extreme biotope on this Earth. Since these organisms have adapted well to their respective biotopes, their cellular machinery is geared to work efficiently in these extreme conditions (Satyanarayana and Noorvez 1996).

In the past few years, surface drillings have expanded the perspective of the Earth's deep biosphere from the terrestrial and oceanic realms to include dark subterranean and subseafloor environments. Conditions in the deep biosphere approach the limits for life, but scientific expeditions undertaken for deep subsurface life suggest that these biospheres are hosting a large fraction of microorganisms; living in water, occupying pore space, or colonizing mineral and rock surfaces. Microbes in the deep biosphere are phylogenetically different from most cultured taxa, and represent novel microbial assemblages with novel metabolic capabilities. Thus, investigating these subsurface terrains, which are the deepest and potentially the largest accessible ecosystem on Earth, provides a unique opportunity for obtaining direct insights into the deep microscopic world. Govil et al. (2019) have described the current understanding of the subsurface depth and environmental limits of our planet's life processes and factors that potentially define the diversity/distribution of life in the deep terrestrial crystal system. The authors have also highlighted microbiological studies from the selected deep terrestrial biospheres such as hydrothermal vents, oil reservoirs, caves, and subsurface mines.

1.2 Classification, Diversity, and Taxonomy

Extremophiles are classified based on the conditions in which they grow optimally. The organisms may be described as acidophilic (optimal growth between pH 1 and pH 5); alkaliphilic (grow optimally at and above pH 9); halophilic (grow optimally at high concentration of salt and require NaCl for growth); thermophilic (optimal growth at temperature between 60 °C and 80 °C), hyperthermophiles (optimal growth above 80 °C temperature); piezophilic or barophilic (grow optimally at high hydrostatic pressure); endolithic (grow within rock or within pores of mineral grains), and xerophilic (grow in dry conditions, with low water availability). Some extremophilic microorganisms are adapted simultaneously to multiple stresses or live in habitats where various physicochemical parameters reach extreme values. For example, many hot springs are acidic and rich in metal content, the deep ocean is cold, has low nutrient content and high hydrostatic pressure. Several hypersaline lakes are very alkaline. Microorganisms living in such environments are termed as polyextremophiles, e.g., thermoacidiphilic, haloalkaliphilic, cryopiezophilic, and oligotrophic (Rampelotto 2013; Gupta et al. 2014).

Extremophiles may be divided into two broad categories: Extremophiles which require one or more extreme conditions in order to grow and extremotolerant organisms which can tolerate extreme values of one or more physicochemical parameters though growing optimally at "normal" conditions.

Extremophiles include members of all three domains of life, i.e., bacteria, archaea, and eukarya. Most extremophiles are microorganisms but also include eukaryotes such as algae, fungi and protozoa, and multicellular organisms (Rampelotto 2013).

The major influence of extremophiles has been the broadening of the classification of the evolutionary development of life on Earth. With the advent of molecular means of comparing genetic sequences of highly conserved regions from various life forms, it became clear that extremophiles were not simple offshoots of bacteria but rather had diverged from both bacteria and eukaryotic cells early in evolutionary history. A new domain called Archaea was created to accommodate extremophiles. Archaea share similarities with bacteria and with eukaryotes (Encyclopedia.com).

Archaea is the main group of microorganisms which thrive in extreme environments and are skilled in adapting to different extreme conditions, holding frequently extremophile records. Some archaea are among the most hyperthermophilic, acidophilic, alkaliphilic, and halophilic microorganisms known. The archaeal *Methanopyrus kandleri* strain 116 grows at 122 °C, the highest recorded temperature, while *Picrophilus torridus* is the most acidophilic organism known, with the ability to grow at pH 0.06. Among bacteria, cyanobacteria are the best-adapted group to various extreme conditions. They often form microbial mats with other bacteria, from Antarctic ice to continental hot springs. They can also develop in hypersaline and alkaline lakes, support high concentration of metals and tolerate xerophilic conditions, forming endolithic communities in desertic regions (Rampelotto 2013).

The phylogenetic diversity of extremophiles is high and very complex to study. Some orders or genera contain only extremophiles and non-extremophiles. There are also groups of organisms belonging to the same phylogenetic family that have adapted to very diverse extreme or moderately extreme conditions (Rampelotto 2013).

There are several environments that are inhospitable to all but the extremophilic bacteria have adapted to live in them. The most studied is at elevated temperature, heat-loving bacteria are referred to as thermophiles. More than 50 species of thermophiles have been discovered to date. Some thermophiles such as *Sulfolobus acidocaldarius* are capable of growth and reproduction in water temperatures that exceed 100 °C. The most heat tolerant thermophile known so far is *Pyrococcus fumarii* that grows in the walls of hydrothermal vents where temperatures exceed 93.33 °C. This bacterium requires a temperature above 90 °C to sustain growth (Encyclopedia.com).

Salt-loving or halophilic bacteria grow in environments where the sodium concentration is extremely high such as in the Dead Sea or Great Salt Lake. Salt-loving bacteria such as *Halobacterium salinarum* content with the sodium discrepancy by increasing the internal concentration of potassium chloride (Encyclopedia.com).

Some extremophiles grow and thrive at very low temperatures. For example, *Palaromonas vacuolata* has an ideal growth temperature of just slightly above the freezing point of water. These bacteria find commercial application in enzymatic processes that operate at refrigeration temperatures.

In 2001, living bacteria were recovered from drill samples kilometers beneath the Earth's crust in an environment where virtually no nutrients were present other than the solid rock surrounding the bacteria. By, yet unknown enzymatic mechanisms, these bacteria are able to extract elemental components including sulfur from the rock and utilize them as nutrients (Encyclopedia.com).

Brooks and Murray (1981) have classified a polyextremophile, radiation-resistant, surviving cold, dehydration, vacuum and acid, world's toughest bacterium, *Deinococcus radiodurans* in Phylum Deinococcus-Thermus, Family Deinococcaceae, Class Deinococci, Genus *Deinococcus*, Domain—Bacteria.

In recent years, culture-dependent and culture-independent (molecular) methods have been employed to study the diversity of extremophilic microorganisms (Satyanarayana et al. 2005).

1.3 Growth and Metabolism of Extremophilic Microorganisms

Extremophilic microorganisms have the capability of using many different types of nutrients for their growth. The utilization of both carbon and nitrogen sources from environments is essential for the thriving of heterotrophic extremophilic microorganisms. Their growth rates are relatively comparable to mesophilic/neutrophilic microorganisms except oligotrophs. They also use novel enzymes for metabolizing carbohydrates including polysaccharides and peptides and amino acids. Conventional and also modified central metabolic pathways including Embden Meyerhof-Parnas (EMP), Entner-Doudoroff (ED) and Pentose phosphate pathways and Tricarboxylic Acid (TCA) cycle are operational. However, which one(s) would be used is dependent on the type of the extremophiles. In general, those that grow on carbohydrates can also grow on peptides and amino acids or both, but some of them can grow on only peptides and amino acids. Tse and Ma (2016) have discussed their growth, substrate specificity, and connections between the carbohydrate metabolism and peptide and amino acids metabolism.

1.4 Physiology of Extremophiles

Extremophilic microorganisms have adapted their lifestyle according to the prevalent conditions in the environment. They adapt themselves to survive in extremities which are important for evolutionary differences across different ecological areas. Diverse groups like psychrophiles, thermophiles, halophiles, and alkaliphiles have adopted several physiological and metabolic adaptations which include high membrane fluidity, increased rate of transcription and translation, heat shock or antifreeze proteins, cytoplasmic pH homeostasis, osmoprotectants, and antiporter system for their survival. Moreover, quorum sensing has also played important role in extremophilic adaptations and regulation of their cell processes such as motility, bioluminescence, and production of metabolites including antibiotics, exopolysaccharides, and enzymes. The advances in molecular tools such as metagenomics and comparative genomics analysis have led to the basic understanding of the physiological adaptations of these organisms (Salwan and Sharma 2020).

Most of the extremophilic microorganisms are bacteria and archaea and possess numerous molecular adaptations that allow them to withstand these conditions. These include production of thermostable or cold-stable proteins, enzymes that are stable at pH extremes, altered membrane composition, production of compatible solutes and secondary metabolites that form radioprotective screens. Extremophiles provide valuable information about the physicochemical boundaries of life (Irwin 2020).

1.5 Biotechnology of Extremophiles

Scientists became interested in extremophiles because of their enzymes. Because extremophiles live in extreme conditions, their enzymes must also work under these conditions. Enzymes extracted from extremophiles have grown into a multibillion-dollar industry. The enzymes are used in industrial and medical applications that range from stone-washed jeans to creating artificial sweeteners to conducting genetic tests. The enzyme-based process known as polymerase chain reaction (PCR) is used to amplify DNA for genetic identification or genetic testing for disease and conditions. The enzyme reactions used in the procedure occur slowly at room temperatures. By using enzyme Taq polymerase from a thermophile, *Thermus aquaticus*, the reactions are performed at a much higher temperature and so at a faster rate (Encyclopedia.com).

The molecules that are produced by extremophiles to withstand extreme conditions, e.g., thermostable or cold stable proteins, enzymes that are stable at pH extremes, have been applied in various industries including food production and secondary metabolites as medicines (Irwin 2020).

Dopson et al. (2015) have reviewed the possibility of the use of extremophilic microorganisms in microbial electrochemical systems. Microbial electrochemical systems exploit the metabolism of microorganisms to generate electrical energy or a useful product. The use of extremophiles in the microbial electrochemical system has opened new possibilities to oxidize substrates in the anode and products in the cathode. Extremophiles can be used to oxidize sulfur compounds in acidic pH to remediate waste waters, generate electrical energy from marine sediment microbial fuel cells at low temperatures, desalinate waste waters, and act as biosensors of low amounts of organic carbon.

Hatti-Kaul and Abouhmad (2020) have described extremophiles as a promising source of novel natural products.

Supuran and Capasso (2020) have reviewed biotechnologically important enzymes including carbonic anhydrases. Extremophiles are a source of biocatalysts, which can be extensively used in biotechnological purposes such as esterases/lipases, glycosidases, aldolases, nitrilases/amidases, phosphatases, racemases, thermostable DNA polymerases, and enzymes used in the biofuel production or mining processes. Carbonic anhydrases (CAs) catalyzing the CO_2 hydration/dehydration reaction have been investigated in extremophiles. CAs have been identified in the genome of *Sulfurihydrogenibium yellowstonense* and *Sulfurihydrogenibium azorense* and investigated for their potential role in biotechnological applications.

Lee (2020) has edited a book on biotechnological applications of extremophilic microorganisms. The book provides a broad overview of how extremophiles can be used in biotechnology including for the production and degradation of compounds. It reviews various recent discoveries and applications related to a large variety of extremophiles considering both prokaryotes and eukaryotes. It is a comprehensive overview of different fields in biotechnology from industrial biotechnology, medical biotechnology, food biotechnology, environmental biotechnology, geobiotechnology to astrotechnology.

Other important biomolecules of extremophiles are the biodegradable plastic, polyhydroxyalkanoates (PHAs). Salgaonkar and Bragança (2019) have reviewed the production of PHAs by extremophilic microorganisms. PHAs are polyesters of hydroxyalkanoic acids synthesized by microorganisms, as reserved material, under excess of carbon and a limited amount of other essential nutrients. PHAs have wide applications in food, medicine, and agriculture as biodegradable materials. Extremophilic microorganisms thrive in unusual habitats and are reported to utilize renewable inexpensive agro-industrial wastes as substrates much more efficiently than their mesophilic counterparts. Extremophilic microorganisms especially halophiles have shown potential as accumulators of PHAs by feeding on readily available waste materials. The use of low-cost carbon-rich wastes as a carbon source for the production of PHAs helps in the valorization of wastes as well as obtaining value-added products.

Basak et al. (2020) have reviewed the exploration of extremophile genomes for biotechnological potential. Metagenomics provides a means of assessing the total genetic pool of the microbes in a particular environment in a culture-independent manner. It has revealed unprecedented diversity in a microbial community composition, which is further reflected in the encoded functional diversity of the genomes, a large proportion of which consists of novel genes. Community metagenomic study also proves the dominating characteristics of sulfur- and nitrogen-cycling genes in the extremophiles. Molecular phylogenetic-type analysis to biosynthesis gene diversity of extremophile environment provides a better understanding of the richness of this hidden biosynthetic reservoir with the discovery of additional novel bioactive natural products.

1.6 Extremophiles and Astrobiology

In recent years, scientists have started searching extreme environments on Earth in hope of discovering clues for finding extraterrestrial life. Prokaryotic life has dominated most of the evolutionary history of our planet, evolving to occupy virtually all available environmental niches. Extremophile research also has implications for the origin of life studies and the search for life on other planetary and celestial bodies. Merino et al. (2019) have reviewed the current state of knowledge for the biospace in which life operates on Earth and discussed it in a planetary context, highlighting knowledge gaps and areas of opportunity. Microorganisms not only thrive under extreme conditions on Earth, but can also survive harsh conditions of space, an environment with extreme radiation, vacuum pressure, extremely variable temperature, and microgravity. The authors have further discussed the parameters that limit life or the "boundary conditions" and explored the possibility of search of life on other planetary bodies.

Seckbach and Stan-Lotter (2020) have edited a book on "Extremophiles as Astrobiological models". The data in the book will serve as the origin of life and evolutionary studies. Endospores of bacteria have a long history of use as model organisms in astrobiology, including survival in extreme environments and interplanetary transfer of life. Numerous other bacteria as well as archaea, lichens, fungi, and algae are now being investigated for the purpose.

1.7 Extremophiles in Sustainability

Arora and Panosyan (2019) have reviewed the applications and roles of extremophiles in environmental sustainability. The extremophiles can be useful to understand the functioning of extreme ecological systems. The information of microbial composition and biogeochemical cycling of biogenic elements of such ecosystems would help to understand global change, threats, and opportunities posed by life in these habitats. Biotechnological tools can further help in identifying and utilizing these amazing microbes in tackling issues such as the restoration of polluted ecosystems and provide increased yields of degraded habitats. There is still a huge diversity of microorganisms particularly from habitats untouched by humans, to be explored.

Many useful species and metabolites will definitely be discovered and utilized for improving the quality of life. Extreme ecosystems are a unique biological resource and a treasure of genetic diversity. Thus, it is necessary to mine this wealth for various green purposes. In near future, these incredible creatures are going to play a crucial role in maintaining environmental sustainability.

1.8 Extremophiles from India

India is endowed with natural habitats like Himalayan Mountain ranges, glaciers, thermal springs, deserts, soda lakes, and above all, a large coastal area. Most of these habitats are extreme and hostile for those organisms that thrive in them. Starting from Jammu, Kashmir at Northern end, Kanyakumari at Southern end, Kachchh, Rajasthan at Western end, and West Bengal at Eastern end, a number of extreme environments are spread over. India is surrounded by Arabian Sea, Indian Ocean and Bay of Bengal. West coast of India is popularly known for biodiversity. Compared to animal and plant diversity, microbial diversity has been paid less attention. However, during the last five decades, Indian researchers have investigated microbial diversity of various ecological niches including extreme environments such as

Hamta, Pindari, Roopkund glaciers in Indian Himalayan ranges; thermal springs in Manikaran, Himachal Pradesh; Sambhar salt lake in Rajasthan; acid mine waste waters rich in metals in Madhya Pradesh, Chhattisgarh; alkaline soda lake in Lonar, Maharashtra, thermal springs at Vajreshwari, Dapoli, Maharashtra, solar saltern pans throughout West coast of India; Andaman and Nicobar Islands and Lakshadweep Islands.

Almost all groups of extremophiles have been studied from India, for example, psychrophiles from glaciers, Indian Himalayan ranges; thermophiles from thermal springs, hydrothermal vents deep in ocean, acidophiles from acid mine waste waters, alkaliphiles from the famous Lonar Lake; halophiles from salt pans and sea water from all the coastal areas; piezophiles from deep ocean floors and xerophiles from Thar desert. Man-made extreme environments such as metal-rich environments are populated by metallophiles and metal-resistant/tolerant microorganisms and nuclear wastes from nuclear reactors by radiophilic, radioresistant, and radiotolerant microorganisms. Polyextremophiles such as acidothermophiles from thermal spring at Manikaran, haloalkaliphilic and alkalithermophilic microorganisms from Lonar lake, halopsychrophiles from high altitude saline lakes in Leh and Ladakh region, and halopiezophiles from deep sea have also been reported.

1.9 Extremophilic Microorganisms in Sustainability

Extremophilic microorganisms have also been found to be associated with plants. Yadav (2017) has reviewed the beneficial role of extremophilic microbes for plant health and soil fertility. The plant microbiomes have the capability of colonizing the rhizosphere, phyllosphere, and internal tissues of different plant parts. The plant microbiomes (epiphytic, endophytic, and rhizospheric) have the ability to produce phytohormones, solubilize nutrients, and be antagonistic against pathogens. The plant-associated microbes have been isolated from plants growing under diverse abiotic stresses conditions and these extremophilic microbes help in plant growth promotion (PGP) and adaptations under harsh environments of temperatures, salinity, pH, and drought stresses. The plant-associated extremophilic microorganisms have been reported from all three domain archaea, bacteria, and eukarya of different phyla/groups, e.g., Actinobacteria, Bacteroidetes, Firmicutes, Proteobacteria, Crenarchaeota, Euryarchaeota, Ascomycota, Basidiomycota, etc. The microbes possess the multifarious plant growth promoting attributes and these efficient and potential microbes may be applied as biofertilizers for crop productivity improvement and soil health for sustainable agriculture. Different extremophilic microorganisms, e.g., psychrophiles, thermophiles, halophiles, acidophiles, alkaliphiles, and xerophiles have been reported from plants. The microbes belonging to different genera such as *Agrobacterium*, *Methylobacterium*, *Pantoea*, and *Pseudomonas* have been reported from the phyllosphere of different crops growing in normal as well as harsh environmental conditions.

Soil salinity is an important limiting factor for agricultural crops especially in arid and semi-arid regions of the world. Although many technologies have been implicated in the improvement of salt tolerance, only PGP microbes-elicited plant tolerance against salt stress has been previously reported. Microbes-associated crops have a high potential for agriculture because they can improve plant growth, under limiting or stress conditions of temperatures. Drought stress limits the growth and productivity of crops, particularly in arid and semi-arid areas. Alkaline/acidic environments are also hot spots for microbial diversity with plant growth attributes. Many acidotolerant bacterial genera have been reported from plants growing in acidic environments. Thus, extremophilic microorganisms with multiferous PGP attributes for plants growth and soil health play an important role in sustainable agriculture (Yadav 2017).

1.10 Contribution to the Knowledge of Extremophilic Microbial Species, New to Science, from India

With the aid of new molecular tools to study microbial diversity, researchers from India have investigated extremophilic novel species from different extreme environments in India. Many microbial species, new to science have been reported from India, as will be seen from the Chapters ahead.

Overall, it can be said that India is a treasure of extremophilic microorganisms with amazing diversity and enormous biotechnological potential.

References

Arora NK, Panosyan H (2019) Extremophiles: applications and roles in environmental sustainability. Environ Sustainability 2:217–218. https://doi.org/10.1007/s42398-019-00082-0

Basak P, Biswas A, Bhattacharyya M (2020) Exploration of extremophiles genomes through gene study for hidden biotechnological and future potential. In: Physiological and biotechnological aspects of extremophiles. Elsevier, London, pp 315–325. https://doi.org/10.1016/B978-0-12-818322-9.00024-1

Brooks BW, Murray RGE (1981) Nomenclature for "*Micrococcus radiodurans*" and other radiation-resistant cocci: deinococcaceae fam. nov. and *Deinococcus* gen. nov. including five species. Int J Syst Bacteriol 31:353–360

Dopson M, Ni G, Sleutels TH (2015) Possibilities for extremophilic microorganisms in microbial electrochemical systems. FEMS Microbiol Rev 40(2):164–181. https://doi.org/10.1093/femsre/fuv044

Govil T, Rathinam NK, Salem DR, Sani RK (2019) Microbial diversity in the genomic era. In: Das S, Dash HR (eds) Taxonomical diversity of extremophiles in the deep biosphere. Academic, New York, pp 631–656. https://doi.org/10.1016/B978-0-12-814849-5.00035-6

Gupta GN, Srivastava S, Khare SK, Prakash V (2014) Extremophiles: an overview of microorganism from extreme environment. Int J Agric Environ Biotechnol 7(2):371–380. https://doi.org/10.5958/2230-732X.2014.00258.7

Hatti-Kaul R, Abouhmad A (2020) Extremophiles: A promising source of novel natural products. In: Biotechnological applications of extremophilic microorganisms. De Gruyter, Berlin. https://doi.org/10.1515/9783110424331-001

Irwin JA (2020) Physiological and biotechnological aspects of extremophiles. In: Salwan R, Sharma V (eds) Overview of extremophiles and their food and medical applications. Academic, New York, pp 65–87. https://doi.org/10.1016/B978-0-12-818322-9.00006-X

Lee NM (2020) Biotechnological applications of extremophilic microorganisms. In: Volume 6 in the series life in extreme environments. De Gruyter, Berlin. https://doi.org/10.1515/9783110424331

Merino N, Aronson HS, Bojanova DP, Feyhl-Buska J, Wong ML, Zhang S, Giovannelli D (2019) Living at the extremes: extremophiles and the limits of life in a planetary context. Front Microbiol 15(10):780. https://doi.org/10.3389/fmicb.2019.00780

Rampelotto PH (2013) Extremophiles and extreme environments. Lifestyles 3(3):482–485. https://doi.org/10.3390/life3030482

Salgaonkar BB, Bragança JM (2019) Advances in biological science research, a practical approach. In: Production of polyhydroxyalkanoates by extremophilic microorganisms through valorization of waste materials. Elsevier, pp 419–443. https://doi.org/10.1016/B978-0-12-817497-5.00026-4

Salwan R, Sharma V (2020) Physiology of extremophiles. In: Physiological and biotechnological aspects of extremophiles. Elsevier, London, pp 13–22. https://doi.org/10.1016/B978-0-12-818322-9.00002-2

Satyanarayana T, Noorwez SM (1996) Extremophiles: biodiversity and biotechnology. The Botanica 46:187–193

Satyanarayana T, Raghukumar C, Shivaji S (2005) Extremophilic microbes: diversity and perspectives. Curr Sci 89(1):78–90

Seckbach J, Stan-Lotter H (2020) Extremophiles as astrobiological models. Scrivener Publishing LLC, Beverly. https://doi.org/10.1002/9781119593096

Supuran CT, Capasso C (2020) Physiological and biotechnological aspects of extremophiles. In: Carbonic anhydrase from extremophiles and their potential use in biotechnological applications. Elsevier, London, pp 295–306

Tse C, Ma K (2016) Growth and metabolism of extremophilic microorganisms. In: Biotechnology of extremophiles. Springer, New York, pp 1–46. https://doi.org/10.1007/978-3-319-13521-2_1

Yadav AN (2017) Beneficial role of extremophilic microbes for plant health and soil fertility. J Agric Sci Bot 1(1):9–12

Halophilic and Halotolerant Microorganisms

2

Abstract

Halophilic microorganisms are salt-loving microorganisms while halotolerant microorganisms grow both at high concentration of salt and without salt. Halophilic microorganisms live in saline environments and are isolated from saltern pans, saline lakes, salted fish, sea water, and salt deposits. Halophilic archaea have been reported from hypersaline habitats. Halophilic microorganisms develop the mechanism of osmoadaptation, wherein they accumulate potassium in the cell; accumulate organic osmolytes like amino acids and sugars to cope up with osmotic stress; and develop compatible solutes that help in stabilization of macromolecular structures and adapt cell proteins to high salinity. Biomolecules produced by halophiles are stable at varied iconic strength and as such many novel enzymes are produced using halophiles. Halophiles play an important role in bioremediation of saline soils and thus contribute to sustainability.

India is blessed with a large coastal line (West coast of India), many saltern pans, salt deposits, salt lake of Sambhar, Rajasthan, Chilika lake, Orissa, and many more from where halophilic microorganisms have been reported by Indian researchers. They were explored for the production of exopolysaccharide, polyhydroxyalkanoate, carotenoid, bacteriorhodopsin, hydrolytic enzymes, and bioremediation of saline soils. Three novel genera and nine novel species of halophilic microorganisms become the contribution, new to Science from India.

Keywords

Halophilic · Halotolerant · Haloarchaea · Compatible solutes · Osmolytes · Saline lakes · Saltern pans · Exopolysaccharides · Carotenoids · Bacteriorhodopsin · Sambhar Lake · West coast of India

P. P. Kanekar, S. P. Kanekar, *Diversity and Biotechnology of Extremophilic Microorganisms from India*, Microorganisms for Sustainability 41,
https://doi.org/10.1007/978-981-19-1573-4_2

Evolution of life forms has extensively given rise to a variety of prokaryotic as well as eukaryotic kingdom. The life originated from unicellular organisms and as the environment changed, went through a series of changes in their morphology, physiology, and behavioral characteristics. Unicellular organisms were evolved into multicellular organisms over the period of time. Due to the variation in their physiological properties, numerous types of prokaryotes and eukaryotes have contributed to the diversity among the species. Different habitats are generated due to geochemical activities which become a treasure of biodiversity. The study of prokaryotic and eukaryotic diversity all over the world has been tremendously expanding over the years. The use of new molecular techniques has been helpful in detecting novel prokaryotic species indicating the richness of the diversity.

Many of the organisms embrace novel mechanisms for survival and thus have an interesting aspect in understanding the physiology. The study of such a variety of microorganisms from an environmental sample encompasses the field of biodiversity. These novel organisms by virtue of their physiological properties may possess the potential to produce a variety of biomolecules. For maintaining the inherent properties of the cultures, one must conserve the microbial cultures and thus help in restoring the biodiversity.

2.1 Origin of Life and Evolution of Halophiles

Life on Earth originated and flourished under primitive extreme conditions like acidic, alkaline, low or high temperatures, high altitude, high salinity, high pressures, etc. Depending upon the type of environment they live in, they are classified as alkaliphiles (soda lakes, alkaline waste waters and optimally growing above pH 9.0), halophiles (saltern pans and ponds, growing in the presence of 2–30% salt), thermophiles (thermal springs, oil wells and optimally growing at temperature 55 °C and above), acidophiles (acid mine water and optimally growing at pH 2–5), psychrophiles (glaciers), barophiles (deep sea, oil wells), etc. (Woese 1993). Such microorganisms capable of surviving and growing in these extreme environments are termed as extremophiles. The study of these extremophiles can lead to understanding the genesis of life on the Earth and interest in astrobiology. Satyanarayana et al. (2005) have reviewed work on the diversity of extremophiles.

Among these extremophiles, organisms which require salt for their growth are known as halophiles. In the earlier period of time, extreme salt-loving organisms, halophilic archaea were dominant in saltern lakes and saltern pans. As the oceans formed, the haloarchaeal community decreased and adaptation of haloarchaea to lower salinity resulted in a different class of halophiles namely moderately halophiles and halotolerant microorganisms. The halophilic archaea are now considered as a purview of primitive microorganisms that existed in hypersaline environments. Moderately halophilic bacteria are distributed among various physiological groups of microorganisms belonging to different genera (Kates 1992). From the perspective of the evolutionary process, it would be interesting to know how these halophiles have adapted to hypersaline conditions. The haloarchaea appeared in the very early stage of evolution and hence are included in the archaeal domain (Dundas 1998).

The halophiles which are salt-loving microorganisms are classified as slight halophiles which grow optimally at 2–5% NaCl; moderate halophiles growing at 5–20% NaCl and extreme halophiles growing at 20–30% NaCl. Halotolerant microorganisms grow both at high concentration and without salt. Halophilic and halotolerant microorganisms grow at a wide range of NaCl concentration in response to physicochemical and growth factors. Halophiles are found all over the globe in hypersaline habitats, in natural hypersaline habitats and artificial salterns. (DasSarma and Arora 2001). Halophilic communities are dense in high salt concentration zones. Kanekar et al. (2012) have reviewed the literature on taxonomy, biodiversity, physiology, and applications of halophiles.

Halophiles are microorganisms which require salt for growth as well as for maintaining their cellular integrity. They can survive and grow in saline environments where most other organisms rarely grow. Halophilic archaea form a significant segment of the prokaryotic microbiota in both aquatic and terrestrial ecosystems, suggesting that the global energy cycles may be influenced by haloarchaea. Halophiles differ from other organisms because of their property of adaptation and survival in hypersaline environments.

Papke (2015) has summarized work presented by researchers all over the globe at the International Congress on Halophilic Microorganisms Halophiles 2013, held at the University of Connecticut in Storrs, USA and emphasized the importance of halophiles in understanding the evolution of life, uncovering basic principles of microbial life, achieve insights into fundamental microbial adaptations, and ecological, biogeochemical, and evolutionary processes.

2.2 Habitats of Halophiles

The halophilic microorganisms are commonly found in saline environments. The metabolic activities of halophilic bacteria and archaea indicate that diverse microbial consortia which are not easily cultured may be present in such saline environments. The extreme hypersaline environments may harbor some novel species. Studies on the diversity of halophiles from such habitats help to understand molecular mechanisms of adaptation to different concentrations of salt. Various saline and hypersaline ecosystems exist on this Earth. Halophiles are found to inhabit mainly saltern pans, soda lakes, salted meat, and fish. They have been isolated from both high- and low-salinity water bodies, salty marshy places, saline lakes, the Dead Sea, etc. (Javor 1989; Oren 2002). Among these, many sites represent ancient environments and consequently the halophiles from these environments as the ancient organisms. Grant et al. (1998) and McGenity et al. (2000) have reviewed the literature on halophiles from ancient salt sediments.

2.2.1 Hypersaline Environments

The hypersaline environments contain salts more than 35 g% and are created by evaporation of sea water thus termed as thalassic. The concentration of salt in sea water brines is over 10–12%. Rodriguez-Valera (1993) has described that the concentration of salt in sea water brines is up to 35%. Only a few extremely halophilic microorganisms such as *Halobacterium, Dunaliella* grow in thalassic brine. Athalassic water bodies contain salts of non-marine proportion, e.g., high concentration of magnesium and potassium salts. Some of the examples of hypersaline environments are subterranean evaporate deposits and deep sea basins.

2.2.2 Saltern Pans, Salt Deposits

Saltern pans are the naturally formed saline bodies wherein the halophiles growing optimally in the salt range of 20–25% are present. Species of *Haloadaptus, Halococcus* were isolated from the Japanese salt field (Fukushima et al. 2007). Haloarchaeal species closely related to genus *Halorubrum* was isolated from Salar de Atacama, Chile (Lizama et al. 2001). Phenotypical characterization of *Haloarcula* and *Halorubrum* species was reported from the salterns of Turkey (Elevi et al. 2004). *Halococcus dombrowskii* sp. nov. was obtained from Permian alpine salt deposit (Stan-Lotter et al. 2002). *Pontibacillus chungwhensis* gen. nov., sp. nov., was identified from a solar saltern in Korea (Lim et al. 2005). Haloarchaea were reported from ancient halite collected from Death Valley, California (Schubert et al. 2010). *Halobacterium, Haloferax*, and *Haloarcula* species were reported from Puerto Rico and the Caribbean (Montalvo-Rodriguez et al. 1997). *Halogeometricum borinquense* gen. nov. sp. nov. was isolated from Puerto Rico. Haloarchaea were isolated from Tuz lake, Turkey as well as Kaldirim and Kayaci salterns (Birbir et al. 2007). New haloarchaeal species *Halococcus hamelinensis* sp. nov. was reported from shark bay, Australia (Goh et al. 2006).

2.2.3 Sea Water

Sea water becomes a good source of microorganisms including halophiles. A moderately halophilic novel species, *Halobacterium volcanii* was isolated from Dead Sea (Mullakhanbhai and Larsen 1975). Extreme halophilic cocci were reported from the sea of Alicante, Spain (Rodriguez-Valera et al. 1979). *Oceanobacillus iheyansis* gen. nov., sp. nov. was reported from the Iheya range (Lu et al. 2001). A novel species of *Pseudomarina* was isolated from shallow coastal water of An-Ping Harbor, Taiwan (Jean et al. 2006). A novel slightly halophile *Erythrobacter flavus* sp. nov. was isolated from the East Sea in Korea (Yoon et al. 2003). A novel species of *Marinococcus* was obtained from China (Li et al. 2005). The abundance of haloarchaeal diversity was revealed in a low-salt, sulfide, and sulfur-rich spring (Elshahed et al. 2004).

2.2.4 Salt/Alkaline Lakes

Soda lakes are naturally formed saline water bodies due to different geochemical activities and thus become favorable ecological niches for halophiles. Species of *Halorubrum* were reported from salt lakes in China (Cui et al. 2006). *Natronobacterium pharaonis* from Wadi Natrun and *Natronococcus occultus* isolated from Lake Magadi grow optimally in the pH range 9.5–10 and but not below pH 8.5. Novel *Bacillus* species from Kenyan soda lake was reported by Vargas et al. (2005). *Alkalimonas* species were studied from soda lakes in China and East Africa (Ma et al. 2004). A novel haloalkaliphilic archaeon *Natronobacterium* was isolated from Lake Magadi, Kenya (Mwatha and Grant 1993). A novel alkalitolerant moderately halophilic *Halomonas* species was isolated from Bolivian lake (Quillaguaman et al. 2004). *Halomonas magadii* sp. nov. was reported from the soda lake of East African rift valley (Duckworth et al. 2000). A moderately haloalkaliphile *Halomonas campisalis* sp. nov. was reported from a saline alkaline lake, Washington by Mormile et al. (1999). A novel haloalkaliphilic archaeon *Natronorubrum sulfidifaciens* sp. nov. was reported from Aiding salt lake in Xin-Jiang, China. A novel haloarchaeon, *Natrinema versiforme* sp. nov. (Xin et al. 2000) and *Haloarcula amylolytica* sp. nov. (Yang et al. 2007) were isolated from Aibi salt lake, China. Birbir and Sesal (2003) have reported haloarchaeal diversity of Sereflikochisar salt lake in Turkey. A novel haloalkaliphilic archaeon *Natronobacterium nitratireducens* sp. nov. was studied from a soda lake in China (Xin et al. 2001). Two novel haloalkaliphilic archaea of genus *Natronorubrum* were isolated from soda lakes of Tibet (Xu et al. 1999). Xue et al. (2005) have reported *Haloalkalicoccus tibetensis* gen. nov., sp. nov. from Tibetan plateau, China. Novel species of *Natrialba* have been reported from soda lakes of China (Xu et al. 2001).

2.3 Halophilic Microorganisms

Halophilic microorganisms are salt-loving microorganisms that live in saline environments and grow optimally in presence of 5–25% NaCl. They include mainly prokaryotic and eukaryotic microorganisms having the ability to balance the osmotic pressure of the surrounding environment and resist denaturing effects of salts. The halophilic bacteria and archaea are classified according to Bergey's Manual of Systematic Bacteriology, Second Edition (Boone and Richard 2001, Vol 1 and Brenner et al. 2005, Vol 2). Grant et al. (2001) have described Class III Halobacteria, class nov. in Bergey's Manual of Systemic Bacteriology (2001).

2.3.1 Halophilic Archaea

The halophilic archaea are the extremely halophilic microorganisms growing optimally at 20–30% NaCl concentration. They are facultative aerobic and impart red color to brines because of their dense growth. They are pleomorphic, rods, cocci, disk-shaped, triangular and square-shaped cells and produce gas vesicles, purple membrane, and red-orange carotenoids. Their cells get lysed in solutions of NaCl having concentration less than 6–9 g% NaCl. Haloarchaea are the members of the family *Halobacteriaceae*. The first haloarchaeon discovered was *Halobacterium* isolated from salted fish and meat from northern Europe and North America (Grant et al. 2001). Many of the haloarchaea have been isolated from saltern pans, salt lakes, salt deposits, brines, and ancient halites. The dominant genera of haloarchaea studied extensively include *Halobacterium, Haloarcula, Halococcus, Haloferax, Halorubrum, Halobaculum, Natronobacterium, Natronococcus, Natrialba, Natronomonas*, *Haloterrigena,* and *Haloadaptus* (Montalvo-Rodriguez et al. 1997; Grant et al. 2001; Lizama et al. 2001; Stan-Lotter et al. 2002; Elevi et al. 2004; Goh et al. 2006; Fukushima et al. 2007).

Quadri et al. (2016) have described halophilic archaea from saline environments (dry salt lakes and salt flat) in Algerian Sahara. Their molecular identification revealed the presence of 10 genera from class *Halobacteria namely Natrinema*, *Natrialba, Haloarcula*, *Halopiger*, *Haloterrigena*, *Halorubrum*, *Halostagnicola*, *Natronococcus*, *Halogeometricum,* and *Haloferax*. The haloarchaeal strains were screened for antimicrobial compounds and hydrolytic enzymes. The strains belonging to genus *Natrinema produced* antimicrobial compounds likely similar to halocin C8.

2.3.2 Aerobic and Facultative Anaerobic Gram-negative Bacteria

Some of the heterotrophic Gram-negative bacteria have been found to be moderately halophilic, e.g., *Halomonas elongata* from solar saltern (Vreeland et al. 1980), *Chromohalobacter marismortui* from Dead Sea (Ventosa et al. 1989), *Salinivibrio costicola* from Australian bacon (Mellado et al. 1996), *Pseudomonas halophila* from Great salt lake (Fendrich 1988), and *Flavobacterium salegens* from Antarctic lakes (Dobson et al. 1993). The genus *Halomonas* has been studied extensively and many new species have been described.

2.3.3 Gram-positive Bacteria

Among Gram-positive bacteria, the *Halobacillus litoralis* from Great Salt lake (Spring et al. 1996), *B. haloalkaliphilus* from Wadi Natrun (Fritze 1996), *Marinococcus* (Li et al. 2005), *Salinococcus* sp. from salterns (Ventosa et al. 1990), and *Tetragenococcus halophilus* from fermented soy sauces (Uchida 2000) have been reported as the moderately halophilic bacteria. Kumar et al. (2012a, b) have studied halophilic bacteria isolated from various saline habitats of India. They were screened for hydrolytic enzymes. Their identification revealed the presence of *Marinobacter*, *Virgibacillus*, *Halobacillus*, *Geomicrobium*, *Oceanobacillus*, *Bacillus, Chromohalobacter, and Halomonas.* They belonged to the moderately halophilic group of bacteria. The solvent stability among halophilic enzymes makes them potentially useful in non-aqueous enzymology.

2.3.4 Anaerobic Bacteria and Archaea

Sediments in hypersaline lakes become the habitats of facultative/strictly anaerobic bacteria and archaea including sulfate-reducing bacteria and methanogenic archaea. *Haloanaerobacter chitinovorans* has been reported from saltern (Liaw and Mah 1992), *Halobacterioides acetoethylicus* from an oil well (Oren et al. 1984), and *Acetohalobium arabaticum* from Lake Sivash (Zhilina and Zavarzin 1990). Among the sulfate-reducing bacteria, moderately halophilic sulfate reducer *Desulfohalobium retbaense* is reported from Lake Retba, Senegal (Ollivier et al. 1991) and *Desulfovibrio halophilus* from Solar Lake (Caumette et al. 1991).

Methanogens are classified as archaea and share some important features with haloarchaea. *Methanohalobium evestigatum,* an extremely halophilic methanogen growing optimally at 4.5 M NaCl and *Methanohalophilus halophilus* have been reported from a microbial mat of Great Salt Lake (Roberts et al. 1992). Moderately halophilic methanogens have been reported from deep-sea brine pools in the Gulf of Mexico.

2.4 Cultivation and Phylogenetic Analysis of Halophiles

For the cultivation of halophiles, special enrichment media are required to be used. Studies have been carried out on different methods for the isolation of halophiles from saline ecosystems. For the cultivation of halophiles, different nutrient media containing high concentration of NaCl have been employed by various researchers. The commonly used media include Sehgal and Gibbons (1960), Mullakhanbhai and Larsen (1975), Tindall et al. (1980), Tindall and Truper (1986), and Rodriguez-Valera et al. (1980). Classification of halophiles is done on the basis of the Bergey's Manual of Systematic Bacteriology (Boone and Richard 2001; Brenner et al. 2005). Gupta et al. (2015a, b) have proposed division of the class *Halobacteria* into additional two new orders viz. *Haloferacales* and *Natrialbales* for better positioning the newly isolated haloarchaea. Gupta et al. (2016a, b) have further proposed division of order *Halobacteriales* into two new additional families *Haloarculaceae* and *Haloco*ccaceae and the order *Haloferacales* into two families *Haloferacaceae* and *Halorubaceae* for better classification of new haloarchaea being isolated.

For 16S rRNA gene sequence identification of extremely halophilic archaea, archaeal primers have to be used rather than the bacterial primers as in the case of moderate halophiles. Based on 16S rRNA gene sequencing, the phylogenetic status of halophiles is determined (Ventosa 1994). For analyzing the information on prokaryotic community structure, 16S rRNA genes extracted from the environmental samples have been characterized. With the advantage of PCR DGGE-based amplification of 16S rRNA genes, a culture-independent evaluation of microbial diversity has been possible suggesting a wide distribution of uncultured haloarchaea in ocean, lake waters, and soil. Using the 16S rDNA clone libraries, the diversity of uncultivated prokaryotes from natural microbiota can be mapped which helps in the interpretation of the evolutionary relationships among the microorganisms. With the help of such molecular surveys, the data on over 20,000 haloarchaeal 16S rRNA gene sequences have been generated from environmental studies. These studies have added to the knowledge of known groups and number of novel lineages. Use of PCR DGGE (denaturing gradient gel electrophoresis) techniques have been used for analyzing the microbial community which revealed the community diversity in the environmental samples like sea water, sand, sediment, and soil samples (Yoshie et al. 2002).

Gudhka et al. (2015) have reported for the first time a haloarchaeon *Halococcus hamelinensis* from Stromatolites. Genome sequencing and analysis of the organism was carried out. The genome consisted of 3,133,046 base pairs with an average G+C content of 60.08% and contained 3,150 predicted coding sequences. Codon usage of the genome was consistent with a highly acidic proteome, a major adaptive mechanism toward high salinity. The genome revealed characteristics reflecting its survival in such extreme environment.

2.5 Mechanisms of Halophily

Studies on halophiles have helped researchers to understand many fundamental biological concepts. The hypersaline conditions are deleterious to cells due to the loss of water to the external medium. The intracellular environments of halophiles are maintained by special mechanisms such as accumulation of compatible solutes, salt in strategy, etc. Halophilic microorganisms accumulate high solute concentrations within the cytoplasm to avoid loss of cellular water (Galinski 1993). The cell volume is maintained after achieving an isoosmotic balance with the medium.

Halophilic nature of any organism is determined by its requirement of high concentration of salt for optimum growth while halotolerance indicates the ability of the organism to grow at higher concentration of salt than the optimum (Imhoff and Thiemann 1991; Ollivier et al. 1994). The salt tolerance of bacteria is influenced by the composition of growth medium and growth temperature (Kushner 1993). One of the restrictive environmental factors is the osmotic stress developed due to hypersaline condition or non-ionic solutes in the surrounding medium resulting in inadequate water content. Halophilic and halotolerant microorganisms follow osmotic adaptation to cope up with such environment (Larsen 1986; Oren 1994; Galinski 1995). The osmoadaptation includes both physiological and genetic adaptation to low water activity. Different features of osmoadaptation have been reviewed (Yancey et al. 1982; Reed and Stewart 1988; Csonka and Hanson 1991; Kinne 1993).

Different mechanisms exist in halophiles which help them to tolerate high osmotic pressure caused by surrounding saline environments. Due to the permeability of the biological membranes to water, halophiles need to maintain their cytoplasm isoosmotic with their surroundings. The cytoplasm should be slightly hyperosmotic to maintain a turgor pressure (Oren 2002). Two different basic mechanisms to overcome the osmotic stress are developed by the halophiles that require more than 0.5 M NaCl for optimal growth.

2.5.1 Osmoadaptation

The halophilic microorganisms adapt the intracellular proteins to high concentrations of salt. The cytoplasmic salt concentration is raised for thermodynamic adjustment of the cell. This mechanism of osmoadaptation was first discovered in haloarchaea namely *Halobacterium, Haloarcula, Haloquadratum, Halorhabdus, Natronobacterium,* and *Natronococcus;* halophilic bacteria from the order *Haloanaerobiales* and the bacterium *Salinibacter ruber* (Oren 2008). Though a high NaCl concentration is present in the surrounding environment, the cytoplasm of haloarchaea is kept relatively free of sodium. They accumulate potassium in the cell and thus K^+ molar concentration becomes 100 times higher inside the cell than the surrounding environment. A part of the proton motive force is used to preserve the ion gradient. In *Haloanaerobium prevalens,* K^+ is the prevalent cation but with also high levels of Na^+ ions (Oren et al. 1997; Mermestein and Zeikus 1998).

The knowledge about the concentrations of Na^+ and K^+ ions inside the cells and their regulations in halophilic microorganisms is now available. The metabolism of anions also appears to be known. In haloadaptaion, the chloride plays an important role (Muller and Oren 2003). Important functions like growth, spore germination, motility, and transport of glycine betaine are found to be dependent on the chloride in *Halobacillus halophilus.* For the growth of *Haloanaerobiales* and *Salinibacter ruber,* a high concentration of chloride is required.

2.5.2 Compatible Solutes

The mechanism for the accumulation of organic osmolytes is known in many microorganisms (Roberts et al. 1992). The microorganisms accumulate amino acids, sugars, polyols, etc., to cope up with osmotic stress. Brown (1976) has defined compatible solutes as the nonionic compounds which are highly soluble in water and do not affect the metabolism at high concentration in cytoplasm. Roberts (2005) has reviewed the different pathways of biosynthesis and the effect of the properties on molecular stability on the compatible solutes. The compatible solutes help in the stabilization of macromolecular structures (Bolen and Baskakov 2008). Recent studies on adaptation to high concentration of salt suggest that halophilic microorganisms either exclude salts by synthesis of osmolytes or contain stable large compounds that resist the likely degradation by salt. Empadinhas et al. (2008) have reviewed the diversity of compatible solutes with their genes involved in their synthesis.

2.5.3 Adaptation of Cell Proteins to High Salinity

Halophilic proteins exhibit different mechanisms of adaptation. The proteins of haloarchaea adapt "salt-in" strategy that imparts them unusual properties as compared to proteins of non-halophiles. Ventosa et al. (1998) and Madern et al. (2000) have reviewed the role of salt on the activity and stability of enzymes in halophiles. Cytoplasmic enzymes, stress proteins, and membrane-bound enzymes have been extensively studied from halophiles. Joo and Kim (2005) have reviewed the recent work on haloarchaeal proteins and useful methods for determining the role of novel halophilic enzymes.

2.5.4 Cell Membrane

The cell membrane protects the cytoplasm of halophilic microorganisms from the high saline environment. The halophiles adjust the membrane according to the surrounding saline environment. Vreeland et al. (1984), Adams et al. (1989), Imhoff and Thiemann (1991), Adams and Russell (1992), and Ritter and Yopp (1993) have extensively studied alterations of membrane composition. The lipid composition in membrane of halotolerant, moderately halophilic and extremely halophilic bacteria changes as per the salt concentration (Kates 1986).

2.5.5 Mechanism of Cytoplasmic Regulation

In extreme halophiles, the intracellular salt concentration is very high and presumably they do not accumulate organic compatible solutes. Accumulation of potassium ions occurs up to 5 mol/L concentration. The concentration of potassium in the cytoplasm is higher than that of sodium. By the arrangement of an electrogenic sodium ion/proton antiporter and a potassium ion uniporter, the potassium gradient is maintained. The sodium ion/amino acid symporter carries out uptake of amino acid (DasSarma and Arora 2001).

2.6 Biotechnology of Halophiles

Halophilic organisms by virtue of their physiological properties bear the biotechnological potential. They produce a number of biomolecules, biomaterials, and secondary metabolites. They also synthesize certain biochemicals known as compatible solutes, e.g., glycine betaine to protect themselves from the high concentration of NaCl. Novel lipids can be derived from the membrane lipids of haloarchaea. Various applications of halophiles have been reviewed by Oren (2010). The enzymes of halophiles function under saline conditions therefore may have industrial applications. Halophiles could also be employed in the bioremediation of saline wastewaters. Rodriguez-Valera (1992) has predicted that the halophiles would have large potential in biotechnological processes in addition to the existing commercial processes like production of β-carotene and fermentation of fish sauces. Halophiles produce biomolecules such as enzymes, e.g., DNAases, lipases, amylases, gelatinases, and proteases which remain active even under saline conditions. Novel biomolecules of halophiles include bacteriorhodopsin having application in biocomputing, pigments in food coloring, and compatible solutes as stress protectants.

As the halophiles thrive and live in high saline environments, the biomolecules they produce are stable at varied ionic strength. Many industries have explored halophiles for various novel biomolecules and enzymes as these are valuable resources for developing new biotechnological processes. With the advantages of sustaining and growing at high ionic strength, halophiles show remarkable properties which are not seen in non-halophiles. Halophiles outgrow the non-halophiles at high salt concentration, thus risk of contamination is minimized (Vreeland 1992). Halophiles can grow easily with simple nutritional requirements (Kushner and Kamekura 1988). Genetic manipulation is simple as the genetic tools applied for non-halophiles are common to that of moderate halophiles. As the haloarchaea grow at high salt concentration, they cannot maintain their cellular integrity at low salt concentration. The biomolecules can be recovered by lysing the cells using water lysis which minimizes the cost of extraction.

2.6.1 Enzymes of Halophiles

Halophilic microorganisms produce stable enzymes many of which are polyextremophilic. Moreno et al. (2009) have described the enzymes of halophiles to be active and stable at high salt concentration, thermotolerant and alkaliphilic. Thus, these enzymes bear a number of biotechnological applications. Applications of extremozymes (extremophilic enzymes) have been reviewed by Gomes and Steiner (2004). The enzymes of halophiles have been applied in food industry, laundry detergents, and textile industries (Setati 2010).

A nuclease enzyme is important as molecular probes and in PCR reactions produced by *Bacillus halophilus* as reported by Onishi et al. (1983). *M. varians* subsp. *halophilus* has been reported for nuclease H production as a flavoring agent by Kamekura et al. (1982). An esterase produced by *Halomonas elongata* was used for the curing of brines by Hinrichsen et al. (1994). Serine proteases produced by *Pseudoalteromonas* sp. strain CP76 useful in the textile industry has been described by Sanchez-Porro et al. (2003). A moderately halophilic *Bacillus* sp. strain TSCVKK was shown to produce surfactant and detergent-stable, halophilic alkalitolerant alpha-amylase (Kiran and Chandra 2008). *Micrococcus varians* has been reported for α-amylase, pullulanases, α and β glucoamylases production for starch bioconversion by Adeleye (2007). Characterization of haloarchaeal strains for esterase and lipase activity was reported by Ozcan et al. (2009). Novel bacillopeptidase F-like proteinase was extracted from moderately halophile *Virgibacillus* sp. (Phrommao et al. 2010).

DasSarma and DasSarma (2015) have updated information on halophilic enzymes that function at very high salinity. Genomic and structural analyses have established that the enzymes of halophilic Archaea and Bacteria are negatively charged due to an excess of acidic over basic residues and altered hydrophobicity which enable the enzymes to function in low water activity. Halophilic enzymes maintain salvation and solubility in low water activity thus useful in industrial processes in organic solvents.

Waditee-Sirisattha et al. (2016) have reviewed work on diverse microbial population from hypersaline environments. The halophiles harbor bioactive compounds, compatible solutes, and novel hydrolytic enzymes. The authors have discussed biotechnological applications of these interesting biomolecules

2.6.2 Production of Exopolysaccharides (EPS)

To protect the cells from harsh environments like high alkalinity, salinity, temperature, etc., the microorganisms synthesize various compounds such as exopolysaccharides (EPS), polyamines, osmolytes, etc. EPS are carbohydrate polymers which are soluble in water and are produced extracellularly by the microbes. The monomers in EPS may be branched/unbranched and interconnected by glycosidic linkages. The applications of microbial EPS have been described by Sutherland (1998), Tombs and Harding (1998) in food, pharmaceutical, petroleum, and other industries. Brown and Lester (1979) have studied the role of microbial EPS in the biosorption of metal wastes. The major applications of EPS are as an emulsifying and gelling agent and in oil recovery. Halophilic bacteria namely *Halomonas* species are studied extensively for the production of EPS. Davis (1974) described EPS as having application as an emulsifier from *H. maura* and *H. eurihalina*. Arias et al. (2003) have reported the application of EPS produced by *H. maura* and named as "mauran" as pseudoplastic. Among the haloarchaea, *Haloferax mediterranei* was the first reported EPS producer from the Mediterranean sea and the EPS produced by *Haloferax mediterranei* is anionic in nature, has high viscosity at low concentration, very good rheological properties, is resistant to extreme temperature, pH and has potential application in oil recovery.

Exopolysaccharide production by *Halomonas* sp. is extensively studied (Pfiffner et al. 1986; Bouchotroch et al. 2001; Martinez-Canovas et al. 2004; Mata et al. 2006). Characterization of nineteen *Halomonas* species for EPS production and application of EPS mauran as immunomodulator was described by Bejar et al. (1998). Novel species of *Halomonas* producing exopolysaccharide were reported by Bouchotroch et al. (2001). *Halomonas eurihalina* capable of producing EPS resulting in hydrocarbon degradation has been reported by Calvo et al. (2002). Poli et al. (2011) have reviewed the production of EPS using halophiles from extreme marine habitats and haloarchaea as the potential biopolymer producers on industrial scale. Sulfated EPS from *Halomonas stenophila* was reported by Ruiz-Ruiz et al. (2011). Production of EPS by marine organisms was reviewed by Satpute et al. (2010) *Alteromonas macleodii* sub sp. *figiensis* isolated from deep sea hydrothermal vent from North Fijian Basin produced sulfated EPS with higher uronic acid content and pyruvate (Costaouec et al. 2012). Biswas and Paul (2017) have reviewed work on the production of exopolysaccharides (EPS) by halophilic microorganisms. EPS is synthesized as an adaptive measure against harsh salt environment by halophiles and exhibits significantly different physicochemical and material properties. *Halomonas* and *Haloferax* have attracted attention in the production of EPS but other genera namely *Idiomarina, Salipiger,* and *Alteromonas also bear interest in EPS production. EPS produced by* halophiles show potential in industrial processes.

EPS exhibits various properties like flocculation, emulsification, biosurfactant, antimicrobial activity, gelling activity, etc. The flocculation activity of microbial EPS has been studied by Bouchotroch et al. (2001), Yun and Perk (2003), Patil et al. (2009), Joshi and Kanekar (2011), and Llamas et al. (2012). Emulsification property of halophilic EPS has been studied by Bouchotroch et al. (2001), Arias et al. (2003), and Kumar et al. (2007). Biosurfactant activity of halophilic EPS has been described by Bouchotroch et al. (2001), Lee et al. (2001), and Poli et al. (2007).

EPS has an important role in enhancing the recovery of oil because of their surfactant and emulsifying activity. Halophilic EPS may be useful as the oil deposits occur in saline environments like deep sea sediments. The EPS exhibits high viscosity at low concentration and resistance to extreme pH and temperature indicating its potential in the recovery of oil especially in oil deposits with high salinity (Anton et al. 1988). Among the halophilic bacteria, *Halomonas* is the potential candidate for the production of EPS with excellent emulsifying activity. EPS of many *Halomonas* species shows potential in oil recovery (Poli et al. 2007; Bouchotroch et al. 2001; Mata et al. 2006; Arias et al. 2003) and also of archaea *Haloferax mediterranei* (Anton et al. 1988; Parolis et al. 1996) and *Hahella chejuensis* (Lee et al. 2001).

2.6.3 Biodegradable Plastic: Polyhydroxyalkanoates (PHA)

Biodegradable plastic has been the most preferred alternative to synthetic plastic as it is biodegradable in nature. It is commonly known as polyhydroxy alkanoate and is accumulated intracellularly in the form of granules by eubacteria as well as archaea as carbon and energy stores. Biodegradable plastic is a polyester of hydroxy acids termed as polyhydroxyalkanoate or PHA. Almost over 300 microbial species have been known to produce PHA. PHA is a biopolymer produced under unbalanced nutritional conditions. In PHA, the carboxyl group of one monomer forms an ester bond with the hydroxyl group of the neighboring monomer. Kanekar et al. (2015a, b, c, 2019) have compiled information on structure, diversity, and groups of PHA produced by microorganisms, their synthesis, production, product development, characterization, their physical properties, biodegradability, etc. PHA is thermoplastic in nature, water insoluble, shows good oxygen barrier capacity and UV resistance. It is biocompatible in nature and partially crystalline polymer. The material properties including Young's modulus (3.5 Gpa) and tensile strength (40 Mpa) are similar to that of synthetic polymer polypropylene but the elongation to break is significantly lower than polypropylene. Density of PHA is higher than polypropylene and thus sink in water facilitating its anaerobic biodegradation. Polypropylene floats on water.

Among the bacterial species, *Halomonas maura* from hypersaline soil and *H. ventosa* from saline soils in Jaen (South eastern Spain) have been reported to produce PHB as described by Bouchotroch et al. (2001) and Martinez-Canovas et al. (2004). Moderately halophilic *Halomonas boliviensis* LC1 isolated from a hypersaline lake located in the Andean region of Bolivia was found to produce PHB (Quillaguaman et al. 2005). A moderate halophile *Halomonas campisalis* isolated from alkaline Lonar Lake, India, produced PHA (Joshi et al. 2007). Improved PHA production by *H. boliviensis* was reported by Quillaguaman et al. (2006, 2007, 2008). Simon-Colin et al. (2008) have discovered that using glucose and glucose plus valeric acid as a carbon source, PHB and PHB-co-PHV production is feasible by *Halomonas profundus*, sp. nov., isolated from a deep sea hydrothermal vent shrimp. Simultaneous synthesis of PHA and ectoine by moderate halophile *H. elongata* was studied by Mothes et al. (2008). Production of PHA by *Halomonas campaniensis* was reported by Strazzullo et al. (2008). A moderate halophile *H. marina* HMA103, isolated from a solar saltern in Orissa, India was found to produce PHB (Biswas et al. 2009). Production of copolymer, PHB-co-PHV by *Halomonas campisalis* from alkaline Lonar Lake, India has been described by Kulkarni et al. (2010). In spite of these efforts, commercial production of PHA using bacterial species like *Alkaligenes* sp., *Halomonas* sp., etc. has not been possible. The reason being the use of costly solvents for extraction increases the overall cost of production.

In recent years, most of the studies have been focused on cost-effective production of PHA using other microbial resources. Extremely halophilic archaea have been explored to some extent for cost-effective production of PHA. PHA-producing haloarchaeal cells get lysed in solutions containing salts $<10\%$ NaCl (DasSarma and Arora 2001; Tindall 1991). The haloarchaeon *Haloferax mediterranei* was found to accumulate 60–65% PHB intracellularly from glucose/starch as sole carbon source in the medium (Fernandez-Castillo et al. 1986; Lillo and Rodriguez-Valera 1990; Huang et al. 2006). This organism accumulates a co-polymer of PHB-co-PHV as the source of carbon as described by Chen et al. (2006). As these PHA-producing cells can be lysed easily by resuspending in distilled water, PHA extraction becomes a much cost-effective process. These organisms can be grown in unsterilized conditions at higher saline condition which inhibits the growth of non-halophilic organisms. Production of PHA by haloarchaeal species was reported by Legat et al. (2010). A review on the production of PHA using haloarchaea enlisted the potential biopolymer producers on industrial scale (Poli et al. 2011). Thus, these haloarchaea are looked upon as future resource microorganisms for commercial PHA production.

PHA offers a wide range of applications including packaging films for food packages, paper coatings, bags and containers, biodegradable carrier for long-term dosage of drugs, medicines, pesticides and fertilizers, disposable items such as cups, utensils, cosmetic containers, diapers, and shampoo bottles. PHA is more popular in biomedical applications such as surgical pins, sutures, staples, swabs, wound dressings, bone replacements, and blood vessel replacements. Applications of PHA have been described at length by Kanekar et al. (2019).

2.6.4 Carotenoids

Large quantities of red-orange carotenoids are produced by halobacteria. Carotenoids are needed to stimulate an active photo repair system which cures thymine dimers generated by ultraviolet radiation. C-50 bacterioruberins are the major carotenoids occurring in halobacteria while ß-carotene and lycopene are present in small amounts. By oxidative cleavage of β-carotene in presence of molecular oxygen, retinal is produced. Rodriguez-Valera (1992) has reported the use of carotenoids produced by *Tetragenococcus muriatianus* as food colors and additives. Kanekar (2014) has studied the production of carotenoids by a haloarchaeal isolate from the Andaman Islands.

2.6.5 Bacteriorhodopsin

A unique feature of haloarchaea is the presence of purple membrane (PM) as a specialized region of the cell membrane. It contains a two-dimensional crystalline lattice of a chromoprotein, Bacteriorhodopsin (BR). Production of BR was first described from *Halobacterium halobium* isolated from salterns of San Fransisco Bay, USA (Oesterhelt and Stoeckenius 1971), further studied by Khorana (1988) and Shand and Betlatch (1991, 1994). BR contains a protein moiety (Bacterio-opsin) and a covalently bound chromophore (retinal) which acts as a light-dependent transmembrane protein membrane (Krebs and Khorana 1993). Only a few halophilic archaea possess BR in their membrane. After *Halobacterium salinarum* (or *H. Halobium*), *Haloarcula japonica* was reported for BR production (Yatsunami et al. 2000), *Natrinema* (Xu et al. 2005), *Halorubrum sodomense* (Mohan and Vatsala 2009), *Haloferax mucosum* (Lynch et al. 2012), *Halostagnicola larsenii* (Kanekar et al. 2015a, b, c), and *Haloferax larsenii* (Kanekar et al. 2017a, b). BR offers a wide variety of applications. BR is used in optical applications as for optical lenses, optical grating, data storage carrier, optical switches, and ultrasensitive optical biosensor. Hampp (2000) has reviewed applications of BR as optoelectric material.

BR has been used as a light-induced electricity generator. Thavasi et al. (2008) reported the feasibility of BR as bio-photosensitizer in extonic solar cells. Yen et al. (2011) have maximized the solar energy conversion of BR by tailoring plasmonic and electrostatic field effects. Shiu et al. (2013) described the procedure for measurement of photocurrent activity of isolated PM solution. Conversion of solar energy into electric energy was observed using liquid culture of BR containing haloarchaea *Halostagnicola larsenii* (2015) and *Haloferax larsenii* (2017). Kanekar et al. (2020a, b) have reviewed work on BR producing haloarchaea, production, properties, and applications.

2.6.6 Other Biotechnological Applications

Amoozegar et al. (2019) have highlighted the role of halophiles in biofuel production. Microbial activities are involved in the degradation of substrates like sugars, starch, agricultural and animal wastes. Many of the industrial production processes are carried out in the presence of high concentrations of NaCl. Hence, attention is focused on halophiles for use in biofuel production. Corral et al. (2020) have reviewed the work on molecules produced by halophilic microorganisms having antimicrobial and anticancer activity. The action mechanisms of these molecules are discussed. Aljohny (2015) has reviewed work on halophilic and halotolerant bacteria, their enzymes, and other biotechnological applications. Thombre et al. (2017) have described pharmaceutical and biotechnological applications.

2.7 Halophiles in Bioremediation and Biodegradation Processes: Role in Sustainability

Saline environments are created by natural geological formations, industrial processes releasing brine effluents and agricultural activities like irrigation. Bioremediation of such saline environments could be possible using halophiles. Halophiles are the best suitable candidates for the bioremediation of hypersaline wastewater as industrial waste waters contain high amounts of chlorides, nitrates, etc.

Food industries which produce pickles and brines, tannery industries that make leather and industries that manufacture dyes generate saline wastewaters. As the halophilic microorganisms have the ability to grow in such hypersaline conditions, research has been done on biological treatment of these types of wastewaters. Hinteregger and Streichsbier (1997) have reported biological treatment of saline phenolic wastewater by moderate halophile *Halomonas* sp. Hypersaline wastewater was treated by salt-tolerant microorganisms as described by Kubo et al. (2001). Using the denitrification process of *Haloferax denitrificans,* nitrates from concentrated wastewater could be removed by anaerobic biodegradation (Cyplik et al. 2007). Bioremediation of saline groundwater containing nitrate and nitrite could be possible by *Haloferax mediterranei* as proposed by Martinez-Espinosa et al. (2007). High saline waste water can be anaerobically treated using halophilic methanogens as described by Riffat and Krongthamchat (2007). Chemical waste waters containing 10% salt was possible using halophilic anaerobe *Haloanaerobium lacusrosei* (Kapdan and Erten 2007; Kapdan and Boylan 2009). Moderately halophilic organisms under saline conditions can degrade phenol as described by Veena and Vasudevan (2010).

The ability of degradation of aromatic compounds by *Halomonas oranivorans* sp. nov. was reported (Garcia et al. 2004). Biosurfactant producing *Acinetobacter* sp. MM74 for diesel and crude oil degradation was reported by Satpute et al. (2008). Production of catechol 1, 2-dioxygenase, muconate cycloisomerase, dienelactone hydrolase produced by *Halomonas halodurans* has been used in bioremediation of hypersaline waters contaminated with herbicides, aromatic compounds (Ward and

Brock 1978; Woolard and Irvine 1992; Maltseva et al. 1996). Phosphorous acid anhydrase production by *Alteromonas haloplanktis* for degradation of toxic organo-phosphorus compounds has been reported by Cheng et al. (1999).

Environmental pollution is caused by heavy metals. EPS can be used in bioremediation of toxic wastewaters containing heavy metals by accumulation (Bitton and Friehofer 1978; Brown and Lester 1979; Loaec et al. 1997, 1998) and biosorption (Brown and Lester 1979) of metals. An anionically charged EPS was found to remove over 99% of Zn and Ag at around pH 8 and ambient temperature (Harvey and Luoma 1985). EPS producing halophiles having role in bioremediation, heavy metal chelation, bio-detoxifier, emulsifier include *Halomonas almeriensis* (Llamas et al. 2012), *H. stenophila* (Llamas et al. 2011), *Alteromonas macleodii* sub sp. *figiensis* (Rougeaux et al. 1998), and *Pseudoalteromonas* sp. (Bouchotroch et al. 2001; Qin et al. 2007; Li et al. 2008). Complexation of EPS with trace metals may influence availability to marine microorganisms. This phenomenon may be important in the downward transport of trace metals and micronutrients in the ocean (Decho 1990).

2.8 Halophiles from India

India is known to be rich in biodiversity including microorganisms. Some geological and geochemical activities have created saline environments like soda lakes. In India, natural saline environments include Sambhar salt lake in Rajasthan, Lonar Lake in North Maharashtra, Great and Little Rann of Kutch, Gujarat. India is blessed with large coastal region across Arabian Sea, Hindi Mahasagar, and Bay of Bengal. Low saline environments include coastal sediments, sand and sea water, from West Coast of India, while manmade habitats are mainly solar salterns from Mumbai, Goa, Kandla Port, Gujarat, Kanyakumari, Tamilnadu, Shiroda, Sindhudurga District, etc. West coast of India popularly known as Western Ghats is recognized as a hot spot of Biodiversity. This coastal region is spread over from Gujarat to Tamilnadu in South India through Maharashtra, Goa, and Kerala and explored for halophilic biodiversity. The halophiles from India are summarized in Table 2.1 and reported below region wise or state wise (Fig. 2.1) so that the interest of researchers if in any particular natural habitat is raised and contributions made to the knowledge of halophiles by researchers in India attract attention of microbiologists over the globe.

2.8.1 Rajasthan

The first report of isolation of haloarchaea from India was presumably from Sambhar salt lake, Rajasthan (Fig. 2.2). The lake is both saline and alkaline and situated at 26° 58′N and 75° 5′ E in the middle of a closed depression in the Aravalli mountain ranges, approximately 65 Km northwest of Jaipur. Isolation of six strains of halophilic archaea from brines from main lake as well as solar evaporation pans have been reported (Upasani and Desai 1990). The salt content of the brines ranged from

Table 2.1 Halophiles reported from India and their biotechnological potential

S. no.	Names of the microorganisms	Source of the microorganisms	Biotechnological potential	Reference
1	*Natronobacterium* sp.	Sambhar salt lake, Rajasthan	Not reported	Upasani and Desai (1990)
2	*Haloferax* MSW 1 and *Halobacterium* MSW 5	Salterns of Mithapur, Gujrat	Cellulolytic activity	Upasani et al. (1994)
3	*Haloarcula quadrata*, *Haloarcula valismortis*	Solar saltern, Kanyakumari coast, Peninsular coast of India	Not reported	Asha et al. (2005)
4	*Natrinema thermotolerant*	Marine salterns, Bhavnagar, Gujrat	Not reported	Dave and Desai (2006)
5	*Marinobacter* sp., *Idiomarina baltica*, *Bacillus* sp., *Vibrio* sp., *Dietzia* sp., *Salinicoccus roseus*, *Pseudomonas* sp., *Ruegeria* sp.	Sponges from sea waters, Tuticorin coast, Gulf of Mannar region, South East India	Antimicrobial activity	Anand et al. (2006)
6	*Halomonas campisalis*	Lonar Lake, North Maharashtra	Polyhydroxyalkanoate (PHA)	Joshi et al. (2007)
7	*Halobacterium* sp.	Salt pan in Kandala, Gujrat	Solvent tolerant protease	Akolkar et al. (2008)
8	*Acinetobacter* sp.	Sea water, sediment, shell samples from West Bengal off Fraserganj, Western (Off Mumbai, Shivree) and Southern (Off Chennai and Calicut) Indian coasts	Emulsification and biosurfactant activity	Satpute et al. (2008)
9	*Bacillus* sp. strain TSCVKK	Soil from salt manufacturing industry, Chennai	Surfactant and detergent stable alpha-amylase	Kiran and Chandra (2008)
10	*Haloferax*, *Halorubrum*, *Haloarcula*, *Halobacterium*, *Halogeometricum*, *Salinibacter*, *Cytophaga*, *Marinococcus*, *Natrinema*	Solar salterns in Kelambakkam, Marakanam and Vedaranyam, Tamilnadu	Not reported	Manikandan et al. (2009)

(continued)

Table 2.1 (continued)

S. no.	Names of the microorganisms	Source of the microorganisms	Biotechnological potential	Reference
11	*Haloarcula amylolytica, Haloadaptus paucihalophilus*	Salt pan at Mumbai, Maharashtra	Not reported	Ahmad et al. (2008)
12	*Halobacterium, Haloarcula*	Water samples from coastal regions of Goa, viz. Tal pona, Vasco, Dona Paula, Miramur, Bainguinim	Not reported	Braganca and Furtado (2009)
13	*Halobacillus, Salinimicrobium halophilum, Halomonas salina, H. shengliensis, H. pacifica, H, aquamarina, H. halophile*	Pulicat brackish water lagoon, Coromandel coast, Andhra Pradesh-Tamilnadu border	Hydrolytic enzymes cellulose, xylanase, amylase, and protease	Sahay et al. (2011)
14	*Marinobacter, Halobacillus, Geomicrobium. Chromohalobacter, Oceanobacillus, Halomonas*	Various saline habitats of India	Solvent stable hydrolytic enzymes, non-aqueous enzymology	Kumar et al. (2012a, b)
15	*Halorubrum* sp.	Solar saltern of Mulund, Mumbai	Carotenoid, Bacterioruberin	Pathak and Sardar (2012)
16	*Bacillus* sp., *Halogeometricum borinquense* strain TSS101	Saurashtra region under Kathiawar peninsula, Gujrat	Solvent stable proteases, non-aqueous peptide synthesis	Singh et al. (2012)
17	*Halobacterium, Natrinema, Haloferax, Haloarcula, Halorubrum, Halogeometricum, Halomicrobium*	Marshy environments of Little and Great Rann of Kutch, Gujrat	Not reported	Thomas et al. (2012)
18	*Halococcus salifodinae, Haloferax volcanii, Haloarcula argentinensis, Haloarcula japonica, Halorubrum, Haloferax alexandrines*	Solar saltern at Ribandar, Goa	Carotenoid, Bacterioruberin, metal tolerance	Mani et al. (2012), Chaudhary et al. (2014)

(continued)

Table 2.1 (continued)

S. no.	Names of the microorganisms	Source of the microorganisms	Biotechnological potential	Reference
19	*Natrinema*	Solar saltern, South India	Halocin	Karthikeyan et al. (2013)
20	*Oceanobacillus, Halomonas, Halostagnicola, Haloferax*	Andaman Islands	Polyhydroxyalkanoate (PHA), Exopolysaccharide (EPS), Bacteriorhodopsin (BR), Carotenoids	Kanekar (2014)
21	*Alkalibacillus haloalkaliphilus*, *Bacillus halodurans*, *Oceanobacillus kimchii*, *Bacillus cohnii.*	Lonar Lake, Buldhana district, North Maharashtra	Hydrolytic enzymes—amylase, protease, caseinase, cellulose	Kanekar (2014)
22	*Domibacillus indicus* sp. nov.	Ocean sediments of Lakshadweep island	Not reported	Sharma et al. (2014)
23	*Staphylococcus* Sp. 13CC	Water samples from sites Barkul, Kalupada Ghat, Parikula and Rambha of Chilika Lake, Odisha	Urease production	Mahapatra and Mishra (2014)
24	*Haliferax larsenii* HA1	Pachpadra salt lake in Rajasthan	Halocin production	Kumar et al. (2015)
25	*Halostagnicola* sp.	Andaman Islands	Genome sequencing revealed the presence of genes encoding enzymes involved in PHA synthesis	Kanekar et al. (2015a, b, c)
26	*Halostagnicola larsenii*	Rock pit sea water, Rock Garden, Malvan, Sindhudurga District, Western Maharashtra	Production of bacteriorhodopsin for conversion of solar energy into electrical energy	Kanekar et al. (2015a, b, c)
27	*Haloferax mediterranei*	Maharashtra surface water of Marine drive area, Mumbai, Western Maharashtra	Stress proteins	Thombre and Oke (2015)
28	*Halobacillus trueperi, Halomonas venusta, Marinomonas* sp.	Lunsu, a natural salt water body in Kangra district, Himachal Pradesh	Salt tolerant protease, amylase and glutaminase useful in bioconversion of organic materials	Gupta et al. (2015a, b, 2016a, b)

(continued)

Table 2.1 (continued)

S. no.	Names of the microorganisms	Source of the microorganisms	Biotechnological potential	Reference
29	*Haloferax marismortui*	Salterns of Bhandup, Mumbai, Western Maharashtra	Not reported	Thombre et al. (2016)
30	*Halobacillus* sp. strain KGW1	Water sample from rhizospheric region of a weed *Phragmites karka* at Kalupara Ghat, Chilika Lake, Odisha	Not reported	Panda et al. (2016)
31	*Haloferax larsenii*	Red rock, Rock Garden, Malvan, Sindhudurga District, Western Maharashtra	Production of bacteriorhodopsin (BR) and conversion of solar energy into electrical energy	Kanekar et al. (2017a, b)
32	*Halostagnicola larsenii* (MCC 2956)	Indian black salt	Production of bacteriorhodopsin	Kanekar and Kulkarni (2017)
33	*Halomonas denitrificans*	Rock pit sea water, Rock Garden, Malvan, Sindhudurga District, Western Maharashtra	Production of exopolysaccharide	Deshmukh et al. (2017a, b)
34	*Halomonas smyrnensis*	Chivla beach water, Deobaug beach water, Malvan, Sindhudurg District, Western Maharashtra	Production of exopolysaccharide	Deshmukh et al. (2017a, b)
35	*Haloferax chudinovii*	Salt pan, Shiroda, Sindhudurga District, Western Maharashtra	Production of exopolysaccharide	Deshmukh et al. (2017a, b)
36	*Haloarcula, Halococcus, Haloferax, Halorubrum, Haloterrigena, Halomonas, Halobacillus, Marinobacter, Oceanobacillus*	Diverse saline habitats	Production of phytohormones, plant growth-promoting compounds to enhance plant growth under saline conditions	Yadav et al. (2018)
37	*Halomonas utahensis*	Puthalam salt pan (Near Kanyakumari), Southern peninsular coast of India	Not reported	Murugan et al. (2018)

(continued)

Table 2.1 (continued)

S. no.	Names of the microorganisms	Source of the microorganisms	Biotechnological potential	Reference
38	*Halobacilus marinus* sp. nov.	Chilika Lake, Odisha	Proteases	Panda et al. (2018)
39	*Salinicoccus roseus*	Karvar mangrove soil, Karnataka	Pigment prodigiosin type	Sahoo et al. (2019)
40	*Halobacillus, Salinicoccus, Virgibacillus, Pontibacillus, Oceanobacillus, Marinobacter*	Coastal regions of Karnataka Malpe, Surathkal, Karwar, Mangalore, Salt pan of Kanasgiri, Karnataka	Amylase, lipase, protease, inulinase, gelatinase, β-galactosidase	Jayachandra et al. (2013, 2014)
41	*Brevibacillus borstelensis*	Marine waters of Bay of Bengal, Rusikonda, Vishakhapatnam	α-amylase	Suribabu and Hemalatha (2014)
42	*Halobacillus, Halomonas, Marinococcus, Marinobacter, Alcaligenes, Bacillus, Micrococcus, Pseudomonas*	Saline soils	Bioremediation of saline soils and improvement of plant growth and crop yield	Arora et al. (2014)
43	*Halomonas aquamarina, Halomonas trueperi*	Sunderban Mangrove soil, North East coast of Bay of Bengal, India	Exopolysaccharide	Halder et al. (2016)
44	*Bacillus cereus, Aeromonas hydrophila, Enterobacter aerogenes, E. coli, Proteus vulgaris, Pseudomonas aeruginosa, Staphylococcus aureus, Salmonella typhimurium*	Marine waters of Vishakhapattanam	Biodegradation of oil and grease	Kiranmai and Anima (2017)
45	*Halomonas* in 90 day cultures of *Dunaliella salina*	Salterns from four sites along coastal belt of eastern coastline of India abutted by Bay of Bengal	Co-habitant of halophilic alga *Dunaliella salina*	Keerthi et al. (2018)
46	*Staphylococcus, Enterococcus, Enterobacter, Proteus*	East coast of India, Odisha, and West Bengal	Salt stress tolerance	Das et al. (2019)

(continued)

Table 2.1 (continued)

S. no.	Names of the microorganisms	Source of the microorganisms	Biotechnological potential	Reference
47	*Natranobacterium* sp-1, *Staphylococcus epidermidis, Staphylococcus intermedius, Staphylococcus citreus, Halobacillus salinus, Halococcus salifodinae, Halobacterium salinarum*	Kovalam, Kanyakumari, Kerala	Amylase, protease, gelatinase	Saju et al. (2011)
48	*Natrialba swarupiae* sp. nov. strain ESP3B-9T	Salt pan of Sambhar lake, Rajasthan	Diversity, New species	Kajale et al. (2020)
49	*Halovivax asiaticus* Sh2.2	Shiroda salt pan, Sindhudurga District, Maharashtra Crude salt from salt pan (Sh.2)	Bacteriorhodopsin (BR) production	Kanekar et al. (2021)
	Halovivax asiaticus Sh2.3	Crude salt from salt pan (Sh.2)	BR production	
	Halorubrum sodomense Sh3.1	Brine from salt crystallizer pond showing pink color (Sh.3)	BR production	
	Halorubrum sodomense Sh3.2	Brine from salt crystallizer pond showing pink color (Sh.3)	BR production	
	Halorubrum californiense Sh4.1	Brine from salt crystallizer pond showing orange color (Sh.4)	–	
	Halorubrum sodomense Sh5.2	Wet soil in vicinity of salt pan showing reddish film on the surface (Sh.5)	BR production	
	Haloferax prahovense Sh6.1	Mud sample in vicinity of salt pan showing reddish film on the surface (Sh.6)	–	
	Haloferax alexandrinus Sh7.1	Dry soil sample in vicinity of salt pan showing reddish film on the surface (Sh.7)	–	

Fig. 2.1 ★ Sites of isolation of halophiles from India (Google image)

12–30% and pH 9.5–10.5. The authors used modified Brown medium having pH 9.5 and varying the content of NaCl and KCl. The strains were studied for their lipid profile and found to contain phosphatidylglycerophosphate (PGP), phosphatidylglycerol (PG), and phosphatidic acid (PA). The haloarchaeal strains were identified as *Natronobacterium* and *Natronococcus*. Kumar et al. (2015) have reported *Haloferax larsenii* strain HA1 from Pachpadra salt lake in Rajasthan. The organism was found to produce halocin.

Vijay et al. (2018) have reported for the first time the use of halophiles isolated from Sambhar Lake, India in microbial fuel fell (MFC). The MFC yielded power output of 207.05 mW/m^2 at 1000 mM NaCl concentration with simultaneous removal of starch (1.83 $kg/m^3/d$) and nitrate (0.83 $kg/m^3/d$ NO^3-N). Maximum (3.2 $kg/m^3/d$) starch degradation was obtained at 3 M.

Fig. 2.2 Sambhar Lake, Rajasthan (Google image)

Kajale et al. (2020) have isolated a haloarchaeal strain (ESP3B_9T) from the salt pan of Sambhar salt lake, Rajasthan, India. Cells were coccoid, non-motile, Gram-stain-negative, and formed reddish-pink pigmented colonies. The strain was aerobic, able to grow at 35–55 °C (optimum, 40 °C), in 20–35% (25%) NaCl and pH 8–10 (pH 9). Mg^{2+} is not required for growth. The cells were lysed in distilled water and the minimum NaCl concentration that prevented cell lysis was 5% w/v. The 16S rRNA gene sequence similarities between strain ESP3B_9T and Natrialba hulunbeirensis JCM 10989T and Natrialba magadii ATCC 43099T were 96.53 and 96.25% respectively. The similarities of the RNA polymerase subunit B gene between strain ESP3B_9T and *N. hulunbeirensis* JCM 10989T and *N. magadii* ATCC 43099T were 84.47 and 84.9% respectively. Genome sequencing revealed a genome size of 4.20 Mbp with DNA G+C content of 62.5 mol%. The major polar lipids were phosphotidylglycerol and phosphatidylglycerol phosphate methyl esters with minor amounts of unidentified lipids. The results of polyphasic analysis determined that strain ESP3B_9T represents a novel species of the genus *Natrialba*, for which the name *Natrialba swarupiae* sp. nov. is proposed. The type strain is ESP3B_9T (MCC 3419T=JCM 33002T=KCTC 4279T=CGMCC 1.16737T).

2.8.2 Gujarat

Gujarat coastal region is extensively explored for halophilic microorganisms. Major salt-producing sites in Gujarat are Saurashtra at Kandala, Jamnagar, Maliya, Mithapur, Porbander, and Bhavnagar which become the natural habitats for haloarchaea. Haloarchaea have been isolated from marine salterns at Mithapur, Gujarat, India (Upasani et al. 1994). The organisms were deposited as *Haloferax* MSW1 (NCMB 2287) and *Halobacterium* MSW 5 (NCMB 2288). These along with the other haloarchaeal strains were studied for their lipid. Microbial diversity of marine salts near Bhavnagar has been studied by Dave and Desai (2006). The sample sites, i.e., marine salterns are located at 22° 10′ N and 72° 15′ E on the Western and Northern coasts of the Gulf of Khambhat, Arabian Sea, about 25–30 km north of Bhavnagar. Out of 73 selected isolates of halophiles, six isolates were found to be highly pleomorphic and halophilic. They were identified as *Natrinema thermotolerant*. Halophilic archaebacteria have been isolated from salt pan situated in Kandla and identified as *Halobacterium* sp. (NCBI Gen Bank Accession No. DQ465014) (Akolkar et al. 2008). The organism was found to produce organic solvent-tolerant enzyme protease. Optimization of nutrient media components for production was studied using the Plackett-Burman method (Akolkar et al. 2009).

Singh et al. (2012) have reviewed work on halophilic bacteria and Actinobacteria with special reference to their enzymes and the role of halophiles in bioremediation. The authors have focused attention on the halophiles from natural and manmade saline environments along the coastal Gujrat (India). Gujrat has long coastal region with intensive industrial activities. The authors have worked on natural microflora occurrence in moderately saline environment of coastal Gujrat. India has unique coastline of two oceans, some gulfs and a bay. Gujrat (Western India) accounts for about 1600 km long shore. Of this, the Saurashtra region under Kathiawar peninsula occupies a total stretch of around 865 km. Exploration of this region has revealed the existence of halotolerant, haloalkalitolerant, and haloalkaliphilic bacteria and actinobacteria. Several microorganisms have been investigated for their ability to secrete enzymes. *Bacillus* species were found to be the key producers of extracellular proteases having application in detergent, food, pharmaceutical, leather, and chemical industries. A novel haloalkaliphilic, thermostable serine protease was purified from the extreme halophilic archeon, *Halogeometricum borinquense* strain TSS101. Many haloalkaliphilic bacterial isolates obtained from saline habitats of coastal region of Gujrat were found to grow and secrete enzymes in the presence of different organic solvents such as butanol, methanol, n-hexane, and propanol indicating their applications in non-aqueous peptide synthesis.

Distribution of haloarchaea in hypersaline marshy environment of Little and Great Rann of Kutch, Gujarat has been investigated (Thomas et al. 2012). The region covered for collection of soil and water samples lies between 23° 10.727′N, 70° 43.733′E of Little Rann and 23° 57.7215′N, 69° 43.7611′E of Great Rann of Kutch (Fig. 2.3). The haloarchaeal population was represented by the genera *Halobacterium, Natrinema, Haloferax, Haloarcula, Halorubrum, Halogeometricum, Halomicrobium*, and some uncultured haloarchaeal clones. Phosphate solubilization activity of haloarchaea isolated from Rann of Kutch, Gujarat has been studied (Yadav et al. 2015). The haloarchaeal genera included *Haloarcula, Halobacterium, Halococcus, Haloferax, Halolamina, Halosarcina, Halostagnicola, Haloterrigena, Natrialba, Natrinema,* and *Natronoarchaeum.*

Joshi and Trivedi (2020) have isolated extreme halophilic archaea, 2 halophilic fungi along with bacteria from soil samples collected from Mandavi beach and Great Rann of Kutch in the Kutch area of Gujrat. Bacteria were explored for the production of different hydrolytic enzymes mainly chitinase, amylase, lipase, cellulose, and protease. *Micrococcus aloeverea* was found to produce all five enzymes and antimicrobial activity.

Halophilic microorganisms identified from Rann of Kutch included *Haloferax* sp., *Haloferax volcanii, Haloferax alexandrines, Haloferax lucentense, Halobacterium salinarum, Halarcula amylolytica, Halomonas elongata, Halobacillus* sp., and *Chromohalobacter salexigens.*

Fig. 2.3 Rann of Kutch, Gujrat (Google image)

2.8.3 Maharashtra

The coastal region of Western Maharashtra is popular for tourist places like Mumbai Fort, Gate Way of India, Shreewardhan and Harihareshwar of Raigad District, Sindhudurga Fort, Rock Garden Malvan, Deobag, Tarkarli beaches of Sindhudurga District, Bhagavati port of Ratnagiri, Thiba palace of Ratnagiri District, and many other places. A number of solar salt pans are also in use, e.g., Mulund, Bhandup, Bhainder of Mumbai, Shiroda of Sindhudurga District, etc. All these saline environments are populated by halophiles but comparatively less explored. Phylogenetic analysis of archaeal Ribosomal DNA sequences from salt pan sediment of Mumbai has been documented (Ahmad et al. 2008). Upper sediment (2–10 cm distant from surface) samples were collected from salt pans of Mumbai, located geographically between 18°53′N and 19°20′N latitude and 72°45′E and 73°00′E longitude. Total genomic DNA was extracted from samples and PCR amplification of 16 S rRNA genes was done using archaea-specific primers. Cloning of environmental PCR products was carried out followed by restriction fragment length polymorphism (RFLP) screening of rDNA clones and nucleotide sequencing of cloned SSU rRNA genes. Phylogenetic analysis of cloned SSU rRNA gene sequences was done by comparing the closest database relatives of all sequences generated to 16S rRNA gene sequences available at NCBI. Based on the similarity values, the closest relatives among Euryarchaeota were found to be *Haloarcula amylolytica*, *Haladaptus pauchihalophilus,* and many uncultured archaea and methanogenic archaea. Satpute et al. (2010) have reviewed work on biosurfactants, emulsifiers, and exopolysaccharides producing marine microorganisms.

Haloarchaea from solar saltern of Mulund (E) 19°10′12″N, 72°57′18″E, Mumbai have been isolated (Pathak and Sardar 2012). Based on biochemical and physiological characterization, the haloarchaeal strains were found to belong to genera

Chromohalobacter, *Haloferax*, *Halobacterium*, and *Halorubrum*. The haloarchaeal strains were explored for their biotechnological potential in the production of carotenoids and optimization of the process parameters. The carotenoid was spectrophotometrically analyzed and found to be bacterioruberin having antioxidant activity. The haloarchaeon *Halorubrum* was proposed as a source of carotenoid.

Isolation of halophilic archaea from sea water and soil samples from salt production sites of Mumbai have been documented (Digaskar et al. 2015). Some strains were found to produce enzyme amylase and caseinase. Studies on stress proteins induced by temperature stress in extremely halophilic archaeon *Haloferax mediterranei* RT 18 isolated from surface water samples of Marine Drive area of Mumbai have been reported (Thombre and Oke 2015). Soluble proteins were isolated, separated by SDS-PAGE and identified by MALDI-TOF MS-MS analysis. The stress protein identified as beta proteosomal subunit is found to play role in maintaining protein homeostatis. Thombre et al. (2016) have studied the mechanism of adaptation in extremely halophilic microorganism *Haloferax marismortui* isolated from hypersaline thalossohaline salterns of Bhandup, Mumbai, India. The intracellular sequestration of K^+ and Cl^- ions in hypo salinity and hyper salinity respectively indicated "salt-in" strategy of osmoadaptation. As a cellular response to salinity stress, small heat shocks like proteins were produced by the organism. This is the first report on the study of the concomitant cellular, molecular and physiological mechanism adapted by the organism when exposed to contrasting salinities in the external environment.

In India, halophiles have been isolated from the West coast of Maharashtra across the Arabian Sea, from places like Mulund and Bhandup of Mumbai, etc. However, coastal areas covering Sindhudurg and Ratnagiri districts are not paid much attention for the isolation of halophiles. Kanekar et al. (2015a, b, c) have isolated halophilic archaea from Rock Garden, Malvan, Sindhudurg District, Western Maharashtra, India. Rock garden is a tourist place in Malvan of Sindhudurg district. It is a rocky beach. Red coloration was observed in some rock pits of this rock garden (Fig. 2.4a). Salinity of the rock pit sea water was found to be 5–6% which is little higher than the sea water. Halophilic archaea were isolated from rock pit sea water using Sehgal and Gibbons (SG) (Sehgal and Gibbons 1960) medium. Pinkish red colonies were studied further and identified as *Halostagnicola larsenii*. The isolates were further studied for the production of Bacteriorhodopsin (BR) which acts as a light-driven proton pump. BR was estimated by BR assay (Lee et al. 1998). Pinkish red colored liquid cultures were exposed to sunlight and electric current generated was measured by immersing electrodes of digital multimeter in liquid culture under sunlight and recorded in millivolts (Fig. 2.4b). The haloarchaeal culture was found to produce electric current of 44 mV in 4 h of exposure to sunlight. This is the first report of production of BR by *Halostagnicola larsenii* and conversion of solar energy into electrical energy by the culture (Kanekar et al. 2015a, b, c). From the Rock garden of Malvan, rock pieces showing reddish coloration were used for the isolation of haloarchaea. Kanekar et al. (2017a, b) reported a haloarchaeon *Haloferax larsenii* from the reddish rock. This isolate also produced BR and generated electric current of 49 mV when the reddish colored liquid culture was exposed to sunlight. This is

A B

Fig. 2.4 (**a**) Red rock from Rock garden, Malvan, Dist. Sindhudurga, Maharashtra (Kanekar et al. 2017a, b). (**b**) Reddish colored liquid culture of *Halostagnicola larsenii* (isolated from rock pit sea water, Rock Garden, Malvan, Dist. Sindhudurga, Maharashtra) exposed to sunlight for conversion of solar energy into electrical energy (Kanekar et al. 2015a, b, c)

also the first report of isolation of haloarchaea from red colored rocks of Rocky beach of Malvan, India. Both the haloarchaeal isolates were found to be resistant to sunlight which is an important feature of haloarchaea.

Kanekar et al. (2017a, b) have also isolated haloarchaea from black salt. The black salt is a type of rock salt commonly called as Himalayan black salt. It is salty, pungent smelling condiment used in South Asia. The black salt contains sodium chloride, sodium bisulfate, sodium bisulfide, iron sulfide, and hydrogen sulfide. Due to the presence of iron sulfide, it forms brownish pink to dark violet translucent crystals when whole. It looks light purple to pink when ground into powder. Black salt is considered as a cooling spice in Indian Ayurvedic preparations. It helps in relieving intestinal gas, heat burn, hysteria, etc. Since haloarchaea have been reported from different rock salts and commercial Indian black salt looks pink in color, an attempt was made to isolate haloarchaea from Indian black salt (IBS). The IBS was procured from local market and used for isolation of haloarchaea using Sehgal and Gibbons (SG) medium with 3.42 M NaCl. Based on morphological, biochemical, physiological characterization, and 16S rRNA gene sequencing, the haloarchaeal isolate was identified as *Halostagnicola larsenii.* The isolate was studied for the production of bacteriorhodopsin (BR) and the effect of different environmental and cultural conditions on the growth and production of BR. BR was estimated by BR assay (Lee et al. 1998). The optimum conditions for the production of BR were NaCl concentration 3.42 M, pH 7–8, incubation temperature of 45 °C and incubation period of 9 days. The maximum yield of BR was 0.288 g/L. This is the first report of optimization of production of BR by *Halostagnicola larsenii.*

Fig. 2.5 Shiroda salt pan, Dist. Sindhudrga, Maharashtra (Kanekar et al. 2021)

Kanekar et al. (2020a, b) have developed bio-sensitized solar cells using cell lysate of haloarchaeon *Halostagnicola larsenii*. Bacteriorhodopsin (BR) present in the cell membrane of haloarchaea and capable of converting light energy into electrical energy is being explored as biosensitizer in bio-sensitized solar cells (BSSC). The haloarchaeon, *Halostagnicola larsenii* RG2.14 (MCC 2809) containing bacteriorhodopsin (BR), was used to prepare BSSC. The cell lysate of the culture containing BR was immobilized on TiO_2 nanocrystalline films. The BSSC was characterized for photoelectric performance, UV–visible spectroscopy, scanning electron microscopy, and chronopotentiometry-chronoamperometry. Studies on the parameters like the use of ethanol for the cell lysate preparation resulted in the maximum open-circuit voltage of 0.52 V, short circuit photocurrent density of 120 $\mu A/cm^2$ fill factor of 0.54 and overall efficiency of 0.11%. The present investigation indicates that the BR containing haloarchaeal culture cell lysate could be used in the preparation of BSSC.

Kanekar et al. (2021) have isolated halophilic archaea from Shiroda solar salt pan ecosystem (Fig. 2.5) situated in District Sindhudurga, West Coast of Maharashtra, India and were explored for production of BR and its direct application in conversion of light energy into electrical energy.

The haloarchaeal strain, Sh2.2 from crude salt of salt pan ecosystem was identified as *Halovivax asiaticus* and found to produce 161 mg/l of BR. Its capability of converting light energy into electric potential (34.6 mV) appears to be the first report. The strains Sh3.1, Sh3.2, and Sh5.2 isolated from brine sample and wet soil in the vicinity of salt pan and identified as *Halorubrum sodomense* were found to produce BR in the range of 26–191 mg/l, which is much higher than the earlier reported one about 12.3 mg/l.

Deshmukh et al. (2017a, b) have isolated halophilic microorganisms from saline environments in the Sindhudurga district of Western Maharashtra namely Malvan, Deobaug, Chivala beaches, and Shiroda salt pan. Sea water and sand samples were employed for isolation of halophiles using Sehgal and Gibbons (SG) medium (Sehgal and Gibbons 1960) with different concentrations of NaCl. Strains of *Halomonas smyrnensis* were obtained from Deobaug and Chivala beach water. A haloarchaeon, *Haloferax chudinovii* was isolated from Shiroda salt pan soil. Production of EPS using these isolates was studied using SG medium supplemented with 3% glucose. EPS was precipitated from cell-free supernatant treated with 20% trichloroacetic (TCA) acid to remove protein followed by using chilled ethanol. Production of EPS ranged between 2 and 10.65 g/l and maximum by *Halomonas smyrnensis*. EPS of *H. smyrnensis* was characterized and found to contain maltose as the constituent sugar and dominant glycosidic linkages, EPS of the group oligosaccharide and type heptasaccharide. The EPS exhibited flocculation activity and emulsification activity indicating its biotechnological potential (Deshmukh et al. 2017a, b).

Biswas and Paul (2017) have reviewed work on halophilic microorganisms from diverse thalassohaline and athalassohaline environments including estuaries, saline and soda lakes and solar salterns. Taxonomic studies indicated that the organisms belong to families *Halobacteriaceae, Methanosarcinaceae,* and the class Gammaproteobacteria. The halophilic microorganisms produce exopolysaccharide (EPS) as an adaptive mechanism, the dominant genera being Halomonas and Haloferax and others as Idiomarina, Salipiger, and Alteromonas. The EPS has possible application in industry and biotechnology.

Satpute et al. (2008) assessed different screening methods to select marine bacteria producing biosurfactant. For isolation of marine bacteria, sea water, sediment, and shell samples were collected from Eastern (West Bengal off Fraserganj), Western (Off Mumbai, Shivree), and Southern (Off Chennai and Calicut) Indian coasts. Marine bacteria were isolated using different nutrient media including Zobell marine medium prepared in artificial sea water containing 3.5% NaCl and natural sea water. Around 18 isolates designated as *Acinetobacter* sp. were screened for detecting biosurfactant using different assays such as oil spread method (OSM), modified drop collapse (MDC) method, blue agar plate (BAP) method, emulsification index (EI), hydrocarbon overlay agar (HOA) method, and emulsification assay (EA). Based on the results obtained, the authors suggest that a combination of various methods is required for effective screening. Biosurfactant/bioemulsifiers play a key role in emulsifying hydrocarbons. Emulsification activity is one of the criteria to support the selection of potential biosurfactant producers.

Lonar Lake is a unique soda lake formed by meteorite impact on basaltic rock. It is situated in the Buldhana District of North Maharashtra. The lake water is alkaline (pH 10) and also contains salts imparting salinity to the lake water. In the past, it was used as a source of washing soda and had high salinity and alkalinity. Salinity of Lonar Lake was recorded as 40.78% in 1910, 31.52% in 1958, 30.87% in 1960, 7.9% in 2000, and 0.38–0.45% in 2008 (Joshi et al. 2008). Thus the salinity of the lake water was found to decrease. Extensive studies have been carried out on alkaliphilic microorganisms from this lake but less attention has been paid to isolate halophiles. Joshi et al. (2007) have reported moderately halophilic *Halomonas campisalis* from Lonar Lake.

Kanekar (2014) has used both culture-dependent and culture-independent methods to study halophiles from the Lonar Lake water. Fourteen samples comprising eight water and six sediment samples were collected from different sites of the lake in 2009. Salinity of the lake water was found to be 0.21–0.67%. Out of total 57 isolates obtained using different nutrient media, 23 were halotolerant alkaliphilic while 34 were haloalkaliphilic. Nineteen isolates required 10% NaCl for growth. Based on morphological, biochemical, and physiological characteristics and 16S r RNA gene sequencing, the 19 isolates were found to belong to *Alkalibacillus haloalkaliphilus*, *Bacillus halodurans*, *Oceanobacillus kimchii,* and *Bacillus cohnii*. The isolates were found to produce hydrolytic enzymes amylase, protease, caseinase, and cellulase. Uncultivable archaeal diversity of the lake was studied by using PCR DGGE (denaturing gradient gel electrophoresis) method. The studies revealed the presence of haloarchaea *Natronococcus occultus* and *Halovivax asiaticus*.

Recently in June 2020, the lake water was found to turn red (Fig. 2.6). The salinity of water was probably increased because of evaporation. Preliminary report based on microbiological investigations carried out at Agharkar Research Institute in Pune revealed the presence of red pigment-producing halophilic microorganisms in the red water (Times of India, July 16, 2020).

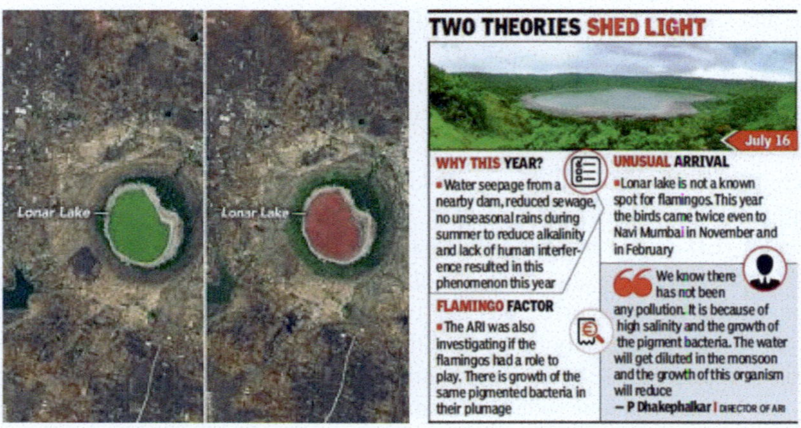

A B

Fig. 2.6 (**a**) Lonar Lake before and after May 2020. (**b**) Times of India News on 16th July 2020

2.8.4 Goa

Goa is a coastal state (15°34′60 N, 74°0′E) and one of the beautiful tourist places of India having natural beaches enjoyed by the tourists. Both the sea water and salt pans in and around Goa have been explored for the isolation of halophiles. Chandramohan (1997) has reviewed work carried out in marine microbiology. Raghavan and Furtado (2004) have described extremely halophilic archaea in sediments from the Continental Shelf of West Coast of India and also production of carotenoid pigment by haloarchaeal cultures exposed to aniline (Raghavan and Furtado 2005). Isolation and characterization of haloarchaea from low salinity coastal sediments and waters of Goa have been reported (Braganca and Furtado 2009). Water and sediment samples were collected from coastal regions of Goa namely Talpona, Vasco, Dona Paula, Miramur, and Bainguinim. The haloarchaeal isolates obtained using nutrient media containing 20% salt were tentatively identified as *Halobacterium* and *Haloarcula* sp. Studies on the isolation of archaea from the natural solar saltern of Goa have been documented (Mani et al. 2012). Salt making is being practiced for many years in Goa. Salt pans are located in Pernem, Bardez, Tisawadi, and Salcete talukas of Goa. The authors selected a single solar saltern at Ribandar (15°30′N, 73°51′E) to examine culturable haloarchaeal diversity. Water and sediment samples were collected 0–10 cm distance from surface twice during the two phases of salt production. In pre-salt harvesting phase/initial stage/salt dilute stage, the haloarchaeal strains of *Halococcus salifodinae* were obtained while in salt harvesting phase/crystallization stage, haloarchaeal strains of *Halococcus salifodinae, Haloferax volcanii, Haloarcula argentinensis, Haloarcula japonica, Haloarcula* sp., *Halorubrum* sp., and *Haloferax alexandrinus* were obtained. The carotenoid pigment was extracted from the orange red growth of the isolates and spectrophotometrically confirmed as bacterioruberin. Metal tolerance of these haloarchaeal strains was investigated (Chaudhary et al. 2014). The strains of *Halococcus salifodinae, Haloferax volcanii, Haloarcula japonica,* and *Halorubrum* sp. could grow and tolerate cadmium concentration ranging from 0.5 to 4 mM.

2.8.5 Tamil Nadu

The peninsular coast of India particularly the Kanyakumari coast is the beautiful tourist place in India. Haloarchaea were isolated from saltern soil samples (Asha et al. 2005). There are over 5000 salterns in the coastal regions of Kanyakumari, thus making the site a congenial environment for the growth of haloarchaea. Based on cultural, morphological, and biochemical characterization, the haloarchaeal isolates were tentatively identified as *Haloarcula vallismortis* (HA3) and *Haloarcula quadrata* (HA9). The isolate HA3 exhibited cellulolytic activity and thus could be a potential candidate for the production of the enzyme cellulase.

Studies on microbial diversity of solar salterns of Tamil Nadu have been reported (Manikandan et al. 2009). Samples from the crystallizer ponds were collected from multipond solar salterns in Kelambakkam, Marakanam, and Vedaranyam of Tamil

Nadu. The haloarchaeal isolates belonged to *Haloferax* sp., *Haloferax larsenii, Halorubrum* sp., *Halogerometricum borinquense, Halobacterium* sp., *Haloarcula* sp., *Halorubrum sodomense, Halobacterium salinarum,* and *Haloferax volcanii.* Thus the microbial community of Tamil Nadu salterns was found to be dominated by haloarchaeal genera *Haloferax* and *Halorubrum.* Production of halocin SH10 by extremely halophilic archaeon *Natrinema* sp. BTSH 10 isolated from salt pans of south India is reported (Karthikeyan et al. 2013). Haloarchaea were isolated from soil samples collected from salt pans of Kanyakumari. Different nutrient media were used for the production of halocin. The haloarchaeal strain BTSH 10 showed strong halocin activity. The production of halocin was optimized with respect to different factors like incubation temperature, NaCl concentration, carbon sources, nitrogen sources, and different inorganic salts. The strain BTSH 10 was identified as *Natrinema* sp. The halocin could be easily produced under submerged fermentation and has applications as preservative against bacteria causing food spoilage, leather processing, control of infectious bacteria, etc.

Sahay et al. (2011) have described halophilic bacteria belonging to genera *Alkalibacillus, Halobacillus, Salimicrobium halophilum, Halomonas salina, H. shengliensis, H. salifodinae, H. pacifica, H. aquamarina,* and *H. halophila* from Pulicat brackish-water Lake of India. Most of the isolates exhibited cellulase, xylanase, amylase, and protease activity. Murugan et al. (2018) have isolated halophilic microorganisms from water samples collected from different sites of Puthalam salt pan, Southern peninsular Coast of India. One halobacterial isolate was identified as *Halomonas utahensis*.

Kiran and Chandra (2008) isolated moderately halophilic alkalitolerant *Bacillus* strain TSCVKK from soil samples obtained from a salt manufacturing industry in Chennai, Tamilnadu, India. The culture produced the hydrolytic enzyme amylase. The enzyme was found to be highly stable in various surfactants and detergents. The end products of starch degradation by the enzyme were glucose, maltose, and maltooligosaccharides indicating that the enzyme is alpha-amylase. Optimization studies indicated that enzyme activity was highest at 10% NaCl in the medium, pH 7.5 and 55 °C. Antimicrobial activity of marine bacteria associated with sponges from the waters off the coast of South East India is reported (Kiran et al. 2018).

Anand et al. (2006) have isolated 75 marine bacterial strains associated with 4 species of sponges from Hare Island, Tuticorin coast, Gulf of Mannar region, and South East coast of India. The bacterial isolates were screened for antimicrobial activity for *Bacillus subtilis, Escherichia coli, Vibrio parahaemolyticus, Vibrio harveyi,* and *Candida albicans.* A potent strain SC3 was explored for the production of antimicrobial compound. The active strains were identified based on 16S rRNA gene sequencing as *Vibrio* and *Bacillus*. This is presumably the first report on the isolation of bacteria associated with sponges from Indian water. Manikandan and Senthilkumar (2017) have reviewed work on halophilic bacteria from hypersaline environments like salt pan. Protease inhibitors find application as antimicrobial agents. Halophiles have great potential in industrial processes for the production of antimicrobial agents.

2.8.6 Karnataka

Halophilic microorganisms have been studied from coastal regions of Karnataka. Jayachandra et al. (2013, 2014) have described halophilic microorganisms isolated from soil samples of coastal regions of Karnataka namely Malpe, Surathkal, Karwar, and salt pan of Kannasagiri. Salinity of the samples was maximum 20% and pH 9.5. Based on the morphological, biochemical, and physiological characteristics and 16S rRNA gene sequencing, 11 isolates capable of growing in presence of 5–25% NaCl were found to belong to genera *Halobacillus, Oceanobacillus, Salinicoccus, Marinobacter, Virgibacillus, Pontibacillus* and *Staphylococcus*. The isolates were screened for hydrolytic enzymes namely amylase, lipase, protease, gelatinase, inulinase, and Beta-galactosidase. Production of Beta-galactosidase by moderately and extremely halophilic bacteria appears to be the first report as per the authors.

Sahoo et al. (2019) have described halophilic bacteria from the soil of Karwar mangrove regions, Karnataka. One green pigmented producer was identified as *Pseudomonas aeruginosa while pink pigment-producing strain belonged to Salinicoccus roseus. The pigments are extractable in chloroform and methanol. Green pigment is suggested to be of phenazine group while the pink pigment of prodigiosin type.*

2.8.7 Kerala

Peninsular coast of India covering Kerala also has been explored for halophiles. Saju et al. (2011) have isolated and characterized halophiles from saltpans of Kovalam (Fig. 2.7) in Kerala. Amylase activity was shown by *Natranobacterium* sp-1, *Staphylococcus epidermidis*, *Staphylococcus intermedius,* and *Staphylococcus citreus* while *Halobacillus salinus, Halococcus salifodinae, Staphylococcus epidermidis,* and *Staphylococcus intermedius* could produce protease. *Halobacillus salinus, Halobacterium salinarum, Natranobacterium* sp-1, *Staphylococcus intermedius,* and *Staphylococcus citreus* showed gelatinase activity.

2.8.8 Andhra Pradesh

Coastal regions of Andhra Pradesh are also explored for halophilic microorganisms. Suribabu and Hemalatha (2014) have isolated *Brevibacillus borstelensis* R1 from marine waters of Bay of Bengal, Rusikonda, Visakhapatnam and tested for production of alpha-amylase. Secondary screening included evaluation of enzyme activity at different salt concentrations in 10 different nutrient media. The organism could produce the enzyme up to 4% salt. Thus the organism was classified as slight halophile. The enzyme has application in starch processing, desizing of textiles, paper sizing, detergent additive, bread improvement, ethanol production, sewage and effluent treatment, and other fermentation processes.

Fig. 2.7 Kovalam salt pan, Kerala (Google image)

Marine waters receive many wastes from different sources and also oil and grease through ship trafficking and oil refinery industries. With the aim to obtain microorganisms capable of degrading oil and grease under saline conditions, Kiranmai and Anima (2017) explored marine waters of the Visakhapatnam coastal area. The isolates were identified as *Aeromonas hydrophila, Bacillus cereus, B. subtilis, Enterobacter aerogenes, Escherichia coli, Proteus vulgaris, Pseudomonas aeruginosa, Staphylococcus aureus,* and *Salmonella typhimurium*. Among these, *Bacillus cereus* was found to be an efficient oil degrader. Thus native marine bacteria could be used in the biodegradation of oil and grease.

Syed et al. (2012) have studied microbial diversity and density from artificial solar salterns in the Nellore district in Andhra Pradesh. Syed et al. (2013a) have isolated amylase-producing bacteria from solar salterns of Nellore district in Andhra Pradesh. Syed et al. (2013b) studied the extracellular enzymatic potential of haloalkaliphiles from solar salterns of Nellore district in Andhra Pradesh.

2.8.9 Chilika Lake, Orissa

Chilika Lake is Asia's largest brackish lake and an estuarine (Fig. 2.8). It is situated between 19° 28′ and19° 54′ N latitude, and 85° 05′ and 85° 38′ E longitude. Its water is subjected to variation in salinity due to the confluence of sea water from Bay of Bengal and fresh water from Daya and Bhargabi rivers and their tributaries.

Mahapatra and Mishra (2014) have reported the presence of extremely halotolerant *Staphylococcus* sp. in water samples from sites Barkul, Kalupada Ghat, Parikula and Rambha of Chilika Lake, Odisha. The organism produces urease enzyme. Panda et al. (2016) have isolated a moderately halophilic *Halobacillus*

Fig. 2.8 Chilika Lake, Orissa (Google image)

sp. strain KGW1 from water sample of rhizospheric region of a weed *Phragmites karka* populating areas at Kalupada Ghat of Chilika Lake. The genome annotation analysis revealed various gene clusters for tolerance to stress such as pH, salt concentration, and toxic metals. Panda et al. (2018) have described *Halobacillus marinus* sp.nov, a novel strain from Chilika Lake. It was found to produce alkaline protease.

Mishra (2019) has studied microbial diversity of brackish water ecosystem of the Chilika Lake, Odisha. The author has expressed that microbes in estuarine conditions constantly undergo tidal flushing which helps them to adapt to changing salt concentration. Such situations make them to combat osmotic stress and tolerance to different environmental factors like pH, temperature, and salinity. By employing classical culture-dependent and modern culture-independent methods, the microbial diversity is evaluated. Behera et al. (2020) have reviewed work on different culture-dependent and culture-independent methods that have been applied to examine microbial communities from the Chilika Lagoon. The authors think that microbial communities in the coastal lagoons are found to have a role in nutrient cycling, mineralization of organic matter, and detoxification of toxic contaminants. The authors have given emphasis on this aspect of Chilika Lagoon.

2.8.10 East coast of India: Orissa and West Bengal

In addition to studies on Chilika Lake, halophiles have also been studied from coastal regions of East coast of India. Sunderban Mangrove Forest is located geographically between 21° 31'N and 22° 30' and latitude 88° 10′E and 89° 51′E along the North East Coast of Bay of Bengal, India and a beautiful natural habitat (Fig. 2.9). Halder et al. (2016) have isolated halophilic microorganisms from soils collected from four different places of Sunderban. The isolates were identified as

Fig. 2.9 Sunderban mangrove forest (Google image)

Halomonas aquamarina and *Halobacillus trueperi*. The organisms were found to produce exopolysaccharide and extracellular enzymes and exhibited multiple heavy metal tolerance and antibiotic resistance.

In the light of soil salinization affecting crop production, Das et al. (2019) thought to study development and inducement of salt stress tolerance in plants to sustain crop production in saline lands. Knowledge about salt stress tolerance genes in natural microbial resources of saline habitats is yet not clear. To assess salt stress tolerance, microorganisms were isolated from water and soil samples along the East coast of India, from Odisha and Wast Bengal. Out of 98 isolates obtained, 63 isolates capable of growing beyond 4% NaCl were studied. Only seven isolates could grow at 8% NaCl. Identification of the isolates revealed the presence of *Staphylococcus, Enterococcus, Enterobacter,* and *Proteus*. The authors think that further studies on salt stress tolerance genes in such microorganisms and development or inducement of salt stress tolerance in crop plants are needed.

The autotrophic alga, *Dunaliella salina* Teodorsco is found in thalassohaline environment worldwide. The studies indicate that the heterotrophic components of such environment vary in different regions. Keerthi et al. (2018) examined waters in salterns from four sites along 650 kM of the coastal belt of Eastern coastline of India abutted by Bay of Bengal, for the presence of microbes. Green colonies represented the alga Dunaliella while pink colonies were of heterotrophic bacteria identified as Halomonas and its pink pigment as a derivative of bacterioruberin. It was observed that cultures of *D. salina* were green in 30 days, orange in 60 days, and pink in 90 days. In 90 days old culture of *Dunaliella*, innumerable numbers of rod-shaped organisms were found. Due to evaporation of water from nutrient media, salt content might have increased leading to growth of halophilic bacteria like *Halomonas*. Thus

Dunaliella and *Halomonas* were found to be co-inhabitants of natural saline environment.

2.8.11 Andaman Islands

Andaman Islands, India are a group of archipelagic islands in the Bay of Bengal and are pristine habitats (6° 45′ and 13° 41′ N and 92° 12′ and 93° 57′ E) for isolation of halophiles from Indian saline environment (Fig. 2.10). So far, isolation of halophiles has not much been reported from Andaman Islands.

Kanekar (2014) has investigated halophilic diversity of Andaman Islands which are archipelagic islands in the Bay of Bengal, India. This pristine habitat has not been much explored for microbial diversity. Twenty-one sea water and sand samples were collected from different sites of Andaman Islands which harbored 62 isolates of halophilic microorganisms. Based on the requirement of salt for growth, 12 isolates representing Wandoor Islands, North Bay, Radhanagar beach of Andaman Islands were selected for growth studies. Ten isolates grew optimally in the range of 5–10% NaCl and 2 isolates grew optimally at 20–22% NaCl indicating their haloarchaeal nature. Based on morphological, biochemical, physiological characteristics, and 16S rRNA gene sequencing, the halophilic isolates were found to belong to genera *Oceanobacillus, Halomonas, Halostagnicola,* and *Haloferax*. The culture-independent diversity of Andaman Islands was studied using molecular technique namely denaturing gradient gel electrophoresis (DGGE). The study revealed the

Fig. 2.10 Andaman Islands, India (Kanekar 2014)

presence of *Natronococcus*, *Methanospirrulum*, *Methanosaetae*, and *Methanomicrobiales*.

The haloarchaeal isolates were explored for their biotechnological potential in the production of polyhydroxyalkanoates (PHA), exopolysaccharide (EPS), carotenoid, and bacteriorhodopsin. Production of PHA was optimized with respect to NaCl concentration, pH, temperature, and inoculum density one factor at a time. Response Surface Methodology and Central Composite Design were used. Maximum PHA was found to be 2.9 g/L under optimum conditions. Rectangular Suspended Cell Reactor (RSCR) was used to produce PHA on large scale. EPS produced by haloarchaeal isolate showed the presence of glucose, galactose, and galactouronic acid. To demonstrate the presence of Bacteriorhodopsin (BR) in the haloarchaeal isolate, pinkish red colored liquid cultures were exposed to sunlight. The electricity generated ranged from 24.3 mV to 34.1 mV in 8 h. The halophilic isolates produced hydrolytic enzymes namely amylase, cellulose, caesinase, protease, and inulinase.

Kanekar et al. (2015a, b, c) have done draft genome sequence of a haloarchaeon *Halostagnicola* sp isolated from Andaman Islands, which contained bases 31, 78, 490 bp in 26 contigs with G + C content 60.8%. The sequence indicated the presence of gene for PHA synthesis. The organism produced PHA substantiating the finding in genome sequencing.

Srinivas et al. (2011) have described a novel, Gram-stain negative, rod-shaped non-motile bacterial strain AMV16T isolated from a soil sample collected from a mud volcano located in Andaman Islands. Based on phenotypic, chemotaxonomic characters, 16 S rRNA gene sequencing, phylogenetic analysis, and DNA-DNA hybridization studies, the strain AMV 16T has been assigned a novel genus and a novel species as *Cesiribacter andamanensis* gen.nov.sp.nov.

Kumar et al. (2012a, b) have described a novel Gram-stain negative, oval to rod-shaped, motile, protobacterim, strain AMV 1T isolated from a soil sample collected from a mud volcano of Andaman Islands. Based on phenotypic, chemotaxonomic characters, 16 S r RNA gene sequencing, phylogenetic analysis, and DNA-DNA hybridization studies, the strainAMV1T was assigned to a novel genus and novel species as *Lutibaculum baratangense* gen.nov.sp.nov.

Srinivas et al. (2015) have described a novel Gram-stain positive, coccoid, non-motile bacterial strain isolated from a soil sample collected from mud volcano located in Andaman Islands. 16 S rRNA gene sequence analysis indicated that the strain AMV4T was a member of the order Actinomycetales and closely related to Aquipuribacter. Based on DNA-DNA relatedness, the strain was assigned a new species as *Aquipuribacter nitratireducens* sp.nov.

2.8.12 Lakshadweep Islands

Lakshadweep is India's smallest Union Territory and is an archipelago consisting of 36 islands with an area of 32 sq. km. It is known for its exotic and sun-touching beaches and lush green landscape. It is comprised of 12 atolls, three reefs, five submerged banks, and 10 inhabited islands. All the islands are 220 to 240 km away from the coastal city of Kochi in Kerala. Sharma et al. (2014) have described the isolation of *Domibacillus indicus* sp. nov. from ocean sediments of Lakshadweep.

2.8.13 Himachal Pradesh

Gupta et al. (2015a, b, 2016a, b) have isolated halophilic bacteria from soil sediments of Lunsu, a salty water body, Himachal Pradesh, India. Their growth at different concentrations of salt, pH, temperature, supplementation of KCL (K^+ ions) was studied. They were analyzed by random amplification of polymorphic DNA (RAPD) and identified based on 16S rRNA sequencing as *Halobacillus trueperi, Shewanella* algae, *Halomonas venusta,* and *Marinomonas* sp. This is the first report on the presence of diverse halophilic bacteria in the foothills of Himalayas. The halophilic bacterial isolates SS1, SS2, SS3, SS5, and SS8 were characterized for the production of industrially important enzymes like amylase, protease, lipase, and glutaminase. Halophilic bacterial isolates SS1 and SS3 exhibited salt-dependent extracellular amylase and protease activities. Both the halophilic isolates SS1 and SS3 exhibited maximum amylase and protease activities in the presence of 1.5 and 1.0 M NaCl, respectively, with the optimum pH 8 and temperature 40 °C. SS2 showed maximum extracellular protease and lipase activities in the presence of 0.75 M NaCl, at optimum pH of 7, and temperature 37 °C. The glutaminase activity of SS3 increased with an increase in the concentration of NaCl up to 2.5 M. The optimum pH and temperature for L-glutaminase activity of SS3 were 8 and 40 °C, respectively. The combined hydrolytic activities of these halophilic bacterial isolates can be used for bioconversion of organic materials to useful products.

Sharma et al. (2016) have isolated a novel moderately halophilic bacterium from a microbial mat deposited at thermal discharges of Manikaran hot spring (with surface water temperature ~95 °C) located in Himachal Pradesh. The bacterial strain AS8T is a Gram-stain-positive, motile, and endospore-forming organism. On the basis of 16 S rRNA gene sequence-based phylogenetic analysis, the strain AS8T was assigned to genus *Fictibacillus*. Based on chemotaxonomic characters and DNA-DNA hybridization studies, the strain AS8T is found to be a novel species for which the name *Fictibacillus halophilus* sp.nov. is proposed. The type strain is AS8T (=MCC 2765T =DSM 100124T=KCT 33758T).

2.9 Role of Halophiles in Sustainability

Environmental pollution is caused to some extent by industrial waste waters. It is estimated that around 5% of industrial waste waters are saline in nature (Singh et al. 2012). Halophilic microorganisms are suitable candidates for bioremediation of such saline waste waters. Moderately halophilic, halotolerant, and extremely halophilic bacteria have a broader catabolic versatility and capability. Singh et al. (2012) have worked on polyaromatic hydrocarbon (PAH) degrading bacteria isolated from Okha-madhi, Porbandar-Madhavpur, and Rajkot. The organisms were found to grow in naphthalene, anthracene, and pyrene as the sole source of carbon and energy. Thus they could be used in the bioremediation of waste waters containing such PAH. Textile industrial wastewaters containing dyes become difficult to treat. The authors isolated moderately halophilic bacteria from the marine site. The isolates were tested for dye degradation and heavy metal tolerance. The isolates showed potential in the treatment of complex dye-containing wastewater.

Yadav and Saxena (2018) have described halophilic and salt-tolerant plant growth-promoting (PGP) microorganisms belonging to genera *Haloarcula, Halobacillus, Halococcus, Haloferax, Halolamina, Halomonas, Halorubrum,* and *Haloterrigena* isolated from diverse saline habitat and characterized for plant growth under the salinity stress. The halophiles have ability to produce phytohormones viz. indole acetic acids, gibberellic acids, and cytokines. They can solubilize and bind nutrients like phosphorus, potassium, and zinc. The halophiles also elicit plant defense reactions against pathogens and help in plant growth under harsh saline environments. Thus, the halophilic PGP-producing microorganisms enhance plant growth, yields, and nutrient uptake. The authors have reviewed work on such PGP-producing halophiles and their role in sustainable agriculture.

In India, salt-affected soils account for 6.7 million ha (2.1%) of geographical area of the country. Saline soils contain salts of sodium, calcium, magnesium and potassium, chlorides, nitrates, sulfates, and carbonates. Arora et al. (2014) have discussed the importance of halophilic microorganisms in the bioremediation of such saline soils. Salt-affected soils become a suitable habitat for halophilic and halotolerant microorganisms. They have a potential to produce plant growth-promoting (PGP) substances and hydrolytic enzymes under salt stress. Dominant halotolerant bacteria in saline soils include *Alkaligenes, Bacillus, Micrococcus,* and *Pseudomonas* while moderately halophilic bacteria are *Halobacillus, Halomonas, Marinococcus,* and *Marinobacter*. These organisms play a role in the remediation of saline soils and thus enhance plant growth and yield. The use of halophiles in bioremediation of saline soils resulting in improvement of plant growth and crop yield would be an eco-friendly approach.

2.10 Contribution to the Knowledge of Halophilic Microbial Species, New to Science, from India

Three novel genera and nine novel species of halophilic microorganism have been added to the global collection of microorganisms and are listed in Table 2.2.

Table 2.2 Halophilic microbial species, new to Science, from India

S. no.	Names of the microorganisms	Source of the microorganisms	Reference
1	*Domibacillus indicus* sp. nov.	Ocean sediments of Lakshadweep island	Sharma et al. (2014)
2	*Halobacilus marinus* sp. nov.	Chilika Lake, Odisha	Panda et al. (2018)
3	*Natrialba swarupiae* sp. nov. strain ESP3B-9T	Salt pan of Sambhar lake, Rajasthan	Kajale et al. (2020)
4.	*Fictibacillus halophilus* sp. nov.	Manikaran hot spring, Himachal Pradesh	Sharma et al. (2016)
5.	*Cesiribacter andamanensis* gen. nov. sp. nov.	Mud volcano soil, Andaman Islands	Srinivas et al. (2011)
6.	*Lutibaculum baratungense* gen. nov.sp.nov.	Mud volcano soil, Andaman Islands	Kumar et al. (2012a, b)
7.	*Aquipuribacter nitratireducens* sp.nov.	Mud volcano soil, Andaman Islands	Srinivas et al. (2015)
8.	*Thalassorhabdus alcalitolerans* gen.nov.sp.nov.	Marine sediment	Sultanpuram and Mothe (2018)
9.	*Bacillus okhensis*, sp. nov. Kh10-101T	Salt pan near Port of Okha, Gujarat	Krishna et al. (2015)

2.11 Future Perspective

Although a number of saline sites have been explored from India for the isolation of halophilic microorganisms, there is still a large scope to study biodiversity and biotechnology of halophiles from India. Moderately halophilic bacteria have demonstrated their versatile metabolic capacity. Exploration of these bacteria for commercial production of industrially important enzymes, exopolysaccharides, polyhydroxyalkanoates (PHA), and carotenoids is awaited. In the light of energy crises, the use of Bacteriorhodopsin (BR) containing haloarchaeal cultures in the conversion of sunlight into electrical energy and in solar cells for absorption of sunlight need to be explored. To make the production of PHA cost-effective, haloarchaeal cultures which are lysed in just water can be employed so that use of organic solvents for extraction of PHA from cell mass is minimized. Since extremely halophilic haloarchaea need high concentration of salt, the spent medium would have high concentration of chlorides leading to waste water pollution. To avoid this, haloarchaea isolated from sea water and able to grow at low concentration of salt can be screened for biomolecules. Due to high concentration of salt, it is likely that the machinery like fermenters, centrifuges, filters, etc. get rusted. Modifications in such equipments need inputs from bioengineers and biotechnologists. Earlier researchers have studied halophiles more with respect to physiology, mechanisms of halophily, and evolution of life. It is now a need of the day that these versatile halophiles are explored for biotechnological and industrial processes.

References

Adams RL, Russell NJ (1992) Interactive effects of salt concentration and temperature on growth and lipid composition in the moderately halophilic bacterium *Vibrio costicola*, 1992. Can J Microbiol 38:823–827

Adams RL, Kogut M, Russell NJ (1989) The effect of salinity on growth and lipid composition of a moderately halophilic Gram-negative bacterium HX. Biochem Cell Biol 68:249–254

Adeleye AI (2007) Purification and properties of α-amylase from *Micrococcus varians*. J Basic Microbiol 30(10):717–722

Ahmad N, Sharma S, Khan FG, Kumar R, Johri S, Abdin MZ, Qazi GN (2008) Phylogenetic analyses of Archaeal ribosomal DNA sequences from salt pan sediments of Mumbai, India. Curr Microbiol 57(2):145–152. https://doi.org/10.1007/s00284-008-9167-z

Akolkar AV, Deshpande GM, Raval KN, Durai D, Nerurkar A, Desai AJ (2008) Organic solvent tolerance of *Halobacterium* sp. 1(1) & its extracellular protease. J Basic Microbiol 48(5): 421–425

Akolkar A, Bharaambe N, Trivedi S, Desai A (2009) Statistical optimization of medium components for extracellular protease production by an extremely haloarchaeon , *Halobacterium* sp. SP1(1). Lett Appl Microbiol 48:77–83

Aljohny BO (2015) Halophilic bacterium - a review of new studies. Biosci Biotechnol Res Asia 12(3):2061–2069. https://doi.org/10.13005/bbra/1874

Amoozegar MA, Safarpour A, Noghabi KA, Bakhtiary T, Ventosa A (2019) Halophiles and their vast potential in biofuel production. Front Microbiol 10:1895. https://doi.org/10.3389/fmicb.2019.01895

Anand TP, Bhat AW, Shouche YS, Roy U, Siddharth J, Sharma SP (2006) Antimicrobial activity of marine bacteria associated with sponges from the waters off the coast of South East India. Microbiol Res 161(3):252–262

Anton J, Meseguer I, Rodriguez-Valera F (1988) Production of an extracellular polysaccharide by *Haloferax mediterranei*. Appl Environ Microbiol 54(10):2381–2386

Arias S, del Moral A, Ferrer MR, Tallon R, Quesada E, Bejar V (2003) Mauran, an exopolysaccharide produced by the halophilic bacterium *Halomonas maura*, with a novel composition and interesting properties for biotechnology. Extremophiles 7:319–326

Arora S, Vanza MJ, Mehta R, Buwa C, Patel CN (2014) Halophilic microbes for bio-remediation of salt affected soil. Afr J Microbiol Res 8(33):3070–3078

Asha KRT, Denslin AJV, Sehgal GK, Manjusha WA, Sukumaran N, Selvin J (2005) Isolation and cultivation of halophilic archaea from solar salterns located in peninsular coast of India. Int J Microbiol 1(2)

Behera P, Mahapatra M, Rastogi G (2020) Microbial ecology of Chilika Lagoon. In: Ecology, conservation and restoration of Chilika Lagoon, India. Springer, Cham, pp 399–414

Bejar V, Llamas I, Calco C, Quesada E (1998) Characterization of exopolysaccharides produced by 19 halophilic strains of the species *Halomonas eurihalina*. J Biotechnol 61:135–141

Birbir M, Sesal C (2003) Extremely halophilic bacterial communities in Sereflikochisar salt lake in Turkey. Turkish J Microbiol 27:7–22

Birbir M, Calli B, Mertoglu B, Bardavid RE, Oren A, Ogmen MN, Ogan A (2007) Extremely halophilic archaea from Tuz lake, Turkey and the adjacent Kaldirim and Kayaci salterns. World J Microbiol Biotechnol 23(3):309–316

Biswas A, Patra AKP (2009) Production of poly-3-hydroxyalkanoic acids by a moderately halophilic bacterium, *Halomonas marina* HMA 103 isolated from solar saltern of Orissa, India. Acta Microbiol Immunol Hung 56:125–143

Biswas J, Paul AK (2017) Diversity and production of extracellular polysaccharide by halophilic microorganisms. Biodiversity Int J 1(2):32–39. https://doi.org/10.15406/bij.2017.01.00006

Bitton G, Friehofer V (1978) Influence of extracellular polysaccharide on the toxicity of copper and cadmium toward *Klebsiella aerogenes*. Microb Ecol 4:119–125

Bolen DW, Baskakov IV (2008) The osmophobic effect: natural selection of a thermodynamic force in protein folding. J Mol Biol 310:955–963
Boone DR, Richard CW (eds) (2001) Bergey's manual of systematic bacteriology, vol 1, 2nd edn. Springer, New York/Berlin/Heidelberg
Bouchotroch S, Quesada E, del Moral A, Llamas I, Bejar V (2001) *Halomonas maura* sp. nov., a novel moderately halophilic, exopolysaccharide-producing bacterium. Int J Syst Evol Microbiol 51(5):1625–1632
Braganca JM, Furtado I (2009) Isolation and characterization of haloarchaea from low salinity coastal sediments and waters of Goa. Curr Sci 96(9):1182–1184
Brenner DJ, Krieg NR, Staley JT (eds) (2005) Bergey's manual of systematic bacteriology, vol 2, 2nd edn. Springer, New York/Berlin/Heidelberg
Brown AD (1976) Microbial water stress. Bacteriol Rev 40(4):803–846
Brown MJ, Lester JN (1979) Metal removal in activated sludge: the role of bacterial extracellular polymers. Water Res 13:817–837
Calvo C, Martínez-Checa F, Toledo FL, Porcel J, Quesada E (2002) Characteristics of bioemulsifiers synthesized in crude oil media by *Halomonas eurihalina* and their effectiveness in the isolation of bacteria able to grow in the presence of hydrocarbons. Appl Microbiol Biotechnol 60(3):347–351
Caumette P, Cohen Y, Matheron R (1991) Isolation and characterization of *Desulfovibrio halophilus* sp. nov., a halophilic sulfate-reducing bacterium isolated from Solar Lake (Sinai). Syst Appl Microbiol 14(1):33–38. https://doi.org/10.1016/S0723-2020(11)80358-9
Chandramohan D (1997) Recent advances in marine microbiology: the Indian scenario. J Mar Biotechnol 5:73–81
Chaudhary A, Pasha IM, Salgaonkar BB, Braganca JM (2014) Cadmium tolerance by haloarchaeal strains isolated from solar salterns of Goa, India. Int J Biosci Biochem Bioinformatics 4(1):1–6
Chen CW, Don TR, Yen HF (2006) Enzymatic extruded starch as a carbon source for the production of poly (3-hydroxybutyrate-co-3-hydroxyvalerate) by *Haloferax mediterranei*. Process Biochem 41:2289–2296
Cheng T-C, DeFrank JJ, Rastogi VK (1999) Alteromonas prolidase for organophosphorus G-agent decontamination. Chem Biol Interact 119-120:455–462
Corral P, Amoozegar MA, Ventosa A (2020) Halophiles and their biomolecules: recent advances and future applications in biomedicine. Mar Drugs 18(1):33. https://doi.org/10.3390/md18010033
Costaouec Le T, Cérantola S, Ropartz D, Ratiskol J, Sinquin C, Colliec-Jouault S, Boisset C (2012) Structural data on a bacterial exopolysaccharide produced by a deep-sea *Alteromonas macleodii* strain. Carbohydr Polym 90(1):49–59. https://doi.org/10.1016/j.carbpol.2012.04.059
Csonka LN, Hanson AD (1991) Prokaryotic osmoregulation: genetics and physiology. Ann Rev Microbiol 45:569–606
Cui H-L, Tohty D, Zhou P-J, Liu S-J (2006) *Halorubrum lipolyticum*, sp. nov. and *Halorubrum aidingense* sp. nov., isolated from two salt lakes in Xin-Jiang, China. Int J Syst Evol Microbiol 56:1631–1634
Cyplik P, Grajek W, Marecik R, Kroliczak P, Dembczynski R (2007) Application of a membrane bioreactor to denitrification of brine. Desalination 207:134–143
Das P, Chatterjee S, Behera BK, Dangar TK, Kumar B, Mohapatra DT (2019) Isolation and characterization of marine bacteria from East Coast of India: functional screening for salt stress tolerance. Heliyon 5(6):e01869
DasSarma S, Arora P (2001) Halophiles. In: Encyclopedia of life sciences, vol 8. Nature Publishing Group, London, pp 458–466
DasSarma S, DasSarma P (2015) Halophiles and their enzymes: negativity put to good use. Curr Opin Microbiol 25:120–126
Dave SR, Desai HB (2006) Microbial diversity at marine salterns near Bhavnagar, Gujarat, India. Curr Sci 90(4):497–500

Davis JS (1974) Importance of microorganisms in solar salt production. In: Coogan AL (ed) Proceedings of the 4th symposium on salt, vol 1. Northern Ohio Geological Society, Cleveland, pp 369–372

Decho AW (1990) Microbial exopolymer secretions in ocean environments: their roles in food webs and marine processes. In: Barnes M (ed) Oceanography and marine biology: an annual review. Aberdeen Univ Press, pp 73–153

Deshmukh SV, Kanekar PP, Bhadekar RK, Dhar SK (2017a) Exopolysaccharide producing halophilic microorganisms from West coast of Maharashtra, India. Int J Pharma Biosci 8(1): 370–375

Deshmukh SV, Kanekar PP, Bhadekar RK (2017b) Production and characterization of exopolysaccharide from marine moderately halophilic bacterium *Halomonas smirnensis* SVD III. Int J Pharma Pharma Sci 9(10):146–151

Digaskar VU, Thombre RS, Oke RS (2015) Screening of extremely halophilic archaea for its biotechnological potential. Int J Pharma Biosci 6(1):811–819

Dobson SJ, Colwell RR, McMeekin TA, Franzmann PD (1993) Direct sequencing of the polymerase chain reaction-amplified 16S rRNA gene of *Flavobacterium gondwanense* sp. nov. and *Flavobacterium salegens* sp. nov., two new species from a hypersaline Antarctic lake. Int J Syst Bacteriol 43(1):77–83. https://doi.org/10.1099/00207713-43-1-77

Duckworth AW, Grant WD, Jones BE, Meijer D, Marquez MC, Ventosa A (2000) *Halomonas magadii* sp. nov., a new member of the genus *Halomonas*, isolated from a soda lake of the East African rift valley. Extremophiles 4(1):53–60

Dundas I (1998) Was the environment for primordial life hypersaline? Extremophiles 2:375–377

Elevi R, Assa P, Birbir M, Ogan A, Oren A (2004) Characterization of extremely halophilic archaea isolated from Ayvalik saltern, Turkey. World J Microbiol Biotechnol 20:719–725

Elshahed MS, Najar FZ, Roe BA, Oren A, Dewers TA, Krumholz LR (2004) Survey of archaeal diversity reveals an abundance of halophilic archaea in a low-salt, sulfide and sulfur rich spring. Appl Environ Microbiol 70(4):2230–2239

Empadinhas N, Milton S, da Costa MS (2008) Osmoadaptation mechanisms in prokaryotes: distribution of compatible solutes. Int Microbiol 11:151–161

Fendrich C (1988) *Halovibrio variabilis* gen. nov. sp. nov., *Pseudomonas halophila* sp. nov. and a new halophilic aerobic coccoid Eubacterium from Great Salt Lake, Utah, USA. Syst Appl Microbiol 11(1):36–43. https://doi.org/10.1016/S0723-2020(88)80046-8

Fernandez-Castillo R, Rodriguez-Valera F, González-Ramos J, Ruiz-Berraquero F (1986) Accumulation of poly (beta-hydroxybutyrate) by halobacteria. Appl Environ Microbiol 51(1): 214–216

Fritze D (1996) *Bacillus haloalkaliphilus* sp. Int J Syst Bacteriol 46(1):98–101 https://doi.org/10.1099/00207713-46-1-98

Fukushima T, Usami R, Kamekura M (2007) A traditional Japanese-style salt field is a niche for haloarchaeal strains that can survive in 0.5% salt solution. Sal Syst 3(2):1–12

Galinski EA (1993) Compatible solutes of halophilic eubacteria: molecular principles, water-solute interactions, stress protection. Experientia 49:487–496

Galinski EA (1995) Osmoadaptation in bacteria. Adv Microb Physiol 37:272–328

Garcia MT, Mellado E, Ostos JC, Ventosa A (2004) Halomonas oranivorans sp. nov., a moderate halophile able to degrade aromatic compounds. Int J Syst Evol Microbiol 54:1723–1728

Goh F, Leuko S, Allen MA, Bowman JP, Kamekura M, Neilan BA, Burns BP (2006) *Halococcus hamelinensis* sp. nov., a novel halophilic archaeon isolated from stromatolites in Shark bay. Australia, Int J Syst Evol Microbiol 56:1323–1329

Gomes J, Steiner W (2004) The biocatalytic potential of extremophiles and extremozymes. Food Technol Biotechnol 42:223–235

Grant WD, Gemmell RT, McGenity TJ (1998) Halobacteria: The evidence for longevity. Extremophiles 2:279–287

Grant WD, Kamemkura M, McGenity TJ, Ventosa A (2001) Class III Halobacteria class, nov. In: Boone DR, Castenholz RW, Garrity GM (eds) Bergey's manual of systematic bacteriology, vol 1, 2nd edn. Springer, New York, pp 294–334

Gudhka RK, Neilan BA, Burns BP (2015) Adaptation, ecology, and evolution of the halophilic stromatolite archaeon *Halococcus hamelinensis* inferred through genome analyses. Archaea 2015:241608. https://doi.org/10.1155/2015/241608

Gupta S, Sharma P, Dev K, Srivastav M, Sourirajan A (2015a) A diverse group of halophilic bacteria exist in Lunsu, a natural salt water body of Himachal Pradesh, India, vol 4. SpringerPlus, p 274. https://doi.org/10.1186/s40064-015-1028-1

Gupta RS, Naushad S, Baker S (2015b) Phylogenomic analyses and molecular signatures for the class Halobacteria and its major clades: a proposal for division of the class Halobacteria into an amended order Halobacteriales and two new orders, *Haloferacales* ord. nov. and *Natrialbales* ord. nov., containing the novel families *Haloferacaceae* fam. Nov. and *Natrialbaceae* fam. nov. Int J Syst Evol Microbiol 65:1050–1069

Gupta S, Sharma P, Dev K, Sourirajan A (2016a) Halophilic bacteria of Lunsu produce an array of industrially important enzymes with salt tolerant activity. Biochem Res Int 2016:9237418. https://doi.org/10.1155/2016/9237418

Gupta RS, Naushad S, Fabros R, Adeolu MA (2016b) Phylogenomic reappraisal of family-level divisions within the class Halobacteria: Proposal to divide the order Halobacteriales into the families *Halobacteriaceae, Haloarculaceae* fam. nov. and *Halococcaceae* fam. nov. and the order Haloferacales into the families *Haloferacaceae* and *Halorubaceae* fam. nov. Antonie Van Leeuwenhoek 109:565–587

Halder D, Dutta P, Mondal A, Basu M (2016) Isolation and characterization of halophilic bacteria from Sundarban soil. IJLSSR 2(4):442–450

Hampp NA (2000) Bacteriorhodopsin: mutating a biomaterial into an optoelectronic material. Appl Microbiol Biotechnol 53:633–639

Harvey RW, Luoma SN (1985) Effect of adherent bacteria and bacterial extracellular polymers upon assimilation by Macoma baltica of sediment bound Cd, Zn and Ag. Mar Ecol Prog Ser 22:281–289

Hinrichsen LL, Montel MC, Talon R (1994) Proteolytic and lipolytic activities of *Micrococcus roseus, Halomonas elongata* and *Vibrio* sp. isolated from Danish bacon curing brines. Int J Food Microbiol 22:115–126

Hinteregger C, Streichsbier F (1997) *Halomonas* sp., a moderately halophilic strain, for biotreatment of saline phenolic waste-water. Biotechnol Lett 19:1099–1102

Huang TY, Duan KJ, Huang SY, Chen CW (2006) Production of polyhydroxyalkanoates from inexpensive extruded rice bran and starch by *Haloferax mediterranei*. J Ind Microbiol Biotechnol 33(8):701–706

Imhoff JF, Thiemann B (1991) Influence of salt concentration and temperature on the fatty acid compositions of Ectothiorhodospira and other halophilic phototrophic bacteria. Arch Microbiol 156:370–375

Javor B (1989) Hypersaline environments, microbiology and biogeochemistry. Springer-Verlag, Berlin, p 384

Jayachandra SY, Kumar AS, Shouche YS, Sulochana MB (2013) Culturable diversity of extremely halophilic bacteria from West Coast of Karnataka, India, JBPAS, 2013. 2(2):391–405

Jayachandra SY, Reddy MK, Parameshwar AB, Sulochana MB (2014) Screening of Halophilic Bacteria for Extracellular Enzymes Production from West Coast of Karnataka, India. Int J Universal Pharma Bio Sci 3(1):31–45

Jean WD, Shieh W-Y, Chiu H-H (2006) *Pseudomarina taiwansesis* gen. nov., sp. nov., a marine bacterium isolated from shallow coastal water of An-Ping Harbor, Taiwan and embedded description of the family Idiomarinaceae. Int J Syst Evol Microbiol 56:899–905

Joo WA, Kim CW (2005) Proteomics of halophilic archaea. J Chrom 815:237–250

Joshi AA, Kanekar PP (2011) Production of EPS by *Vagococcus carniphilus* MCMB-1018 isolated from alkaline Lonar Lake. Ann Microbiol 61:733–740

Joshi CM, Trivedi NS (2020) Isolation, identification and characterization of microorganisms isolated from unexplored saline regions of Kutch, Gujarat, India. Acta Sci Microbiol 3(7):3–9

Joshi AA, Kanekar PP, Kelkar AS, Sarnaik SS, Shouche Y, Wani A (2007) Moderately halophilic, alkalitolerant *Halomonas campisalis* MCM B-365 from Lonar Lake, India. J Basic Microbiol 47(3):213–221

Joshi AA, Kanekar PP, Kelkar AS, Shouche YS, Wani AA, Borgave SB, Sarnaik SS (2008) Cultivable bacterial diversity of alkaline Lonar lake, India. Microbiol Ecol 55:163–172

Kajale S, Deshpande N, Pali S, Shouche Y, Sharma A (2020) *Natrialba swarupiae* sp. nov., a halophilic archaeon isolated from a hypersaline lake in India. Int J Syst Evol Microbiol 70(3): 1876–1881

Kamekura M, Hamakawa T, Onishi H (1982) Application of halophilic nuclease H of *Micrococcus* varians subsp. *halophilus* to commercial production of flavoring agent 59-GMP. Appl Environ Microbiol 44:994–995

Kanekar SP (2014) Biodiversity and biotechnological exploration of halophiles isolated from Andaman Islands and Lonar lake, India. Ph. D. thesis, Savitribai Phule Pune University, Pune, Maharashtra, India

Kanekar PP, Kulkarni SO (2017) Optimization of Bacteriorhodopsin production by *Halostagnicola larsenii* IBS (MCC 2956). Int J Pharma Bio Sci 8(2):618–621

Kanekar PP, Kanekar SP, Kelkar AS, Dhakephalkar PK (2012) Halophiles- taxonomy, diversity, physiology and applications. In: Satyanarayana J, Prakash A (eds) Microorganisms in environmental management. Springer Business, B.V., pp 1–34

Kanekar PP, Kulkarni SO, Nilegaonkar SS, Sarnaik SS, Kshrsagar PR, Ponraj M, Kanekar SP (2015a) Environmental friendly microbial polymers, polyhydroxyalkanoates (PHA's) for packaging and biomedical applications. In: Alavi S, Thomas S, Sandeep KP, Kalarikkal N, Verghese J, Yaragalla S (eds) Polymers for packaging applications. Apple Academic Press, Toranto, pp 197–226

Kanekar PP, Kulkarni SO, Kanekar SP, Shouche Y, Jani K, Sharma A (2015b) Exploration of a haloarchaeon, *Halostagnicola larsenii*, isolated from rock pit sea water, West Coast of Maharashtra, India, for production of bacteriorhodopsin. J Appl Microbiol 118:1345–1356. https://doi.org/10.1111/jam.12784

Kanekar SP, Saxsena N, Pore SD, Arora P, Kanekar PP, Dhakephalkar PK (2015c) Draft genome sequence of *Halostagnicola* sp. A56, an extremely halophilic archeon isolated from the Andaman islands. Genome announcements 3(6):e01332–e01315

Kanekar PP, Kulkarni SO, Dhakephalkar PK, Kulkarni KG, Saxsena N (2017a) Isolation of a halophilic bacteriorhodopsin – producing archaeon, *Haloferax larsenii* RG3D.1 from the rocky beach of Malvan, West coast of India. Geomicrobiology 34(3):242–248

Kanekar PP, Deshmukh SV, Kanekar SP, Dhakephalkar PK, Ranjekar PK (2017b) Exopolysaccharides of halophilic microorganisms: an overview. In: Thangadurai D, Sangeetha J (eds) Industrial biotechnology – sustainable production and bioresource utilization. Apple Academic Press, Toronto, pp 1–27

Kanekar PP, Kulkarni SO, Kanekar SP (2019) Packaging: Polyhydroxyalkanoate (PHA's) in. In: Mishra M (ed) Encyclopedia of polymer applications, vol III, 1st edn. CRC Press, Taylor and Francis Group, Boca Raton, FL, pp 2010–2020

Kanekar PP, Kulkarni SO, Kanekar SP (2020a) Production of Bacteriorhodopsin, an industrially important membrane protein of halophilic archaea: Biotechnological challenge. In: Sangeetha J, Thangadurai D, Tanasupawat S, Kanekar PP (eds) Biotechnology of microorganisms – diversity, improvement and application of microbes for food processing, healthcare, environmental safety and agriculture. Apple Academic Press, Toranto, pp 123–145

Kanekar PP, Kulkarni SO, Jagtap CV, Kadam VS, Pathan HM (2020b) A novel approach for the development of bio-sensitized solar cell using cell lysate of a haloarchaeon *Halostagnicola larsenii* RG2.14 (MCC 2809) containing bacteriorhodopsin. Solar Energy 212:326–331

Kanekar PP, Kulkarni SO, Dhakephalkar PK, Saxena N, Pathan HM (2021) Bacteriorhodopsin producing halophilic archaea isolated from solar salt pan saline environment for conversion of light energy into electrical energy. ES Energy Environ 13:57–64

Kapdan IK, Boylan B (2009) Batch treatment of saline wastewater by *Halanaerobium lacusrosei* in an anaerobic packed bed reactor. J Chem Technol Biotechnol 84(1):34–38

Kapdan IK, Erten B (2007) Anaerobic treatment of saline wastewater by *Halanaerobium lacusrosei*. Process Biochem 42:449–453

Karthikeyan P, Bhat SG, Chandrasekaran M (2013) Halocin SH10 production by an extrme haloarchaeon *Natrinema* sp. BTSH10 isolated from salt pans of South India. Saudi J Biol Sci 20:205–212

Kates M (1986) Influence of salt concentration on membrane lipids of halophilic bacteria. FEMS Microbiol Lett 39:95–101

Kates M (1992) Archaebacterial lipids: structure, biosynthesis and function. Biochem Soc Symp 58: 51–72

Keerthi S, Koduru UD, Nittal SS, Parine NR (2018) The heterotrophic eubacterial and archaeal co-inhabitants of the halophilic Dunaliella salina in solar salterns fed by Bay of Bengal along south eastern coast of India. Saudi J Biol Sci 25(7):1411–1419

Khorana HG (1988) Bacteriorhodopsin, a membrane protein that uses light to translocate protons. J Biol Chem 263:7439–7442

Kinne RKH (1993) The role of organic osmolytes in osmoregulation: from bacteria to mammals. J Exp Zool 265:346–355

Kiran KK, Chandra TS (2008) Production of surfactant and detergent-stable, halophilic and alkalitolerant alpha-amylase by a moderately halophilic *Bacillus* sp. strain TSCVKK. Appl Microbiol Biotechnol 77:1023–1031

Kiran GS, Priyadharsini S, Sajayan A, Ravindran A, Selvin J (2018) An antibiotic agent pyrrolo [1,2-a]pyrazine-1,4-dione,hexahydro isolated from a marine bacteria *Bacillus tequilensis* MSI45 effectively controls multi-drug resistant *Staphylococcus aureus*. RSC Adv 8:17837–17846. https://doi.org/10.1039/C8RA00820E

Kiranmai MR, Anima SD (2017) Isolation of marine bacteria from Visakhapatnam coast for degradation of oil. J Bioremediat Biodegrad 8:392. https://doi.org/10.4172/2155-6199.1000392

Krebs MP, Khorana HG (1993) Mechanism of light-dependent proton translocation by bacterio-rhodopsin. J Bacteriol 175:1555–1560

Krishna PS, Sreenivas A, Singh DK, Shivaji S, Prakash JSS (2015) Draft genome sequence of *Bacillus okhensis* Kh10-101T, a halo-alkali tolerant bacterium from Indian saltpan. Genomics Data 6:283–284. https://doi.org/10.1016/j.gdata.2015.10.019

Kubo M, Hiroe J, Murakami M, Fukami H, Tachiki T (2001) Treatment of hypersaline-containing wastewater with salt-tolerant microorganisms. J Biosci Bioeng 91:222–224

Kulkarni SO, Kanekar PP, Nilegaonkar SS, Sarnaik SS, Jog JP (2010) Production and characterization of a biodegradable poly (hydroxybutyrate-co-hydroxyvalerate) (PHB-co-PHV) copolymer by moderately haloalkalitolerant *Halomonas campisalis* MCM B-1027 isolated from Lonar Lake, India. Biores Technol 101:9765–9771

Kumar AS, Mody K, Jha B (2007) Evaluation of biosurfactant / bioemulsifier production by a marine bacterium. Bull Environ Contam Toxicol 79:617–621

Kumar PA, Srinivas TNR, Manasa P, Madhu S, Shivaji S (2012a) *Lutibaculum baratangense* gen. nov., sp. nov., a proteobacterium isolated from a mud volcano. Int J Syst Evol Microbiol 62 (Pt_8):2025–2031

Kumar S, Karan R, Kapoor S, Singh SP, Khare SK (2012b) Screening and isolation of halophilic bacteria producing industrially important enzymes. Braz J Microbiol 43(4):1595–1603

Kumar V, Saxena J, Tiwari SK (2015) Description of a halocin-producing *Haloferax larsenii* HA1 isolated from Pachpadra salt lake in Rajasthan. Arch Microbiol 198(2):181–192. https://doi.org/10.1007/s00203-015-1175-3

Kushner DJ (1993) In: Vreeland RH, Hochstein LI (eds) The biology of halophilic bacteria. CRC Press, Inc, Boca Raton, FL, pp 87–103

Kushner DJ, Kamekura M (1988) Physiology of halophilic eubacteria. In: Rodriguez-Valera F (ed) Halophilic bacteria, vol I. CRC Press, Inc., Boca Raton, FL, pp 109–138

Larsen H (1986) Halophilic and halotolerant microorganisms – an overview and historical perspective. FEMS Microbiol Rev 39:3–7

Lee SY, Chang HN, Um YS, Hong SH (1998) Bacteriorhodopsin production by cell recycle culture of *Halobacterium halobium*. Biotechnol Lett 20(8):763–765

Lee HK, Chun J, Moon EJ, Ko SH, Lee DS, Lee HS, Bae KS (2001) *Hahella chejuensis* gen. nov. sp. nov., an extracellular-polysaccharide producing marine bacterium. IJSEM 51:661–666

Legat A, Gruber C, Zangger K, Wanner G, Stan-Lotter H (2010) Identification of polyhydroxyalkanoates in *Halococcus* and other haloarchaeal species. Appl Microbiol Biotechnol 87:1119–1127

Li WJ, Schumann ZY-Q, Chen G-Z, Tian X-P, Xu L-H, Stackerbrandt E, Jiang C-L (2005) *Marinococcus halotolerans* sp. nov. isolated from Qinghai, north-west China. Int J Syst Evol Microbiol 55:1801–1804

Li WW, Zhou WZ, Zhang YZ, Wang J, Zhu XB (2008) Flocculation behavior and mechanism of an exopolysaccharide from the deep-sea psychrophilic bacterium *Pseudoalteromonas* sp. SM 9913. Bioresour Technol 99(15):6893–6899

Liaw HJ, Mah RA (1992) Isolation and characterization of *Haloanaerobacter chitinovorans* gen. nov., sp. nov., a halophilic, anaerobic, chitinolytic bacterium from a solar saltern. Appl Environ Microbiol 58(1):260–266. https://doi.org/10.1128/AEM.58.1.260-266.1992

Lillo JG, Rodriguez-Valera F (1990) Effects of culture conditions on poly beta- hydroxybutyric acid production by *Haloferax mediterranei*. Appl Environ Microbiol 56:2517–2521

Lim J-M, Jeon C-O, Song S-M, Kim C-J (2005) *Pontibacillus chungwhensis* gen. nov., sp. nov., a moderately halophilic Gram-positive bacterium from a solar saltern in Korea. Int J Syst Evol Microbiol 55:165–170

Lizama C, Monteoliva-Sanchez M, Prado B, Ramos-Cormenzana A, Weckesser J, Campos V (2001) Taxonomic study of extreme halophilic archaea isolated from the "Salar de Atacama", Chile. Syst Appl Microbiol 24:464–474

Llamas I, Bejar V, Martinez-Checa F, Martinez-Canovas MJ, Molina I, Quesada E (2011) *Halomonas stenophila* sp. nov., a halophilic bacterium that produces sulphate exopolysaccharides with biological activity. Int J Syst Evol Microbiol 61:2508–2514

Llamas I, Amjres H, Mata JA, Quesada E, Bejar V (2012) The potential biotechnological applications of the exopolysaccharide produced by the halophilic bacterium *Halomonas almeriensis*. Molecules 17:7103–7120

Loaec M, Olier R, Gueennec J (1997) Uptake of lead, cadmium and zinc by a novel bacterial exopolysaccharide. Water Res 31:1171–1179

Loaec M, Olier R, Guezennec J (1998) Chelating properties of bacterial exopolysaccharides from deep-sea hydrothermal vents. Carb Polym 35:65–70

Lu J, Nogi Y, Takami H (2001) *Oceanobacillus iheyansis* gen. nov., sp. nov., a deep-sea extremely halotolerant and alkaliphilic species isolated from a depth of 1050 m on the Iheya range. FEMS Microbiol Lett 205(2):291–297

Lynch EA, Langille MGI, Darling A, Wilbanks EG, Haltiner C, Shao KSY, Starr MO, Teiling C, Harkins TT, Edwards RA, Eisen JA, Facciotti MT (2012) Sequencing of seven haloarchaeal genomes reveals patterns of genomic flux. PLoS One 7:e41389

Ma Y, Xue Y, Grant WD, Collins NC, Duckworth AW, van Steenbergen RP, Jones BE (2004) *Alkalimonas amylotytica* gen. nov., sp. nov. and *Alkalimonas delamerensis* gen. nov., sp. nov., novel alkaliphilic bacteria from soda lakes in China and East Africa. Extremophiles 8:193–200

Madern D, Ebel C, Zaccai G (2000) Halophilic adaptation of enzymes. Extremophiles 4:91–98

Mahapatra PD, Mishra BB (2014) Isolation and molecular identification of extreme an halotolerant *Staphylococcus* Sp. 13cc from brackish water of Chilika Lake. Biosci Biotechnol Res Asia 11(1):277–282. dx.doi.org. https://doi.org/10.13005/bbra/1268

Maltseva O, McGowan C, Fulthorpe R, Oriel P (1996) Degradation of 2, 4-dichlorophenoxyacetic acid by haloalkaliphilic bacteria. Microbiol 142:1115–1122

Mani K, Salgaonkar BB, Braganca JM (2012) Culturable halophilic archaea at the initial crystallization stages of salt production in a natural solar saltern of Goa, India. Aquatic Biosystem 8:15

Manikandan P, Senthilkumar PK (2017) An overview of saltpan halophilic bacterium. J Antimicrob Agents 2(4):28–33

Manikandan M, Kannan V, Pasic L (2009) Diversity of microorganisms in solar salterns of Tamil Nadu, India. World J Micro Biotech 25:1001–1017

Martinez-Canovas MJ, Quesada E, Llmas I, Bejar V (2004) *Halomonas ventosae* sp. nov., a moderately halophilic, denitrifying, exopolysaccharide-producing bacterium. Int J Syst Evol Microbiol 54(3):733–737

Martinez-Espinosa RM, Zafrilla B, Camacho M, Bonete MJ (2007) Nitrate and nitrite removal from salted water by *Haloferax mediterranei*. Biocatal Biotransform 25:295–300

Mata JA, Béjar V, Llamas I, Arias S, Bressollier P, Tallon R, Urdaci MC, Quesada E (2006) Exopolysaccharides produced by the recently described bacteria *Halomonas ventosae* and *Halomonas anticariensis*. Res Microbiol 157:827–835

McGenity TJ, Gemmell RT, Grant WD, Stan-Lotter H (2000) Origins of halophilic microorganisms in ancient salt deposits. Env Microbiol 2:243–250

Mellado E, Moore ERB, Nieto JJ, Ventosa A (1996) Analysis of 16S rRNA Gene Sequences of *Vibrio costicola* Strains: Description of *Salinivibrio costicola* gen. nov., comb. nov. Int J Syst Bacteriol 46(3):817–821

Mermestein LD, Zeikus JG (1998) In: Horikoshi K, Grant WD (eds) Extremophiles: microbial life in extreme environments. Wiley-Liss, New York, p 322

Mishra SR (2019) Microbial biodiversity study of a brackish water ecosystem in Eastern India: the Chilika Lake. In: Microbial diversity in the genomic era. Academic Press, London, pp 47–63. https://doi.org/10.1016/B978-0-12-814849-5.00004-6

Mohan RS, Vatsala TM (2009) Tryptophan inhibits bacteriorhodopsin formation in *Halorubrum sodomense* A01. J Basic Microbiol 49:304–309

Montalvo-Rodriguez R, Ruiz-Acevedo A, Lopez-Garriga J (1997) New isolates of extremely halophilic archaebacteria (halobacteria) from Puerto Rico and the Caribbean. Caribbean J of Sci 33(1-2):98–104

Moreno MDL, Garcia MT, Ventosa A, Mellado E (2009) Characterization of *Salicola* sp. IC10, a lipase- and protease-producing extreme halophile. FEMS Microbiol Ecol 68:59–71

Mormile MR, Romine MF, Garcia MT, Ventosa A, Bailey TJ, Peyton BM (1999) *Halomonas campisalis* sp. nov., a denitrifying, moderately haloalkaliphilic bacterium. System Appl Microbiol 22:551–558

Mothes G, Schubert T, Harms H, Maskow T (2008) Biotechnological coproduction of compatible solutes and polyhydroxyalkanoates using the Genus Halomonas. Eng Life Sci 8:658–662

Mullakhanbhai MF, Larsen H (1975) *Halobacterium volcanii* sp. nov., a Dead sea halobacterium with a moderate salt requirement. Arch Microbiol 104(1):207–214

Muller V, Oren A (2003) Metabolism of chloride in halophilic prokaryotes. Extremophiles life under extreme conditions. Extremophiles 7(4):261–266

Murugan S, Subha T, Asha KRT (2018) Isolation and characterization of haloarchaeal strain from Puthalam Salt Pan located in the Southern Peninsular Coast of India. J Microbial Biochem Technol 10:3. https://doi.org/10.4172/1948-5948.1000400

Mwatha WE, Grant WD (1993) *Natronobacterium vacuolata* sp. nov., a haloalkaliphilic archaeon isolated from Lake Magadi, Kenya. Int J Syst Bacteriol 43(3):401–404

Oesterhelt D, Stoeckenius W (1971) Rhodopsin-like protein from the purple membrane of *Halobacterium halobium*. Nat New Biol 233(39):149–152

Ollivier B, Hatchikian CE, Prensier G, Guezennec J, Garcia J-L (1991) *Desulfohalobium retbaense* gen. nov., sp. nov., a halophilic sulfate-reducing bacterium from sediments of a hypersaline lake in Senegal. Int J Syst Evol Microbiol 41(1):74–81. https://doi.org/10.1099/00207713-41-1-74

Ollivier B, Caumette P, Garcia JL, Mah RA (1994) Anaerobic bacteria from hypersaline environments. Microbiol Rev 58:27–38

Onishi H, Mori T, Takeuchi S, Tani K, Kobayashi T, Kamekura M (1983) Halophilic nuclease of a moderately halophilic *Bacillus* sp.: production, purification and characterization. Appl Environ Microbiol 45(1):24–30

Oren A (1994) The ecology of extremely halophilic Archaea. FEMS Microbiol Rev 13:415–440

Oren J (2002) Diversity of halophilic microorganisms: environments, phylogeny, physiology, and applications. J Ind Microbiol Biotechnol 28:56–63

Oren A (2008) Microbial life at high salt concentrations: phylogenetic and metabolic diversity. Saline Syst 4:2–14

Oren A (2010) Industrial and environemental applications of halophilic microorganisms. Environ Biotechnol 31:825–834

Oren A, Weisburg WG, Kessel M, Woese CR (1984) *Halobacteroides halobius* gen. nov., sp. nov., a moderately halophilic anaerobic bacterium from the bottom sediments of the Dead Sea. Syst Appl Microbiol 5:58–69

Oren A, Heldal M, Nordland S (1997) X-ray microanalysis of intracellular ions in the anaerobic halophilic eubacterium *Haloanaerobium praevalens*. Can J Microbiol 43:588–592

Ozcan B, Ozyilmaz G, Cokmus C, Caliskan M (2009) Characterization of extracellular esterase and lipase activities from five halophilic archaeal strains. J Ind Microbiol Biotechnol 36(1):105–110

Panda AN, Mishra SR, Ray L, Sahu N, Acharya A, Jadhao S, Suar M, Adhya TK, Rastogi G, Pattnaik AK, Raina V (2016) Draft genome sequence of *Halobacillus* sp. strain KGW1, a moderately halophilic and alkaline protease-producing bacterium isolated from the rhizospheric region of Phragmites karka from Chilika Lake, Odisha, India. Genome Announc 4(3):e00361–e00316. https://doi.org/10.1128/genomeA.00361-16

Panda AN, Mishra SR, Ray L, Das S (2018) Taxonomic description and genome sequence of *Halobacillus marinus* sp. nov., a novel strain isolated from Chilika Lake, India. J Microbiol 56(4):223–230. https://doi.org/10.1007/s12275-018-7387-x

Papke RT (2015) Preface to the proceedings of halophiles 2013, editorial article. Front Microbiol 6. https://doi.org/10.3389/fmicb.2015.00341

Parolis H, Parolis LAS, Boan IF, Valera FR, Widmalm G, Manta MC, Jansson PE, Sutherland IW (1996) The structure of the exopolysaccharide produced by the halophilic archaeon *Haloferax mediterranei* strain R4 (ATCC 33500). Carbohydr Res 295:147–156

Pathak AP, Sardar AG (2012) Isolation and characterization of carotenoid producing haloarchaea from solar saltern of Mulund, Mumbai, India. Ind J Nat Products Res 3(4):483–488

Pfiffner SM, McInerney MJ, Jenneman GE, Knapp RM (1986) Isolation of halotolerant, thermotolerant, facultative polymer-producing bacteria and characterization of the exopolymer. Appl Environ Microbiol 51:1224–1229

Phrommao E, Rodtong S, Yongsawatdigul J (2010) Identification of novel halotolerant bacillopeptidase F-like proteinases from a moderately halophilic bacterium, *Virgibacillus* sp. SK37. J Appl Microbiol 110:191–201

Poli A, Esposito E, Oriando P, Lama L, Giordano A, de Appolonia F, Nicolaus B, Gambacorta A (2007) *Halomonas alkaliantarctica* sp. nov. isolated from saline lake Cape Russell in Antarctica, an alkalophilic moderately halophilic, exopolysaccharide-producing bacterium. Syst Appl Microbiol 30:31–38

Poli A, Di-Donato P, Abbamondi GR, Nicolaus B (2011) Synthesis, production and biotechnological applications of exopolysaccharides and polyhydroxyalkanoates by archaea. Archaea 2011: 693253. https://doi.org/10.1155/2011/693253

Qin K, Zhu L, Chen L, Wang PG, Zhang Y (2007) Structural characterization and ecological roles of a novel exopolysaccharide from the deep-sea psychrotolerant bacterium *Pseudoalteromonas* sp. SM 9913. Microbiology 153:1566–1572

Quadri I, Hassani II, l'Haridon S, Chalopin M, Hacène H, Jebbar M (2016) Characterization and antimicrobial potential of extremely halophilic archaea isolated from hypersaline environments of the Algerian Sahara. Microbiol Res 186-187:119–131. https://doi.org/10.1016/j.micres.2016.04.003

Quillaguaman J, Hatti-Kaul R, Mattiasson B, Alvarez MT, Delgado O (2004) *Halomonas boliviensis* sp. nov., an alkalitolerant, moderate halophile isolated from soil around a Bolivian hypersaline lake. Int J Syst Evol Microbiol 54:721–726

Quillaguaman J, Hasher S, Bento F, Mattiasson B, Hatti-Kaul R (2005) Poly (b-hydroxybutyrate) production by a moderate halophile, *Halomonas boliviensis* LC1 using starch hydrolysate as substrate. J Appl Microbiol 99:151–157

Quillaguaman J, Delgado O, Mattiasson B, Hatti-Kaul R (2006) Poly (β-hydroxybutyrate) production by a moderate halophile, *Halomonas boliviensis* LC1. Enzyme Microb Technol 38:148–154

Quillaguaman J, Munoz M, Mattiasson B (2007) Optimizing conditions for poly (b-hydroxybutyrate) production by *Halomonas boliviensis* LC1 in batch culture with sucroseas carbon source. Appl Microbiol Biotechnol 74:981–986

Quillaguaman J, Doan-Van T, Guzman H, Guzman MJ, Everest A, Hatti-Kaul R (2008) Poly (3-hydroxybutyrate) production by *Halomonas boliviensis* in fed-batch culture. Appl Microbiol Biotechnol 78:227–232

Raghavan TM, Furtado I (2004) Occurrence of extremely halophilic archaea in sediments from the Continental Shelf of West Coast of India. Curr Sci 86(8):1065–1067

Raghavan TM, Furtado I (2005) Expression of carotenoid pigments of haloarchaeal cultures exposed to aniline. Int J Environ Toxicol 20(2):165–169

Reed RH, Stewart WDP (1988) The responses of cyanobacteria to salt stress. In: Rogers LJ, Gallon JR (eds) Biochemistry of the algae and cyanobacteria. Claredon Press, Oxford, p 217

Riffat R, Krongthamchat K (2007) Anaerobic treatment of high-saline wastewater using halophilic methanogens in laboratory-scale anaerobic filters. Water Environ Res 79:191–198

Ritter D, Yopp JH (1993) Plasma membrane lipid composition of the halophilic cyanobacterium *Aphanothece halophytica*, 1993. Arch Microbiol 159:435–439

Roberts MF (2005) Organic compatible solutes of halotolerant and halophilic microorganisms. Saline Syst 1:5–35

Roberts MF, Lai MC, Gunsalus RP (1992) Biosynthetic pathways of the osmolytes N epsilon-acetyl-beta-lysine, beta-glutamine, and betaine in *Methanohalophilus* strain FDF1 suggested by nuclear magnetic resonance analyses. J Bacteriol 174:6688–6693

Rodriguez-Valera F (1992) Biotechnological potential of halobacteria. In: Danson MJ, Hough DW, Lunt GG (eds) The archaebacteria: biochemistry and biotechnology. Portland Press, London, pp 135–147

Rodriguez-Valera F (1993) In: Vreeland R, Hochstein L (eds) Biology of halophilic bacteria. CRC Press, Boca Raton, FL, pp 1–23

Rodriguez-Valera F, Ruiz-Beeraquero FF, Ramos-Cormenzana A (1979) Isolation of extreme halophiles from sea water. Appl Environ Microbiol 38:164–165

Rodriguez-Valera F, Ruiz-Berraquero F, Ramos-Cormenzana A (1980) Isolation of extremely halophilic bacteria able to grow in defined inorganic media with single carbon sources. J Gen Microbiol 119:535–538

Rougeaux H, Talaga P, Carlson RW, Guerzennec J (1998) Structural studies of an exopolysaccharide produced by *Alteromonas macleodii* subsp. *fijiensis* originating from a deep-sea hydrothermal vent. Carbohydr Res 312:53–59

Ruiz-Ruiz C, Srivastava GK, Carranza D, Mata JA, Llamas I, Santamaría M, Quesada E, Molina IJ (2011) An exopolysaccharide produced by the novel halophilic bacterium *Halomonas stenophila* strain B100 selectively induces apoptosis in human T leukaemia cells. Appl Microbiol Biotechnol 89(2):345–355. https://doi.org/10.1007/s00253-010-2886-7

Sahay H, Singh S, Kaushik R, Saxena A, Arora D (2011) Characterization of halophilic bacteria from environmental samples from the brackish water of Pulicat Lake, India. Biologia 66(5): 741–747. https://doi.org/10.2478/s11756-011-0094-2

Sahoo SR, Goli D, Maringanti B (2019) Studies of two pigment producing halophilic bacteria from Karnataka Mangrove Soil (Karnataka). Indian J Pharm Sci 81(5):892–899. https://doi.org/10.36468/pharmaceutical-sciences.584

Saju KA, Babu MM, Murugan M, Raj ST (2011) Survey on Halophilic microbial diversity of Kovalam Saltpans in Kanyakumari District and its industrial applications. J Appl Pharm Sci 01 (05):160–163

Sanchez-Porro C, Mellado E, Bertoldo C, Antranikian G, Ventosa A (2003) Screening and characterization of the protease CP1 produced by the moderately halophilic bacterium *Pseudoalteromonas* sp. strain CP76. Extremophiles 7:221–228

Satpute SK, Bhawsar BD, Dhakephalkar PK, Chopade BA (2008) Assessment of screening methods for selecting biosurfactant producing marine bacteria. Ind J Mar Sci 37(3):243–250

Satpute SK, Banat IM, Dhakephalkar PK, Banpurkar AG, Chopade BA (2010) Biosurfactants, bioemulsifiers and exopolysaccharides from marine microorganisms. Biotechnol Adv 28:436–450

Satyanarayana T, Raghukumar C, Shivaji S (2005) Extremophilic microbes: diversity and perspectives. Curr Sci 89(1):78–90

Schubert BA, Lowenstein TK, Timofeeff MN, Parker MA (2010) Halophilic archaea cultured from ancient halite, Death valley, California. Environ Microbiol 12(2):440–454

Sehgal SN, Gibbons NE (1960) Effect of some metal ions on the growth of Halobacterium cutirubrum. Can J Microbiol 6:165–169

Setati ME (2010) Diversity and industrial potential of hydrolase producing halophilic/halotolerant eubacteria. Afr J Biotechnol 9(11):1555–1560

Shand RF, Betlatch MC (1991) Expression of the bop gene cluster of *Halobacterium halobium* is induced by low oxygen tension and by light. J Bacteriol 173:4692–4699

Shand RF, Betlatch MC (1994) Bop gene cluster expression in bacteriorhodopsin overproducing mutants of *Halobacterium halobium*. J Bacteriol 176:1655–1660

Sharma A, Dhar S, Prakash O, Vemuluri VR, Thite V, Shouche Y (2014) Description of Domibacillus indicus sp. nov., isolated from ocean sediments of Lakshadweep and emended description of the genus *Domibacillus*. Int J Syst Evol Microbiol 64:3010–3015

Sharma A, Kohli P, Singh Y, Schumann P, Lal R (2016) *Fictibacillus halophilus* sp. nov., from a microbial mat of a hot spring atop the Himalayan Range. Int J Syst Evol Microbiol 66(6): 2409–2416. https://doi.org/10.1099/ijsem.0.001051

Shiu PJ, Ju YH, Chen HM, Lee CK (2013) Facile isolation of purple membrane from *Halobacterium salinarum* via aqueous-two-phase-system. Protein Expr Purif 89:219–224

Simon-Colin C, Raguenes G, Cozien J, Guezennec JG (2008) *Halomonas profundus* sp. nov., a new PHA-producing bacterium isolated from a deep-sea hydrothermal vent shrimp. J Appl Microbiol 104:1425–1432

Singh SP, Rawat VH, Purohit TJT, Gohal SD, Pandey S, Akbari VG, Rawal CM (2012) Haloalkaliphilic bacteria and actinobacteria from the saline habitats: new opportunities for biocatalysis and bioremediation. In: Satyanarayana T et al (eds) Microorganisms in environmental management: microbes and environment, Chapter 19. Springer science + business media B. V., London, pp 415–425

Spring S, Ludwig W, Marquez MC, Ventosa A, Schleifer K-H (1996) *Halobacillus* gen. nov., with Descriptions of *Halobacillus litoralis* sp. nov. and *Halobacillus trueperi* sp. nov., and Transfer of *Sporosarcina halophila* to *Halobacillus halophilus* comb. nov. Int J Syst Evol Microbiol 46(2):492–496. https://doi.org/10.1099/00207713-46-2-492

Srinivas TNR, Kumar PA, Madhu S, Sunil B, Sharma TVRS, Shivaji S (2011) *Cesiribacter andamanensis* gen. nov., sp. nov., isolated from a soil sample from a mud volcano. Int J Syst Evol Microbiol 61(7):1521–1527. https://doi.org/10.1099/ijs.0.025429-0

Srinivas TNR, Kumar PA, Tank M, Sunil B, Poorna M, Zareena B, Shivaji S (2015) *Aquipuribacter nitratireducens* sp. nov., isolated from a soil sample of a mud volcano. Int J Syst Evol Microbiol 65(8):2391–2396. https://doi.org/10.1099/ijs.0.000269

Stan-Lotter H, Pfaffenhuemer M, Legat A, Busse HJ, Radax C, Gruber C (2002) *Halococcus dombrowskii* sp. nov., an archaeal isolate from a Permo-Triassic alpine salt deposit. Int J Syst Evol Microbiol 52:1807–1814

Strazzullo G, Gambacorta A, Vella MF, Immirzi B, Ronmano I, Calandrelli V, Nicolaus B, Lama L (2008) Chemical-physical characterization of polyhydroxyalkanoates recovered by means of a simplified method from cultures of *Halomonas campaniensis*. World J Microbiol Biotechnol 24(8):1513–1519

Sultanpuram VR, Mothe T (2018) *Thalassorhabdus alkalitolerans* gen. nov., sp. nov., a novel *Bacillaceae* member isolated from marine sediment. Int J Syst Evol Microbiol 68(9): 2969–2976. https://doi.org/10.1099/ijsem.0.002931

Suribabu K, Hemalatha KPJ (2014) Optimization of slight halotolerant α-amylase producing *Brevibacillus borstelensis* R1 in various media by submerged fermentation. Int J Eng Sci Res Technol 3(9):289–295

Sutherland IW (1998) Novel and established applications of microbial polysaccharides. Trends Biotechnol 16:41–46

Syed S, Prasada BG, Paramgeetham CH (2012) A major of soil microbial diversity and density from artificial solar salterns of Nellore district in Andhra Pradesh, India. Int J Res Biol Sci 2(2):83–86

Syed S, Prasada BG, Paramgeetham CH (2013a) Isolation of amylase producing bacteria from solar salterns of Nellore district in Andhra Pradesh, India. Res Rev J Microbiol 2(1):1

Syed S, Prasada BG, Paramgeetham CH (2013b) Extracellular enzymatic potential of haloalkaliphiles from solar salterns of Nellore district in Andhra Pradesh, India. Asian J Biol Sci 4(2):302–305

Thavasi V, Lazarova T, Filipek S, Kolinski M, Querol E, Kumar A, Ramakrishna S, Padros E, Renugopalakrishnan V (2008) Study on the feasibility of bacteriorhodopsin as bio-photosensitizer in excitonic solar cell: a first report. J Nanosci Nanotechnol 8:1–9

Thomas M, Pal KK, Dey R, Saxena AK, Dave SR (2012) A novel haloarchaeal lineage widely distributed in the hypersaline marshy environment of Little and Great Rann of Kutch in India. Curr Sci 103(9):1078–1084

Thombre RS, Oke RS (2015) Study of stress proteins induced by temperature stress in extremely halophilic archaea, *Haloferax mediterranei* RT 18. Int J Curr Microbiol Appl Sci Special issue-2:199–209

Thombre RS, Shinde VD, Oke RS, Dhar SK, Shouche YS (2016) Biology and survival of extremely halophilic archaeon *Haloarcula marismortui* RR12 isolated from Mumbai salterns, India in response to salinity stress. Sci Reports 6:25642

Thombre RS, Joshi VS, Oke RS (2017) Halophiles: pharmaceutical potential and biotechnological applications. In: Thangadurai D, Jeyabalan S (eds) Industrial biotechnology, sustainable production and bioresource utilization. Apple Academic Press, Burlington ON, pp 111–140

Tindall BJ (1991) In: Balows AJ (ed) The prokaryotes, vol 1, 2nd edn. Springer-Verlag, New York, pp 754–808

Tindall BJ, Truper HG (1986) Ecophysiology of the aerobic halophilic archaebacteria. Syst Appl Mcrobiol 7:202–212

Tindall BJ, Mills AA, Grant WD (1980) An alkaliphilic red halophilic bacterium with low magnesium requirement from a Kenyan soda lake. J Gen Microbiol 116:257–260

Tombs MP, Harding SE (1998) An introduction to polysaccharide biotechnology. Taylor & Francis, London, pp 1–176

Uchida K (2000) Diversity and ecology of salt tolerant lactic acid bacteria: *Tetragenococcus halophilus* in soy sauce fermentation. Jpn J Lactic Acid Bact 11(2):60–65

Upasani V, Desai S (1990) Sambhar salt lake chemical composition of the brines and studies on haloalkaliphilic archaebacteria. Arch Microbiol 154:589–593

Upasani VN, Desai SG, Moldveanu N, Kates M (1994) Lipids of extremely halophilic archaeobacteria from saline environments in India: a novel glycolipid in *Natronobacterium* strains. Microbiol 140:1959–1966

Vargas VA, Delgado OD, Hatti-Kaul R, Mattiasson B (2005) *Bacillus bogoriensis* sp. nov., a novel alkaliphilic, halotolerant bacterium isolated from Kenyan soda lake. Int J Syst Evol Microbiol 55:899–902

Veena GK, Vasudevan N (2010) Enrichment of phenol degrading moderately halophilic bacterial consortium from saline environment. Int J Biotech Biochem 6(5):783–791

Ventosa A (1994) Taxonomy and phylogeny of moderately halophilic bacteria. In: Priest FG (ed) Bacterial diversity and systematics. Plenum Press, New York, NY, pp 231–242

Ventosa A, Gutierrez MC, Garcia MT, Ruiz-Berraquero F (1989) Classification of "*Chromobacterium marismortui*" in a new genus, *Chromohalobacter* gen. nov., as *Chromohalobacter marismortui* comb. nov., nom. rev. Int J Syst Evol Microbiol 39(4): 382–386. https://doi.org/10.1099/00207713-39-4-382

Ventosa A, Márquez MC, Ruiz-Berraquero F, Kocur M (1990) *Salinicoccus roseus* gen. nov., sp. nov., a new moderately halophilic gram-positive coccus. Syst Appl Microbiol 13(1):29–33. https://doi.org/10.1016/S0723-2020(11)80177-3

Ventosa A, Nieto JJ, Oren A (1998) Biology of moderately halophilic aerobic bacteria. Microbiol Mol Biol Rev 62(2):504–544

Vijay A, Arora S, Gupta S, Chhabra M (2018) Halophilic starch degrading bacteria isolated from Sambhar Lake, India, as potential anode catalyst in microbial fuel cell: a promising process for saline water treatment. Bioresour Technol 256:391–398

Vreeland RH (1992) The family *Halomonadaceae*. In: Balows A, Truper HG, Dworkin M, Harder W, Schleifer KH (eds) The prokaryotes. A handbook on the biology of bacteria: ecophysiology, isolation, identification, applications, vol IV, 2nd edn. Springer-Verlag, New York, N.Y., pp 3181–3188

Vreeland RH, Litchfield CD, Martin EL, Elliot E (1980) *Halomonas elongata,* a new genus and species of extremely salt tolerant bacteria. Int J Syst Bacteriol 30:485–495

Vreeland RH, Anderson R, Murray RG (1984) Cell wall and phospholipid composition and their contribution to the salt tolerance of *Halomonas elongate*. J Bacteriol 160(3):879–883

Waditee-Sirisattha R, Kageyama H, Takab T (2016) Halophilic microorganism resources and their applications in industrial and environmental biotechnology. AIMS Microbiol 2(1):42–54. https://doi.org/10.3934/microbiol.2016.1.42

Ward DM, Brock TD (1978) Hydrocarbon biodegradation in hypersaline environments. Appl Environ Microbiol 35:353–359

Woese C (1993) The archaea: their history and significance. In: Kates M, Kushner D, Matheson A (eds) The biochemistry of archaea (Archaebacteria). Elsevier, Amsterdam, pp vii–xxix

Woolard CR, Irvine RL (1992) Biological treatment of hypersaline wastewater by a biofilm of halophilic bacteria, 1992. Syst Appl Microbiol 15:530–534

Xin H, Itoh T, Zhou P, Suzuki K-I, Kamekura M, Nakase T (2000) *Natrinema versiforme* sp. nov., an extremely halophilic archaeon from Aibi salt lake, Xinjiang, China. Int J Syst Evol Microbiol 50:1297–1303

Xin H, Itoh T, Zhou P, Suzuki K, Nakase T (2001) *Natronobacterium nitratireducens* sp. nov., a haloalkaliphilic archaeon isolated from a soda lake in China. Int J Syst Evol Microbiol 51(5): 1825–1829

Xu Y, Zhou P, Tian X (1999) Characterization of two novel haloalkaliphilic archaea *Natronorubrum bangense* gen. nov., sp. nov., and *Natronorubrum tibetense* gen. nov., sp. nov. Int J Syst Evol Microbiol 49:261–266

Xu Y, Wang Z, Xue Y, Zhou P, Ma Y, Ventosa A, Grant WD (2001) *Natrialba hulunbeirensis* sp. nov. and *Natrialba chahannaoensis* sp. nov., a novel haloalkaliphilic archaea from soda lakes in Inner Mongolia autonomous region, China. Int J Syst Evol Microbiol 51:1693–1696

Xu XW, Wu M, Huang WD (2005) Isolation and characterization of a novel strain of *Natrinema* containing a bop gene. J Zhejiang Univ Sci B 6:142–146

Xue Y, Fan H, Ventosa A, Grant WD, Jones BE, Cowan DA, Ma Y (2005) *Haloalkalicoccus tibetensis* gen. nov., sp. nov., representing a novel genus of haloalkaliphilic archaea. Int J Syst Evol Microbiol 55:2501–2505

Yadav AN, Saxena AK (2018) Biodiversity and biotechnological applications of halophilic microbes for sustainable agriculture. J Appl Biol Biotechnol 6(1):48–55. https://doi.org/10.7324/JABB.2018.60109

Yadav AN, Sharma D, Gulati S, Singh S, Dey R, Pal KK, Kaushik R, Saxena AK (2015) Haloarchaea endowed with phosphorus solubilization attribute implicated in phosphorus cycle. Sci Reports 5:12293. https://doi.org/10.1038/srep12293

Yadav AN, Verma P, Sachan SG, Kaushik R, Saxena AK (2018) Psychrotrophic microbiomes: molecular diversity and beneficial role in plant growth promotion and soil health. In: Microorganisms for green revolution. Springer, Singapore, pp 197–240

Yancey PH, Clark ME, Hand SC, Bowlus RD, Somero GN (1982) Living with water stress: the evolution of osmolyte systems. Science 217:1212–1222

Yang Y, Cui H-L, Zhou P-J, Liu S-J (2007) *Haloarcula amylolytica* sp. nov., an extremely halophilic archaeon isolated from Aibi salt lake in Xin-Jiang, China. Int J Syst Evol Microbiol 57:103–106

Yatsunami R, Kawakami T, Ohtani H, Nakamura SA (2000) Novel bacteriorhodopsin like protein from *Haloarcula japonica* strain TR–1: Gene cloning, sequencing, and transcript analysis. Extremophiles 4:109–114

Yen CW, Heyden SC, Dreaden EC, Szymanski P, El-Sayed MA (2011) Tailoring plasmonic and electrostatic field effects to maximize solar energy conversion by bacteriorhodopsin, the other natural photosynthetic system. Nano Lett 11:3821–3826

Yoon J-H, Kim H, Kim I-G, Kang KH, Park Y-H (2003) *Erythrobacter flavus* sp. nov., a slight halophile from the East Sea in Korea. Int J Syst Evol Microbiol 53:1169–1174

Yoshie S, Noda N, Miyano T, Tsuneda S, Hirata A, Inamori Y (2002) Characterization of microbial community in nitrogen removal process of metallurgic Wastewater by PCR-DGGE. Water Sci Technol 46(11-12):93–98

Yun UJ, Perk HD (2003) Physical properties of extracellular polysaccharide produced by *Bacillus* sp. Lett Appl Microbiol 36:282–287

Zhilina TN, Zavarzin GA (1990) A new extremely halophilic homoacetogen bacteria *Acetohalobium arabaticum*, gen. nov. sp. nov. Dokl Akad Nauk SSSR 311:745–747

Alkaliphilic, Alkalitolerant Microorganisms 3

Abstract

Alkaliphilic microorganisms are those microbes that require alkaline pH for growth with optimum growth at pH 9.0. They cannot grow or grow only slowly at near neutral pH. Alkalitolerant microorganisms can grow at alkaline pH at 9–10 but can also grow at neutral pH. They occur in soda lakes, alkaline wastewaters, commercial process effluents, etc. To cope up with the alkaline environment, they maintain internal pH 7.0–8.5, possess acidic polymers in cell wall components, and use sodium ions for solute transport through membrane. Alkaliphilic bacteria are widely distributed in nature and represented by *Bacillus alcalophilus*, cyanobacteria, Proteobacteria, and archaea. Alkaliphilic bacteria are more explored for their alkaline enzymes, exopolysaccharides (EPS), polyhydroxyalkanoate (PHA), and bioremediation of alkaline wastewaters.

In India, alkaliphilic microorganisms have been extensively studied from the alkaline soda lake of Lonar and explored for production of EPS, PHA, hydrolytic enzymes and antimicrobial compounds, and bioremediation of alkaline wastewaters. Alkaliphilic microorganisms have also been reported from alkaline soils, alkaline wastewaters from chemical industries, etc. from different parts of India. Three new genera and nine novel species of alkaliphilic microorganisms have been described, thus contributing to the knowledge of alkaliphiles, new to Science, from India.

Keywords

Alkaliphilic · Alkalitolerant · Soda lakes · Alkaline wastewaters · pH homeostasis · Acidic polymers · *Alkalibacillus* · Cyanobacteria · Alkaline enzymes · Lonar Lake · Polyhydroxyalkanoate · Exopolysaccharide

P. P. Kanekar, S. P. Kanekar, *Diversity and Biotechnology of Extremophilic Microorganisms from India*, Microorganisms for Sustainability 41,
https://doi.org/10.1007/978-981-19-1573-4_3

There are no precise definitions of what characterizes an alkaliphile or alkalitolerant organism. Several microorganisms exhibit more than one pH optimum for growth depending on the growth conditions, particularly nutrients, metal ions, and temperature. The term alkaliphile is used for microorganisms that grow optimally or very well at pH values above 9, often between 10 and 12, but cannot grow or grow only slowly at the near-neutral pH value of 6.5 (Horikoshi 1999). Alkaliphiles are widely distributed in deep sea, alkaline lake, sub-ground soils all over the world. Naturally occurring alkaline environments include carbonate springs, alkaline soils, and soda lakes which are characterized by their high pH values (pH 8–11). Alkaline pH is mainly due to the presence of high concentrations of sodium carbonate salts which are generally formed by evaporative concentration. Alkaliphiles are classified into two main groups, alkaliphilic and alkali tolerant. Alkaliphiles actually require alkaline pH for growth with optimum growth at pH 9. Some microorganisms can grow at alkaline pH 9 or 10 but they can also grow at neutral or less pH and hence termed as alkalitolerant. Alkaliphilic microorganisms could be both aerobic and anaerobic. Alkaliphiles consist of two main physiological groups of microorganisms namely alkaliphiles and haloalkaliphiles. Alkaliphiles require pH 9 or more for growth while haloalkaliphiles require both alkaline pH (pH 9) and high salinity. Alkaliphiles have been isolated from alkaline environments as well as neutral environments, sometimes even acidic environments. Haloalkaliphiles are isolated from alkaline saline environments like soda lakes. Joshi et al. (2007) has compiled basic information as origin of life and evolution of alkaliphiles, special physiological features, mechanism of cytoplasmic pH regulation, habitats of alkaliphiles, genesis of soda lake environments, microorganisms in alkaline environments, and many more aspects of alkaliphiles and extensively studied alkaliphilic bacteria.

3.1 Origin and Evolution of Alkaliphilic Bacteria

Although alkaliphiles are widespread, it seems probable that these organisms especially those unique to hypersaline lakes, evolved separately within an alkaline environment. Although present-day soda lakes are geologically quite recent, they have probably existed since archaean times, permitting the evolution of independent communities of alkaliphiles from an early period in Earth's history (Jones et al. 1998). Many of nonexacting alkaliphiles have spread beyond the borders of their soda lake environments due to grazing flamingoes, migratory birds, and dust storms in arid regions.

The 2.3 billion years old Venters drop formation in South Africa, which dates from the late Archaean to the early Proterozoic era, shows evidence of process similar to those presently going on in Lake Magdi (Karpeta 1989). One of the largest fossil soda lakes is the Green River formation in Utah and Wyoming, USA, created from two lakes (Lake Gosutue, Lake Uinta) covering an area of 100,000 km^2 during the Eocene (55–36 million years ago) (Eugster and Hardie 1978). Over a period of more than 12 million years, sediments were laid down together with trona in vast quantities. Mining of the trona beds and mine brine samples have yielded alkaliphilic and haloalkaliphilic bacteria and archaea. Zavarzin (1993) has argued convincingly that alkaline soda lakes are ancient prokaryotic communities and also speculated that continental soda lakes are the origin of prokaryotic diversity. The apparent uniqueness of some of the soda lake prokaryotes implies that they must have evolved in an alkaline, saline environment. These extreme conditions are reflected in the adaptation of microorganisms found in the lakes, many of which are both alkaliphilic and halophilic or extremely halotolerant species (Sorokin et al. 2002, 2003; Hoover et al. 2003; Ma et al. 2004a, b).

3.2 Special Physiological Features of Alkaliphiles

3.2.1 Internal pH

Most alkaliphiles have an optimal growth at pH around 10, which is the most significant difference from well-investigated neutrophilic microorganisms. It suggests that there is difference between physiological and structural aspects of neutrophilic and alkaliphilic bacteria. Studies on the internal pH of a number of alkaliphiles favor the concept that it does not exceed pH 9.0–9.5, despite a high external pH of 8–11 (Guffanti et al. 1978; Horikoshi and Akiba 1982; Krulwich and Guffanti 1983). Therefore, one of the key features in alkaliphily is associated with the cell surface, which discriminates and maintains the intracellular neutral environment separate from the extracellular alkaline environment.

3.2.2 Na^+ Ions and Membrane Transport

A feature of certain alkaliphiles is their absolute requirement for Na^+ ions (Horikoshi and Akiba 1982). The presence of sodium ions in the surrounding environment has proved to be essential for effective solute transport through the membranes of alkaliphilic *Bacillus* spp. In Na^+-dependent transport systems, the H^+ is exchanged with Na^+ by Na^+/H^+ antiporter systems, thus generating a sodium motive force, which drives substrates accompanied by Na^+ into the cells. The presence of sodium ions significantly enhanced the incorporation of the substrates. Instead of Na^+, other cations including K^+, Li^+, NH_4^+, Cs^+, and Rb^+ showed no any effect.

3.2.3 Cell Surface and Alkaliphily

Cell wall composition of alkaliphilic bacteria differs from that of neutrophilic bacteria. Cell wall is directly exposed to high alkaline environments. The intracellular pH of alkaliphilic bacteria is neutral (pH 7–8.5). The pH difference might be due to cell surface components. It has been observed that protoplasts of alkaliphilic *Bacillus* strains lose their stability against alkaline environments. This suggests that the cell wall may play some role in protecting the cell from alkaline environments (Horikoshi 1991).

3.2.4 Peptidoglycan

Major constituents detected commonly in the hydrolysates of the peptidoglycans of alkaliphilic bacteria are glucosamine, muramic acid, D and L alanine, D-glutamic acid, meso-diaminopimelic acid, and acetic acid (Horikoshi 1991). The culture pH has no effect on the composition of peptidoglycan.

3.2.5 Acidic Polymers

Non-peptidoglycan acidic components found in the cell walls of alkaliphilic *Bacillus* sp. No. C-125, were galacturonic acid, gluconic acid, glutamic acid, aspartic acid, and phosphoric acid (Aono and Horikoshi 1983). The negative charges on the acidic nonpeptidoglycan polymer give the cell surface its ability to adsorb sodium and hydronium ions and repulse hydroxide ions, and as a result may enable the cells to grow in alkaline environments.

3.3 Adaptation Process in Alkaliphiles

Aono (2011) has described the adaptation process in alkaliphilic strains of *Bacillus* spp. The cell walls of alkaliphilic strains of *Bacillus* spp. consist of peptidoglycans and acidic polymers linked through acid-labile linkages. Three types of polymers namely teichoic acids, teichuronic acids, and teichuronopeptides are found in alkaliphiles. Teichuronopeptides are formed by covalent combination of two acidic polymers, polyuronates and acidic polypeptides. Alkaliphilic strains of *Bacillus* species contain teichuronopeptides which are not found in neutrophilic strains. Amino acid composition of teichuronopeptide is specific for each strain. The facultative alkaliphilic strain (C-125) produces teichuronic acid and teichuronopeptide in the cell walls. The strain C-125 teichuronic acid is composed of glucuronic acid, galacturonic acid and N-acetylfucoseamine while teichuronopeptide of this strain is composed of poly-gamma-L-glutamic acid and poly-alpha- and beta-1,4-glucuronic acid. A mutant defective in its synthesis of teichuronopeptide was constructed from the strain C-125. This mutant showed sensitivity to high pH (alkaline conditions). The gene *tup* A cloned from C-125 chromosome directs teichuronopeptide synthesis in the mutant. The mutant was found to restore the teichuronopeptide synthesis and alkaliphily upon the introduction of *tup* A gene. Anionic charges present in the cell walls aid intracellular pH homeostasis in the strain exposed to alkaline pH.

3.4 Lipids in Alkaliphiles

Lipids are a major component of the cell membrane. The cell membrane is a pH barrier to keep intracellular pH lower than the extracellular pH. The first studies on lipids in alkaliphilic bacteria were conducted by Koga et al. (1982). Major polar lipids in alkaliphilic *Bacillus* sp. No. A-007 are phosphatidylethanolamine, phosphatidic acid phosphatidylglycerol, cardiolipin, monoglycosyldiacylglycerol, diglycosylglycerol and one of the important lipid components is bis-(monoacylglycero)-phosphate (BMP) found only in the alkaliphilic bacteria. A characteristic distribution of fatty acids is observed in alkaliphilic strains. These all contribute to the alteration of the membrane properties of alkaliphilic bacteria.

3.5 Polyamines in Alkaliphilic Bacteria

Spermidine, which stabilizes the cell membrane, is the major polyamine in alkaliphilic *Bacillus* sp. No. 25 cultivated at pH 10 (Yonezawa and Horikoshi 1978).

3.6 Mechanism of Cytoplasmic pH Regulation

The cell walls containing negatively charged acidic polymers may reduce the pH values at the cell surface. The surface of the plasma membrane must presumably be kept below pH 9.0, because it is very unstable at alkaline pH values (8.5–9.0). Plasma membranes may also maintain pH homeostasis by using the Na^+/H^+ antiporter system, the K^+/H^+ antiporter, and ATPase-driven H^+ expulsion. It was reported that membrane-associated proteins must catalyze the inward movement of proton. One such protein is Na^+/H^+ antiporter, which enables cells to adapt to a sudden upward shift in pH and to maintain cytoplasmic pH 2–2.3 units below the external pH in the most alkaline range of pH for growth. Another is the proton-translocating ATP synthase that catalyzes the production of ATP under conditions in which the external proton concentration and the bulk chemiosmotic driving force are low. These mechanisms indicated that the membrane proteins play a key role in keeping the intracellular pH values in the range of 7.0–8.5.

Padan et al. (1976) and Schalachev (1978) suggested that prokaryotic cation/proton antiporters could serve to regulate cytoplasmic pH. Most of the obligately alkaliphilic bacteria show an absolute requirement of Na^+. The cells or membrane vesicles from these alkaliphilic bacteria show a typical proton motive force pattern if Na^+ is present (Mandel et al. 1980; Krulwich 1982). When cells of alkaliphilic *Bacillus firmus* RAB were incubated in the absence of Na^+ at pH 10.5, the cytoplasmic pH immediately rose from less than 9.5 to 10.5; and they lost viability very rapidly. On the other hand, cells incubated at pH 10.5 in the presence of Na^+ regained greater viability, and internal pH was maintained below 9.0 (Kitada and Horikoshi 1982). These data indicate the presence of Na^+/H^+ antiporter activity at alkaline pH, which catalyzes energy-dependent acidification of the cytoplasmic space. Thus it is quite evident that alkaliphilic bacteria possess a Na^+/H^+ antiporter that is very active in the alkaline range of pH and has a crucial role in pH homeostasis. *E. coli* has three distinct antiporter systems Ca^{2+}/H^+, Na^+/H^+, and K^+/H^+. K^+/H^+ antiporter plays an important role in pH homeostasis in alkalitolerant marine bacteria (Nakamura et al. 1984).

3.7 Habitats of Alkaliphilic Bacteria

Alkalinity in the terrestrial and aquatic environments may be generated by a number of factors, either naturally occurring or brought about by man's activities. Commercial processes ranging from cement manufacturing to the preparation of indigo dye, mining operations, paper and pulp production and food processing effluents all generate examples of highly alkaline environments.

Naturally occurring alkaline environments may arise through biological activities such as ammonification, sulfate reduction, and oxygenic photosynthesis (Abd-el-Malek and Rizk 1963a, b, c; Brewer and Goldman 1976; Langworthy 1978). In these instances, the alkalinity may be localized or relatively short-lived or even diurnal. Natural geochemical processes such as the weathering of silicate minerals can lead to alkaline water about pH 11 because of $Ca\ (OH)_2$ (Grant 1992; Grant et al. 1990; Bath et al. 1987) but they have extremely low buffering capacity. The pH stability is limited by the low solubility of $Ca\ (OH)_2$ (about 10 mM) and exposure to atmospheric CO_2. Carbonate ions are rapidly removed from solution as insoluble calcium or magnesium minerals. The most stable and most significant naturally occurring environments are caused by a combination of geological, geographical, and climatic conditions. Soda lakes and soda deserts represent the major type of naturally occurring highly alkaline environments on the Earth where large amounts of carbonate minerals can generate pH values >11.5. Such sites are widely distributed throughout the world to name some soda lakes, Manito in Canada, Alkali valley, Albert lake, Mono lake and many more in the USA, Langunilla valley in Venezuela, Antofagasta in Chile, Kalunda Steppe, Tanatar lakes, and many more in Russia, Qinhgai Hu, Chahar, and many more in China, Wadi Natrun in Egypt, Lake Bogoria, Lake Magadi, Lake Elementieta in Kenya, Lake Natron, Lake Magad and many more in Tanzania, Lake Corangamite and Red Rock lake in Australia (Joshi 2007). Although these environments are widely distributed, they are often located in inaccessible continental interiors, which have hindered their detailed and systematic study (Jones et al. 1998). The best-studied soda lakes are those of East African Rift valley, where detailed limnological and microbiological investigations have been carried out over many years (Jones et al. 1994; Grant et al. 1990; Rich 1933; Beadle 1932; Jenkin 1932). Microbiological studies of central Asian soda lakes have also been well documented (Zhilina and Zavarzin 1994; Issatchencko 1951).

In volcanic areas, alkaline hot springs have been reported with pH 9.5 (Hensel et al. 1997) where alkalinity is probably also generated by decomposition of silicates. In this case, the maintenance of stable alkaline pH is limited probably because of exposure to O_2 and outgassing of CO_2.

3.8 The Genesis of Soda Lake Environment

Soda lakes are highly alkaline extreme environments that are formed in closed drainage basins exposed to high evaporation rates. The alkaline conditions in soda lakes were known to be derived from the presence of high levels of sodium carbonate. A number of theories concerning the source of sodium carbonate have been proposed (Abd-el-Malek and Rizk 1963a, b, c; Baker 1958; Eugster and Hardie 1978). The simplest and most favored explanation was that this was due to a geological anomaly, where the rocks contained high levels of sodium carbonate, washed out by ground water, and accumulated in the lake basin (Baker 1958). The factors leading to the formation of alkaline saline brines may be divided into climatic, geological, and topographical (Tindall 1988). The influence of climate controls the amount of water entering the system as rainfall or surface runoff and the amount leaving by evaporation, favoring the formation of a saline lake. Geological conditions determine which ions enter the system. One of the most significant factors governing the evolution of saline brines is the relative concentration of divalent cations (Mg^{++} and Ca^{++}) to anions (CO_3^{2-}, SO_4^{2-}, and SiO_2). In case of alkaline environment, the most important factor is a lack of magnesium and calcium in the surrounding strata, which essentially means an absence of rocks of sedimentary origin.

Under the prevailing conditions, groundwater is saturated with atmospheric CO_2, which results in the formation of a weak acid, carbonic acid, which undergoes ion exchange with the surrounding rock, leaching the minerals. The geology of the region directly affects the ionic composition of the groundwater, although this may be further modified by precipitation of insoluble salts or replenishment of lost bicarbonate or carbonate by further dissolution of CO_2. Concentration of the ions in the groundwater through evaporation leads to a shift in the CO_2/bicarbonate/carbonate equilibrium in favor of carbonate.

An increase in carbonate ion concentration results in the precipitation of insoluble carbonates of calcium and magnesium, removing these divalent cations from solution and leaving more soluble carbonates of sodium and potassium to accumulate. The formation of an alkaline saline environment is also dependent on the topography, which allows the salts to concentrate in a shallow depression, often a closed basin with a high marginal relief, having sufficient rainfall to sustain streams entering the basin to produce a standing water body. It restricts surface outflow from the drainage basin to produce a standing body of water. There is generally little or no outflow and water loss is by evaporation. In arid zones with high rates of evaporation exceeding inflow, salts accumulate. Depending upon the salinities, the soda lakes are divided into diluted lakes with NaCl concentrations about 5% and hypersaline lakes where NaCl concentrations are >15%.

3.9 The Soda Lake Aerobic Environment

Soda lakes are probably the most productive naturally occurring environments in the world with gross primary productivities more than any aquatic environment. This remarkable productivity is the consequence of relatively high temperatures (30–45 °C), high light intensities, and unlimited supply of CO_2 for photosynthesis.

One of the noticeable features of many soda lakes is their color. The water may be of various shades of green or red because of the massive blooms of microorganisms. This coloring is a reflection of the very high primary productivities associated with them. Depending on a variety of conditions related to water chemistry, dense populations of microorganisms may color the lakes green, orange, purple, or claret. Most of the autotrophic activity in soda lakes is due to purple sulfur bacteria and various cyanobacteria (Jones et al. 1998). The microorganisms are mainly related to blooms of algae, cyanobacteria, eubacteria, or archaebacteria (Tindall 1988). The most striking feature of the Lonar Lake is the occurrence of *Spirulina*, imparting "Pea soup" green color to the lake brine (Badve et al. 1993). Lonar Lake water also has flora comprising of blue green algae viz. *Arthrospira*, *Oscillatoria*, diatoms, and marshy macrophytes. The formation of thick blooms of *Oscillatoria aghardii*, *Arthospira*, and *Spirulina parentensis* in the Lonar Lake water and their role in nitrogen economy have been reviewed (Singh 1961).

3.10 Microorganisms in Alkaline Environments

Several authors have isolated a number of alkaliphiles from soda lakes showing considerable phylogenetic diversity. Physiological types include proteolytic, cellulolytic, saccharolytic, acetogenic, and sulfate reducing bacteria, methanogenic and halophilic archaea (Zavarzin et al. 1999; Jones et al. 1998; Duckworth et al. 1996; Zhilina and Zavarzin 1994).

3.10.1 Aerobic Alkaliphiles

A variety of heterotrophic organisms have been isolated from a number of alkaline environments. The majority of organisms studied were aerobic; spore-forming Gram-positive, motile rods belonging to the genus *Bacillus*. Alkaliphilic members of the genus *Bacillus* include *Bacillus alcalophilus*, *Bacillus alcalophilus* sub. sp. *halodurans*, *Bacillus pasteurii*, *Bacillus firmus*, *Bacillus lentus*, *Bacillus rotans*, *Bacillus sphaericus*, and *Bacillus pantothenticus* (Grant and Tindall 1986). An alkaliphilic bacterium *Exiguobacterium auranticum* has been isolated from the waste of potato lye treatment factory (Gee et al. 1980; Collins et al. 1983). A number of other bacteria from alkaline environments have also been described and include members of the genera *Rhizobium* (Allen and Holding 1974), *Thiobacillus* (Rupela and Tauro 1973a, b), and *Flavobacterium* (Souza and Deal 1977). Some alkalitolerant organisms include the genera *Pseudomonas*, *Sarcina*, *Micrococcus*, *Proteus*, and *Bacillus* (Horikoshi and Akiba 1982).

Tindall et al. (1984) have isolated and classified haloalkaliphilic archaebacteria as *Natronobacterium* and *Natronococcus* from Lake Magadi. Other organisms present in visible accumulations in alkaline environments are *Cynobacteria–Chloroflexus* mats characteristically associated with alkaline hot springs throughout the world (Brock 1978; Tindall 1980). Takahar and Tanabe (1962) isolated an alkaliphilic *Bacillus alkaliphilus* from a vat of fermenting indigo leaves used for reducing indigo dye. Akiba and Horikoshi (1980) isolated strains of alkaliphilc *Corynebacteria* from soils (7–11). Strains of alkaliphilic *Pseudomonas* were isolated from soil (Watanabe et al. 1977). Souza et al. (1974) reported *Flavobacterium* sp. isolated from highly alkaline springs (pH 10.8–12.1) in California. Alkaliphilc *Streptomyces* (*Streptomyces alkaliphilus* and *Nocardiopsis* sp.) were isolated from soils by Mikami et al. (1982) and Nakanishi (1975).

Examinations of the number of alkaline and highly saline lakes from different geological locations including Lake Magadi (Kenya), the lakes of Wadi natrum (Egypt), and Owens Lake (California) indicated the presence of haloalkaliphilic archaea of the family Halobacteriaceae (Tindall 1988). Halobacteria are the most halophilic microorganisms and they dominate the microbial population when hypersaline waters become saturated, imparting red coloration to liquor because of the presence of C_{50} carotenoids in their core lipids (Rodriguez-Valera et al. 1981). Thus, highly saline soda lakes harbor alkaliphilic halobacteria of the various genera such as *Natronobaterium* and *Natronococcus Natronomonas*, *Natrialba*, *Natronorubrum*, and *Halorubrum.* They flourish on the organic matter arising from the evaporation of brine and the death of the microbial population (Zavarzin et al. 1999). *Roseinatronobacter monicus*, new alkaliphilic bacteriochlorophyll-a containing bacterium was isolated from the hypersaline soda Mono Lake California, U.-S. (Boldareva et al. 2007). A haloalkaliphilic Gram-positive bacterium, *Bacillus haloalkaliphilis* (renamed as *Alkalibacillus haloalkaliphilus*) was isolated from Lake Gabara in the Wadi Natrun, Egypt (Fritze 1996).

A number of alkaliphilic bacteria were reported from different soda lakes. Duckworth et al. (1998) reported alkaliphilic aerobic bacteria, *Dietzia natronolimnaios* from moderately saline and alkaline little Lake Naivasha in East Africa. An alkaliphilic and moderately halophilic *Salinicoccus alkaliphilus* was isolated from Baer soda lake in Inner Mongolia, China (Zhang et al. 2002). *Rhodobaca bogoriensis*, an alkaliphilic, purple nonsulfur bacterium was isolated from water and sediment samples of Lake Bogoria and Crater Lake, East Africa (Milford et al. 2004). *Alkalimonas amylolytica* and *Alkalimonas delamerensis*, novel aerobic heterotrophic alkaliphilic and slightly halophilic bacteria, were isolated from Lake Chahannor in China and Lake Elementeita in East Africa respectively (Ma et al. 2004a, b).

Another group of microorganisms inhabiting the alkaline saline environments such as soda lakes are halophilic and halotolerant bacteria. A number of moderately halophilic bacteria especially *Halomonas* sp. were isolated from saline lakes. Moderately haloalkaliphilic *Halomonas campisalis* was isolated from the salt plain of

Alkali lake in Washington State (Mormile et al. 1999). *Halomonas magadii* sp. nov., haloalkalitolerant bacterium, was isolated from a soda lake of the East African Rift Valley (Duckworth et al. 2000). *Halomonas boliviensis* sp. nov., an alkalitolerant moderate halophile, was isolated from soil around a Bolivian hypersaline lake (Quillaguman et al. 2004).

3.10.2 Anaerobic Alkaliphiles

In contrast to aerobic alkaliphiles, a very few anaerobic alkaliphilic bacteria were reported from the soda lakes. Sulfate-reducing bacteria (SRB) and haloalkaliphilic anaerobic bacterium *Clostridium thermoautotrophicum* were isolated from the brines of Lake Magadi. *Thermopallium natrnophilum* was reported from hot spring at Lake Bogoria (Jones et al. 1998). Few methanogenic archaea belonging to the genus *Methanosarcina* have been isolated from Lonar Lake (Thakker and Ranade 2002). Wani et al. (2006) have isolated *Methanoculleus* and *Methanocalculus* from Lonar Lake sediment samples.

3.11 Phylogenetic Diversity of Alkaliphilic Bacteria from Soda Lakes

The most extensively studied soda lakes are those of the East African rift valley (Grant et al. 1990; Jones and Grant 1999). Phylogenetic diversity of East African soda lake (pH 8 to >12) alkaliphiles has been investigated by isolating alkaliphilic bacteria from different soda lakes in Kenyan Tanzanian Rift Valley (Duckworth et al. 1996). Their results indicated the presence of Gram-negative bacteria related to γ^3 subdivision of Proteobacteria, Gram-positive isolates of % G + C and Low % G + C division with a majority of bacteria belonging to *Bacillus* group, others to *Arthrobacter* and an anaerobic thermophilic isolate related to *Thermotogales* from samples collected from lake Bogoria, having a temperature of 66–96 °C.

Diversity and activity of microorganisms in hyperalkaline spring waters (pH 12.9) in Maquarin, Jordan was studied (Pedersen et al. 2004) using cultivation-based methods and identified by 16S rDNA sequencing. Most of the populations were found to be related to Proteobacteria with very few to Firmicutes, Actinobacteria, Candidate division, *Deinococcus thermus*, and Archaea.

Diversity of aerobic heterotrophic populations from nonsaline alkaline environment (pH 11) from the ground water of Cabeco de Vide in south Portugal was assessed (Tiago et al. 2004) by isolating those using different media of varying pH. The selected isolates were subjected to the phenotypic characterization and 16S rDNA sequencing which has revealed the vast majority of the isolates of Gram-positive with High G+C group with some of the isolates related to Low G+C group and only five isolates with Proteobacteria. Despite the pH of the environment, the vast majority of the populations were alkalitolerant; only two strains could grow at pH 11.

Microbial diversity of Kenyan Soda Lake was assessed by molecular methods (Rees et al. 2003) by extracting genomic DNA from water and sediment samples collected from different soda lakes in the Kenyan Tanzanian Rift valley and their amplification by PCR. The diversity studies were done using denaturing gradient gel electrophoresis (DGGE) of 16S rDNA amplicons. The results of the phylogenetic analysis indicated that the sequences were related to the haloarchaea, *Bacillus/Clostridium*, *Methylobacter*, and *Cytophaga-Flavobaterium-Bacteroids* (CFB) groups and the Enterobacteriaceae/*Aeromonas/Vibrio* part of the γ subdivision of the Proteobacteria. The presence of CFB group of microorganisms was found to be an important finding due to their role in nutrient recycling.

Horikoshi (2006) has done the pioneering work on the discovery, research, and applications of alkaliphilic bacteria. The isolation, distribution, and taxonomy of alkaliphilic microorganisms as well as their cell structure and physiology are discussed for a basic understanding of these entities in this book. The molecular biology and genome sequencing of some alkaliphilic bacterial strains are also presented. Part II of the book focuses on enzymes of alkaliphiles and their applications. The enzymes include alkaline proteases, starch degrading enzymes, and many more. Some of these enzymes are currently in wide commercial as laundry detergent additives and in waste treatment. Alkaliphiles are a relatively recent field of research that will no doubt provide avenues to a wide range of further discoveries and applications for a new generation of workers in bioscience and technology.

Wani et al. (2006) reported the prokaryotic diversity associated with Lonar Lake, India using culture-independent approach. The studies were done by extracting community DNA directly from four sediment samples collected from the lake followed by PCR amplification of 16S rDNA, cloning, RFLP analysis for screening the clones and sequencing. Diversity analyses demonstrate greater bacterial diversity with a skewed representation of species of Archaea.

Kulkarni et al. (2019) have reviewed the diversity of alkaliphiles. The authors have described alkaline environments all over the world, soda lakes, their genesis, examples of soda lakes such as Kenyen Tanzanian Rift valley, Wadi Natrun lakes, Kulunda Steppe Altai, Russia, Mono Lake, California, USA, Lonar Lake, India and Sambhar lake, India. They have discussed adaptation strategies of alkaliphiles. Microbial diversity is described with respect to aerobic alkaliphilic bacteria, anaerobic alkaliphilic diversity, methanogenic archaea, haloalkaliphilic archaea, chemolithotrophs, and phototrophic bacteria. Phylogenetic diversity of alkaliphiles determined using different molecular techniques is also described. Alkaliphiles are more popularly studied for their biotechnological potential in the production of industrial enzymes mainly alkaline proteases, antimicrobial compounds, polyhydroxyalkanoate (PHA), and exopolysaccharide (EPS). Alkaliphiles have also application in the bioremediation of alkaline wastewater.

Pathak and Rathod (2019) have reviewed the distribution of alkaliphiles, their habitats like soda lakes and diversity of alkaliphiles. The review also covers applications of these microorganisms in the production of alkalozymes like protease, amylase, xylanase, carotenoids, siderophores, organic acids, antibiotics, and chalic acid derivatives.

3.12 Biotechnology of Alkaliphiles

Alkaliphilic microorganisms offer a multitude of actual or potential applications in various fields of biotechnology. Not only do many of them produce compounds of industrial interest, but they also possess useful physiological properties which can facilitate their exploitation for commercial purposes (Ulukanli and Digrak 2002). Microbial communities in natural alkaline environments have attracted attention because of the possible biotechnological use of enzymes and metabolites from such organisms.

3.12.1 Alkaline Enzymes

Alkaliphilic bacteria are of considerable interest particularly for the production of enzymes such as alkaline protease, amylase, cellulase, lipase, xylanase, catalase, pectinase, pullulanase, chitinase which have potential applications in various fields of industrial biotechnology like detergent additive, hide dehairing process, gelatin removal from X-ray films, detergent formulations, waste water treatment, pulp and paper manufacture, food and dairy industries.

Soda lakes contain dense populations of aerobic organotrophic and alkaliphilic bacteria, some of which are potential sources of alkali-stable enzymes (Gessesse and Gashe 1997). Studies on alkaline enzymes have been concentrated largely on alkaliphilic *Bacillus* sp. (Horikoshi 1999). Starch hydrolyzing bacteria were isolated from soil and sediment samples obtained from Ethiopian soda lakes in the Rift Valley area—Lake Shala (pH 9.65), Lake Abijata (9.85), Lake Arenguadi (pH 10.3) (Martins et al. 2001). Most of these isolates were found to be closely related to the *Bacillus* cluster. Hashim et al. (2004) reported obligately alkaliphilic and halotolerant *Bacillus halodurans* from hot springs around Lake Bogoria, Kenya. The organism exhibited extracellular enzyme activity at 55 °C and at pH 10. Lipase-producing microorganisms were isolated and characterized from Lake Bogoria, Kenya (Vargas et al. 2004). It was observed that members of *Bacillus* and *Halomonas* obtained in their study produced lipolytic activity under alkaline conditions. Alkali and thermostable protease from *Bacillus* sp. and *Arthrobacter* sp. from Lonar Lake was studied (Kanekar et al. 2002) and characterized (Nilegaonkar et al. 2002).

Sarethy et al. (2011) have reviewed applications of alkaliphiles in industrial biotechnology. Enzymes of alkaliphiles mainly proteases find application in industries primarily in the detergent and laundry, in cheese production, processing of meat products, for recovery of silver from photographic films, silk degumming, leather dehairing, cosmetics, and pharmaceuticals. Amylases find utility in biocatalysis of starch-based products, cellulases in facilitating usage of cellulose-based substrates as renewable source of energy. Lipases cause hydrolysis of triglycerides releasing fatty acids and glycerol, are involved in catalysis of trans-esterification reactions and enantioselective hydrolyses. Xylanases, pectinase, chitinase, and catalase are other important enzymes. Alkaliphiles have also been studied as source of antibiotics and carotenoids. The organisms have been used in Indigo blue dye production process and iron reduction. Alkaliphiles are known to produce siderophores having application in therapeutics, agriculture, and environment. Alkaliphilic cyanobacteria can become a food supplement. Thus, alkaliphiles yield a rich array of products.

Chinnathambi (2015) has reviewed work on enzymes from alkaliphilic microorganisms which thrive at pH 9.0 and above, represent a variety of taxa and which can be isolated also from normal environments. The enzymes have potential application in detergent and laundry industries. Their other products include antibiotics and carotenoids.

Borkar (2015) has reviewed diversity, physiology, and industrial applications of alkaliphilic bacteria. It was interesting to know that Gram-positive alkaliphilic bacteria were present in both alkaline and non-alkaline environments. These isolates produced enzymes such as amylase, protease lipase, and cellulase having activity at alkaline pH. Two isolates identified as *Bacillus lehensis* strain SB-D and *Bacillus halodurans* strain SB-W required sodium for their growth and their internal pH was two units less as compared to external pH. These isolates possessed high cytoplasmic buffering capacities absent in neutrophilic cultures. *B. lehensis* produced a viscous exopolysaccharide (EP) having string metal adsorption and adhesive property for glass and paper.

Syed (2016) has reviewed biotechnological potential of haloalkaliphilic *Bacillus* species particularly from solar salterns. Haloalkaliphiles require both an alkaline pH (>9.0) and high salinity (up to 33% w/v NaCl). The hypersaline alkaline brines provide a suitable environment having pH around 12 for haloalkaliphilic microorganisms namely *Natronococcus*, *Natronobacterium*, *Natrialba*, *Halorubrum*, *Natronorubrum*, and *Natronomonas*. The only other aerobic haloalkaliphilic microorganism cultivated from such saline alkaline environment is *Bacillus* sp. Haloalkaliphilic *Bacillus* sp. has potential applications in the production of enzymes, organic acids, lipopeptides (Lantibiotics), biosurfactants, exopolysaccharides, bioplastics and in biodegradation of phenol, petroleum hydrocarbons, biosorption of heavy metals, and degradation of reactive dyes.

Sharma et al. (2019) have reviewed work on microbial proteases. They have applications in leather industry, detergent, silver recovery, and other industrial sectors. The authors have highlighted their application in food industry in the preparation of fish hydrolysate, whey hydrolysate, and soy hydrolysate. These alkaline proteases are produced mostly using alkaliphilic microorganisms since the optimum pH required for production is alkaline (9–11). The alkaliphilic bacterial strains producing alkaline protease include *Bacillus firmus*, *B. licheniformis*, *B. megaterium*, *B. clausii*, *B. pumilus*, *B. horikoshi*, *B. amovivorus*, *B. cereus*, *Micrococcus* sp., *Virgibacillus pantothenticus* (MTCC6729), *Stenotrophomonas maltophilia*, and *Elizabethkingia meningoseptica*.

3.12.2 Production of Polyhydroxyalkanoates (PHAs)

Polyhydroxyalkanoates (PHAs) are a group of bacterial polyesters that have received much attention in recent times due to their biodegradable nature and properties close to plastics from petrochemical routes (Anderson and Dawes 1990). PHAs are synthesized by several bacterial species and accumulated intracellularly in large quantities due to the limitation of one or more environmental or nutritional factors such as nitrogen, phosphorus, potassium, or oxygen in the presence of excess carbon (Anderson and Dawes 1990). Practical applications of PHAs include packaging films (for food packages), bags, containers, paper coatings, biodegradable carrier for long-term dosage of drugs, medicines, insecticides, herbicides, fertilizers, disposable items such as razors, utensils, diapers, cosmetic containers, shampoo bottle, cups and use in the production of surgical pins, sutures, staples, swabs, wound dressings, bone replacement and plates and blood vessel replacement. Ponraj (2007) has described the production of PHA by alkaliphilic strain of *Bacillus cereus* isolated from Lonar Lake. Joshi et al. (2007) have reported the formation of PHA granules in cells of haloalkaliphilic strain of *Halomonas campisalis* isolated from Lonar Lake.

3.12.3 Exopolysaccharide Production from Alkaliphilic Microorganisms

Extreme environments are proving to be a valuable source of microorganisms that secret interesting new molecules. It has been hypothesized that the synthesis of exo-cellular polysaccharides in microorganisms plays a major role in protecting the cells from stress in extreme habitats (Nicholaus et al. 1999a, b). Microbial polysaccharides are water-soluble polymers and may be ionic or non-ionic in nature. The repeating units of these polysaccharides are very regular, branched or unbranched, and interconnected by glycosidic linkages. Microbial polysaccharides have found a wide range of applications in food, pharmaceutical, petroleum, and other industries (Sutherland 1990, 1998; Tombs and Harding 1998). It has crucial role in metal biosorption process (Brown and Lester 1979) and thus could be used in heavy metal bioremediation. Some alkaliphilic and halophilic exopolysaccharide (EPS) producing bacteria include alkaliphilic *Bacillus* sp. (Carsaro et al. 1999), *Halomonas maura* (Soledad et al. 2003), *Halomonas eurihalina* (Martinez-Checa et al. 2002), and haloalkaliphilic *Bacillus* sp. (Ganeshkumar et al. 2004).

There are few reports on the production of exopolysaccharides by alkaliphilic bacteria. Carsaro et al. (1999) has reported characterization and structural determination of exopolysaccharide produced by an alkaliphilic *Bacillus* isolated from Lake Natron. Haloalkaliphilic *Bacillus* isolated from soil samples of heavily polluted tidal mudflats of the Korean Yellow Sea produced an exopolysaccharide with very good flocculating activity (Ganesh Kumar et al. 2004). Moderately halophilic, *Halomonas eurihalina* strain H-28 isolated from a solar saltern at Alicante, Spain produced EPS in hydrocarbon-containing media and thus could be useful for biotreatment of saline soils and in oil recovery process (Martinez-Checa et al. 2002). Bouchotroch et al.

(2001) reported EPS producing *Halomonas maura*, isolated from soil samples from saltern at Asilah (Morocco), which is of potential industrial use as a viscosifier and emulsifier. Exopolysaccharide produced by the same organism namely, Mauran has novel composition and interesting properties for biotechnology (Arias et al. 2003). There is still hope for the development of new polysaccharides with superior properties than the existing polymers because of the diversity offered by microorganisms.

3.13 Role of Alkaliphiles in Sustainability

Many of the waste waters are alkaline wherein alkaliphilic microorganisms play an important role in their treatment. Bioremediation of alkaline waste waters has been attempted using alkaliphilic microorganisms (Kanekar et al. 1999; Kumar et al. 2008; Kulshreshtha et al. 2012; Tambekar et al. 2014a, b; Zampieri et al. 2019).

Wormald et al. (2020) have reported hydrogenotrophic methanogenesis (by hydrogen utilizing methanogenic archaea) under alkaline conditions at pH >9.0. The authors investigated a range of alkaline and neutral pH sediments for methanogenesis. This finding has application in the disposal of intermediate-level radioactive wastes by using a cement-based geological facility (GDF). GDF having both organic and metallic materials become a suitable substrate for methanogenesis. Since GDF has alkaline environment, methanogenesis under alkaline condition is desirable. *Methanobacterium* sp. dominated at pH 10.0.

3.14 Alkaliphilic Microorganisms from India

Studies on alkaliphilic microorganisms from India are mainly carried out on the unique ecosystem on the Earth, the Lonar Lake.

3.14.1 Lonar Lake

Lonar crater is situated in a village Lonar in the Buldhana district of Maharashtra State, India (Latitude 19°58′, Longitude 76°36′) in formerly volcanic Deccan-trap geological region. It is an almost circular depression (Fig. 3.1) with its longest and shortest diameters being 1875 m and 1787 m, respectively, with a raised rim of about 30 m and a depth of 135 m (La Touche and Christie 1912).

Based on geological studies, it is postulated that the lake was originated as a meteorite impact crater about 50,000–60,000 years ago (Fredriksson et al. 1973). Through research, it is believed that the meteor was approximately 60 m in diameter and weighed million tones in form of stone, which struck the Earth at a velocity creating tremendous energy and heat and a cloud of molten ash and dust. It resulted in the destruction of neighboring flora and fauna. The destruction was an estimated 6 megatons bomb explosion. The meteor here is believed to have been buried 600 m

Fig. 3.1 Alkaline saline Lonar Lake, village Lonar, District Buldhana, Maharashtra State, India (Joshi 2007)

below the crater level. Till a couple of decades ago, the crater site was looked as a result of volcanic activity until investigations by Indian and American geologists were carried out. An important conclusion was that the geologist discovered mineral maskelynite, which is formed under tremendous impact and heat, which is not found in volcanic impact. Fredriksson et al. (1973) have conducted geologic mapping, drilling, trenching, and geochemical and geophysical studies of Lonar Lake. Three diamond drill holes were drilled from a barge in the lake. Discovery of shock-metamorphosed material establishes the impact origin of Lonar crater. Their results offered definitive evidence that Lonar crater is indeed an impact crater. Lonar Lake is a unique basaltic rock meteorite impact crater, ranking third in the world based on the area (1800–2000 m), first being the Ashanti Lake in Ghana, Africa (10,000 m), second New Quebec, Labrador in Canada (3500 m), and fourth being Canyan diable, Arizona, North America (1300 m).

Lonar Lake, also known as Lonar crater is a notified National Geo-heritage monument located at Lonar in Buldhana District of Maharashtra, India (now designated as Ramsar Wetland since 22 July 2020). It is a saline soda lake created by a meteorite collision impact during the Pleistocene Epoch. It is one of the four known hyper velocity, impact craters in basaltic rock anywhere on Earth. Lonar Lake has a mean diameter of 1.2 km (3900 ft.) and is about 137 m (449 ft.) below the crater rim. The meteor crater rim is about 1.8 km (5900 ft.) in diameter. Lonar crater sits inside the Deccan Plateau—a massive plain of volcanic basalt rock created by eruptions some 65 million years ago. The formation of the Lonar crater is widely accepted as a result of a meteorite impact that occurred between 35,000 and 50,000 years ago. Lonar Lake lies within the only known extraterrestrial impact crater found within the great Deccan Traps, a huge basaltic formation in India (Pittarello et al. 2010). The crater has an oval shape. The meteorite impact came from the east, at an angle of 35–40°. There are various estimates of the age of the crater. Earlier thermoluminescence analyses gave a result of 52,000 years while recent argon-argon dating suggests that the crater is much older (Jourdan et al. 2011). It could be 570,000 ± 47,000 years old. The greater age is in line with the degree of erosion of the crater rim.

The saline deposits obtained from the lake are of exceptional nature and mainly composed of the carbonates of soda. Lonar salts are derived from an unknown source in the bed of the lake. Water enters the lake through rain, ground water seepage and springs situated in the cliffs at the edge of the lake. Uniqueness of the lake water is its salinity and high alkalinity. The water is alkaline with pH of 10–10.5; this high alkalinity is due to the high concentration of sodium carbonate (Jhingram and Rao 1954) and had been exploited for manufacture of soda since fairly ancient days. The lake does not receive any industrial seepage. The lake has no outlet at all, and has in the main only one feeder stream, which drains into it at its northern-eastern end. Water is continually flowing into the lake and there is no loss except for by evaporation. The main feeder stream could not however supply this amount of alkali. The salts collected from this lake vary in their nature and composition depending upon the period and condition of crystallization. With the process of crystallization, sodium chloride or common salt is formed along with the carbonates of soda. Commercial exploitation of the salts from the lake is recorded from 1842 until 1903. Gaylussite is the mineral reported from drill core in Lonar Lake (Anoop et al. 2013). Gaylussite is a carbonate mineral, a hydrated sodium calcium carbonate ($Na_2Ca\ (CO_3)_2{\cdot}5\ H_2O$). The alkalinity of Lonar Lake is thus attributed to the presence of carbonates in the lake water.

Jhingram and Rao (1954) investigated the exact nature of salinity and its source in the Lonar Lake. They reported that the Lonar Lake water is characterized by a high concentration of chloride, carbonate, and sodium radicals. Besides, it contains minor quantities of SiO^2, SO^4, Ca, K, Mg, and B. From a discussion of the data, it is suggested that the alkalinity of the water is the result of a number of reactions, as chief of them being the conversion of sulfate to carbonate through intermediate sulfide formation and the concentration of the lake water.

Nandy and Deo (1961) reported that the alkalinity of the lake water is due to the meteoric waters feeding the lake. These influents while passing through the volcanic rocks which are the source of the sodium salt dissolve the soluble salt and bring them into the lake. Evaporation and interactions of the salts in the lake water aided by the inorganic compounds added by the rain and the organisms growing in the lake have resulted in the concentration of the salts in the lake.

Tambekar et al. (2010) have compared chemical characteristics particularly pH, chlorides, salinity, alkalinity of the Lonar Lake water from 2008 to that in the past. It is observed that the alkalinity and salinity of the Lonar Lake is decreasing day by day and the authors express that the uniqueness of this national treasure should be preserved.

Review of literature reveals that the alkalinity of the lake water in terms of carbonate was 15, 22.26, and 7.52 % while salinity was 40.78, 31.52, and 30.87 % in 1910, 1958, and 1960 respectively (Nandy and Deo 1961). The salinity of lake is now lowered down to 7.9% (Malu et al. 2000). It is clear from the literature survey that most of the analyses of the Lake have been limnological rather than microbiological. Eutrophication and presence of blue green algae in Lonar Lake have been described (Badve et al. 1993).

Surve et al. (2021) have reviewed the chemical and ecological status of Lonar lake. Alkalinity of the lake is reported to decrease nearly by 19%, salinity by 7% and pH dropped down. This decrease is mainly due to the process of remediation of salt at bottom of the lake. Consequently, the microbial flora is affected. Conservation of Lonar Lake is therefore necessary.

It is no wonder having such a unique extreme environment, a number of researchers are fascinated to explore this amazing ecosystem for microbial diversity.

Alkaliphilic microorganisms reported from Lonar Lake, India are enlisted in Table 3.1.

As seen from Table 3.1, extensive studies have been carried out on diversity and biotechnology of alkaliphilic microorganisms from Lonar Lake.

Wani et al. (2006) investigated microbial diversity associated with Lonar Lake using culture-independent approach. Community DNA was extracted from four sediment samples and 16S rDNA amplified by PCR using primers for both bacteria and archaea. By employing the restriction fragment length polymorphism (RFLP) technique, a total of 44 unique phylotypes were obtained within the domain Bacteria. 34% of the clones were classified as firmicutes, 29.5% as proteobacteria, 6.8% actinobacteria, 4.5% *Deinococcus thermus*, 13.3% cytophages-flavobacterium-bacteroidetes, 6.8% planctomycetes, 4.5% cyanobacteria, and 2.27% spirochetes. In case of archaea, 13 distinct phylotypes were obtained among which 5 phylotypes were associated with Crenarchaeota and 8 with Euryarchaeota most of which were related to methanogens. This molecular study of Lonar Lake was presumably the first report revealing the presence of highly diverse bacterial population and comparatively less diverse archaeal population.

Surakasi et al. (2007) carried out a phylogenetic analysis of methanogenic enrichment cultures from Lonar Lake. The diversity of methanogenic archaea was investigated using 16S rDNA molecular phylogenetic approach. Archaeal 16S rRNA clone libraries were generated from enrichment cultures and 13 RFLP groups were obtained. Representative sequence analysis of each RFLP group indicated that the majority of the 16S rRNA gene sequences were phylogenetically affiliated to uncultured archaea. Three RFLP groups were related to *Methanoculleus* sp., while two related to *Methanocalculus* sp. The studies indicated Lonar Lake harbors unexplored methanogens.

Joshi et al. (2008) investigated the bacterial diversity of Lonar Lake using different nutrient media with pH adjusted to 10 since pH of Lonar Lake was found to be 10.5. The total number of microorganisms present in water was 10^2 to 10^4 cfu (colony forming units)/ml and 10^2 to 10^6 cfu/g sediment sample. Total 196 strains isolated were screened for pH and salt tolerance and 64 out of them were further

Table 3.1 Reports of alkaliphilic microorganisms from Lonar Lake, Buldhana district, Maharashtra state, India

Name of the microorganism/ group	Biotechnological potential/ diversity	Reference
Bacillus cereus	Bioremediation of phenol containing alkaline industrial waste water	Kanekar et al. 1999
Bacillus laterosporus		
Arthrobacter sp.		
Micrococcus agilis		
Pseudomonas putida		
Pseudomonas pseudoalkaligenes		
Citrobacter freundii		
Arthrobacter ramosus	Protease activity	Kanekar et al. 2002, Nilegaonkar et al. 2002
Bacillus alcalophilus		
Methanosarcina sp.	Methane production	Thakker and Ranade 2002
Alkaliphilic bacteria	Chitin degradation	Nawani and Kapadnis 2003
44 unique phylotypes obtained 34% firmicutes, 29.5% proteobacteria, 6.8% actinobacteria, 4.5% deinococcus–thermus, 13.3% cytophages–flavobacterium–Bacteroidetes, 6.8% planctomycetes, 4.5 % cyanobacteria, 2.27% spirochetes, 13 distinct phylotypes of archaea–5 Crenarchaeota, 8 Euryarchaeota (mostly related to methanogens)		Wani et al. 2006
Alkaliphilic bacteria	Chitinase	Bansode and Bejakal 2006
Methanoculleus sp. IIE1	Methane production	Surakasi et al. 2007
Methanocalculus sp. 01F97C		
Halomonas campisalis	Production of polyhydroxyalkanoate (PHA)	Joshi et al. 2007; Kulkarni et al. 2010, 2011, 2014, 2015; Kshirsagar et al. 2012, 2013; Kanekar et al. 2014, 2018
Magnetotactic bacteria	Diversity	Chavdar and Bajekal 2009
Alkaliphilic actinomycetes	Nitrogen-fixing bacteria	Raut and Bejakal 2009
Actinomycetes	Chitinase production	Loni and Bajekal 2009
Actinomycetes	Nitrogen fixation	Raut and Bejakal 2009
Non-symbiotic nitrogen-fixing microaerophilis	Nitrogen-fixing diversity	Raut and Bejakal 2010
Arthrospira fusiformis		Surakasi et al. 2010
Planococcus rifientensis		
Methylobacterium variabile		
Bordetella hinzii		
Tindella megadiensis		

(continued)

Table 3.1 (continued)

Name of the microorganism/ group	Biotechnological potential/ diversity	Reference
Planococcus rifietensis		
Beijerinckiaceae family members		
Methylomicrobium buryaticum	Methylotrophic	Kumar et al. 2010a
Alteromonadales bacterium		
Methylophaga marina		
Nitritalea haloalkaliphilia gen. nov. sp. nov.		
Indibacter alkaliphilus gen. nov. sp. nov.		Kumar et al. 2010b
Vagococcus carniphilus	Production of exopolysaccharide	Joshi and Kanekar 2011
Bacillus fusiformis	Amylase having optimum enzyme activity at pH 10	Ahirrao et al. 2011
Bacillus firmus SBPL-05	Alkaline chitinase	Loni and Bajekal 2011
Diazotrophic actinobacteria	Agar degrading	Raut and Bejakal 2011
Pseudomonas stutzeri	Phenol degradation bacterium	Tambekar et al. 2012
Synechocystis sp.	Phycocyanin production	Deshmukh and Puranik 2012a, b
Halomonas campisalis	Antimicrobial activity for human pathogenic bacteria and phytopathogenic fungi	Borgave et al. 2012
Planococcus maririmus		
Paenibacillus sp. L55		
Bacillus badius Di	Biodegradation of anthracene	Ahmed et al. 2012
Bacillus sp.	Antibacterial substance	Tambekar and Dhundale 2012a, b
Synechocystis sp.	Phycocyanin production	Deshmukh and Puranik 2012a, b
Cecembiua lonarensis gen. nov. sp. nov.		Kumar et al 2008
Methylophaga lonarensis	Methylotroph	Antony et al. 2012a
Methanolobus sp.	Methanogens	Antony et al. 2012b
Stenotrophomonas sp. DL18	ATP synthase	Jadhav and Gawai 2013
Paenibacillus sp. L55	Production of cyclodextrin glycosyl transferase (CGTase) having application in the production of cyclodextrins	Thombre et al. 2013
Natrinema sp.	Xylanase and cellulase	Patil and Bajekal 2014
Proteus mirabilis	Bioremediation of chromium (VI)	Tambekar et al. 2014a, b
Paenibacillus pasadenesis	Alkaline chitinase	Loni et al. 2014
Pseudomonas aeruginosa DHT2 *Enterobacter cloacae* DHT8	Bioremediation of methanol	Tambekar and Rajgire 2015

(continued)

Table 3.1 (continued)

Name of the microorganism/ group	Biotechnological potential/ diversity	Reference
Nitrincola alkalisediminis sp. nov.	Diversity	Joshi et al. 2016
Bacterial community structural composition 28 bacterial phyla, Proteobacteria 30%, Actinobacteria 24%, Firmicutes 11%, cyanobacteria 5%	Microbial diversity analysis	Dhiraj et al. 2016
Halophilic archaea	Siderophore production	Patil et al. 2016
Phormidium sp.	Alkaliphilism and carbonate transport	Deshmukh and Puranik 2017
Rhizobium sp.	Biotreatment of paper mill effluent	Raut et al. 2018
Nitrincola tapani sp. nov.	A novel bacterium	Joshi et al. 2020
Spirulina platensis		Deshmukh and Puranik 2020
Synechocystis aquatilis		
Oscillatoria minimus		
Oscillatoria amphibian		
Phormidium laminosum		
Phormidium tenue		
Phormidium fragile		
Cyanobacteria		
Alkaliphilic bacteria	Lipase production	Dhundale 2021

subjected to phenotypic, biochemical characterization, and 16 S r RNA gene sequencing. Phylogenetic analysis (Figs. 3.2, 3.3, and 3.4) (Joshi 2007) carried out for 31 isolates selected based on their enzyme profile indicated dominance of strains belonging to Phylum Firmicutes and genera *Bacillus*, *Paenibacillus*, *Alkalibacillus*, *Exiguobacterium*, *Planococcus*, *Enterococcus*, and *Vagococcus*. Seven strains constituted Proteobacteria with genera *Halomonas*, *Stenotrophomonas*, *Providencia*, *Alkaligenes*, and *Paracoccus*. Among Actinobacteria were *Cellulosimicrobium*, *Dietzia*, *Arthrobacter*, and *Micrococcus*. Most of the strains were alkalitolerant and only two strains were obligately alkaliphilic. Many of the isolates produced biotechnologically important enzymes at alkaline pH, one isolate produced polyhydroxyalkanoate (PHA) and one produced exopolysaccharide. Joshi et al. (2007) reported moderately halophilic, alkalitolerant *Halomonas campisalis* from Lonar Lake. Two out of seven strains produced polyhydroxyalkanoate.

Kanekar et al. (2008) have extensively studied alkaliphilic and halophilic bacteria from Lonar Lake. Culture-dependent phenotypic characterization and 16S rDNA-based phylogenetic analysis were applied. 196 bacterial strains were obtained using six different enrichment media. Isolation and identification of alkaliphilic and halophilic bacteria namely *Halomonas campisalis*, *Alkalibacillus haloalkaliphilus*,

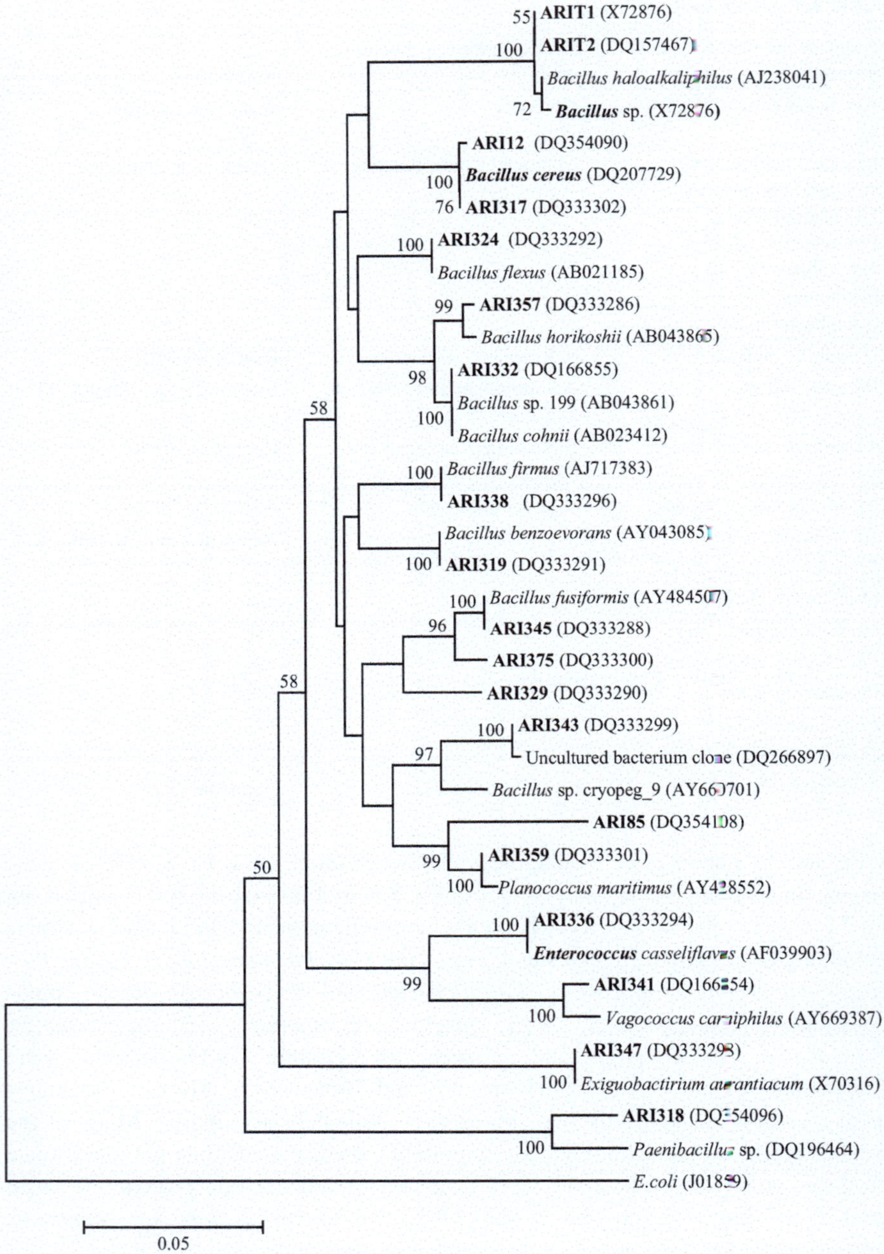

Fig. 3.2 Phylogenetic tree showing the relationship among bacterial 16S rDNA gene sequences of Low G+C containing bacteria from Lonar Lake with reference sequences obtained through BLAST analysis. Numbers in the bold are sequences obtained in the present study. Bar, inferred nucleotide substitutions per nucleotides. *E. coli* was used as an outgroup taxa (Joshi 2007)

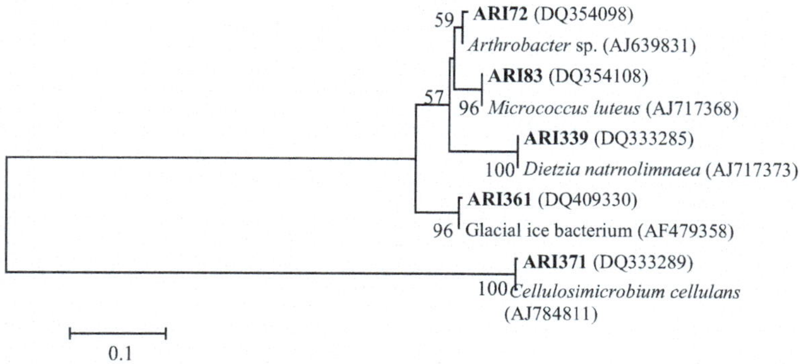

Fig. 3.3 Phylogenetic tree showing the relationship among bacterial 16S rDNA gene sequences of High G+C containing bacteria from Lonar Lake with reference sequences obtained through BLAST analysis. Numbers in the bold are sequences obtained in the present study. Bar, inferred nucleotide substitutions per nucleotides (Joshi 2007)

Dietzia natronolimnaea, *Vagococcus carniphilus*, *Exiguobacterium aurantiacum*, *Bacillus horikoshi*, *Cellulosimicrobium cellulans*, *Thermoactinomyces thalpophilus*, *Paenibacillus* sp., *Marinobacter excellens*, *Marinobacter alkaliphilus*, *Roseinatronobacter monicus*, *Rhodobacteriaceae bacterium*, *Alkalimonas delamerensis*, and *Rhodobaca bogoriensis* appear to be the first report from Lonar lake. Production of polyhydroxyalkanoate (PHA) by *Halomonas campisalis* and exopolysaccharide by *Vagococcus carniphilus* also become the first report.

Kumar et al. (2010a, b) have described a novel Gram-stain negative, rod-shaped, non-motile bacterium, strain LW1T isolated from a water sample collected at a depth of 3.5 m, from Lonar Lake, Buldhana District, Maharashtra, India. Based on phenotypic, chemotaxonomic characters, 16 S r RNA gene sequencing, phylogenetic analysis, and DNA-DNA hybridization studies, the strain LW1T was assigned a novel species as *Indibacter alkaliphilus* gen. nov. sp. nov.

Surakasi et al. (2010) investigated the phylogenetic diversity of bacterial communities in microbial mats from Lonar Lake in two different seasons, using 16S rRNA gene library analysis, Arthrospira (Cyanobacteria)-related clones dominated libraries of both the seasons. The operational taxonomic units (OTUs) related to Fusibacter and *Tindallia magadiensis* were found in post-monsoon but not in pre-monsoon library. The OTUs related to *Planococcus rifietensis*, *Bordetella hinzii*, and *Methylobacterium variable* found in the pre-monsoon were not found in post-monsoon. Methanotrophs in the surface mat were related to *Methylomicrobium buryaticum*.

Raut and Bejakal (2010) isolated nitrogen-fixing bacteria from soil samples collected from the perimeter of Lonar Lake using Ashby's nitrogen-free mannitol medium by the standard technique. Out of 32 isolates obtained, four isolates were able to grow on inorganic medium containing Martian soil simulant components and free of nitrogen and carbon. One isolate was Gram-negative rod while three were

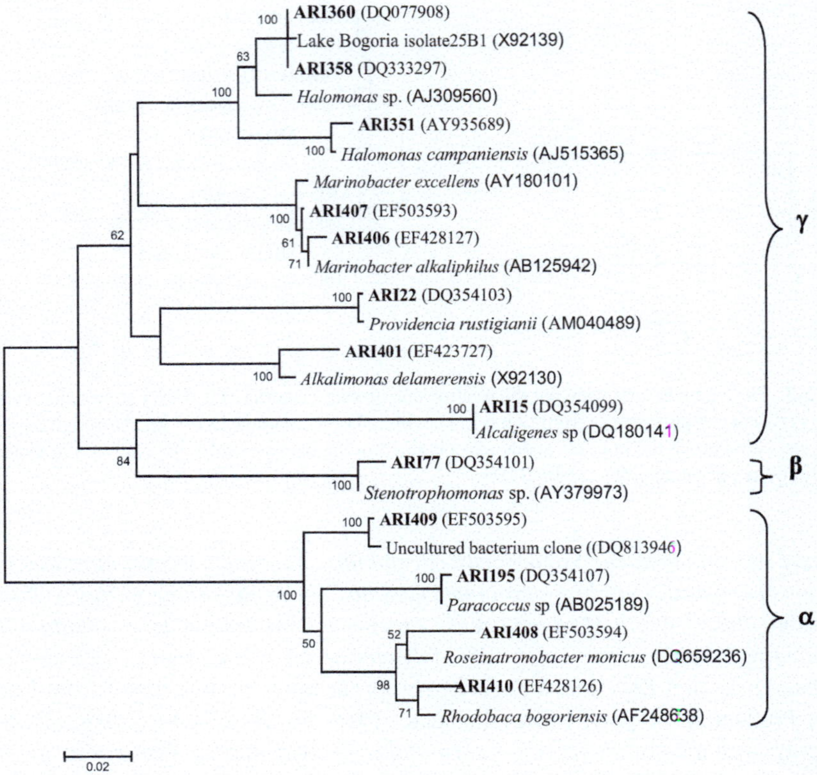

Fig. 3.4 Phylogenetic tree showing the relationship among bacterial 16S rDNA gene sequences of Proteobacteria from Lonar Lake with reference sequences obtained through BLAST analysis. The numbers on the tree indicate the percentages of bootstrap sampling derived from 1000 replications. Numbers in bold are sequences obtained in the present study. Bar, inferred nucleotide substitutions per nucleotides (Joshi 2007)

actinomycetes. Nitrogen fixed in the medium ranged between 0.029% and 0.215% (0.009% being in the uninoculated control medium).

Ahirrao et al. (2011) have isolated haloalkaliphilic bacteria from Lonar Lake water samples having pH 10.5. One of the bacterial strains, identified as *Bacillus fusiformis*, exhibited optimum growth at pH 12.5 and 18% salt concentration. The organism produced enzyme amylase having optimum enzyme activity at pH 10.

Shivaji et al. (2012) have reported draft genome sequence of *Cecembia lonarensis* strain LW9, isolated from Lonar Lake water sample collected at 4.5 m depth.

Antony et al. (2012b) have studied molecular diversity of methanogens and methylotrophic archaea in Lonar Lake sediments.

Joshi et al. (2016) isolated two Gram-stain negative (strains MEB087T and MEB142) aerobic alkaliphilic bacteria from sediment and water samples, respectively, collected from the alkaline Lonar Lake in Maharashtra, India. The 16S rRNA gene sequences of both strains showed close relationship with the genus *Nitrincola*. The authors have described phenotypic, biochemical, and chemotaxonomic characteristics of these two strains. Based on phylogenetic analysis and chemotaxonomic characteristics, strains MEB087T and MEB142 represent a novel species in the genus *Nitrincola*, for which the name *Nitrincola alkalisediminis* sp. nov. is proposed. The type strain is MEB087T (=KCTC42948T) with MEB142 (=KCTC42949) as an additional strain.

Paul et al. (2016) investigated microbial community structure associated with Lonar Lake sediment and water samples using high-throughput sequencing. Microbial diversity analysis revealed the existence of diverse, consistent communities. The sequencing survey indicated predominance of proteobacteria (30%), Actinobacteria (24%), Firmicutes (11%), and cyanobacteria (5%). Microbial communities of Lonar lake formed a distinct cluster, suggesting a different microbial community composition and structure. The authors report for the first time, the difference in composition of indigenous microbial communities between the sediment and water samples of Lonar Lake.

Deshmukh and Puranik (2017) have investigated the mechanism developed by alkaliphilic cyanobacteria in the alkaline lake of Lonar. The cyanobacterium, *Phormidium* sp. isolated from Lonar Lake was able to tolerate 1M Na_2CO_3. The authors isolated cyanobacterial sheath under carbonate stress and analyzed by FTIR. The studies revealed that the cyanobacterium possessed two carbonate transport systems operational at pH 9 and 10. This suggests the adaptation of the cyanobacteria to seasonal variation in the alkaline—saline lake Lonar. Thus, an attempt was made to understand the alkaliphilism or adaptation of cyanobacteria in the alkaline Lonar Lake.

Joshi et al. (2020) reported a novel Gram-negative bacterial strain MEB193T from Lonar Lake sediment sample. The organism grew optimally at pH 9.0 and 1% (w/v) NaCl. It was found to belong to genus *Nitrincola*. Based on draft genome, the isolate was found to be a novel species of the genus *Nitrincola*. The name *Nitrincola tapanii* sp. nov. is proposed for the novel strain isolated from Lonar Lake.

Deshmukh and Puranik (2020) have extensively studied the diversity of alkaliphilic cyanobacteria from Lonar Lake. Based on morphological characteristics, seven species belonging to four genera and two orders (listed in Table 3.1) were identified. A significant correlation was observed between chlorophyll a content and cyanobacterial density.

3.15 Biotechnology

Alkaliphilic microorganisms are studied all over the world more for their biotechnological potential. Many of the industrial processes are carried out under harsh alkaline conditions wherein alkaline enzymes catalyze the reactions. Many alkaliphilic microorganisms have been explored for their alkaliphilic enzymes.

3.15.1 Enzymes

Chaphalakar and Dey (1994) have worked on alkaline metalloprotease from the alkaline strain of *Streptomyces* sp. isolated from Lonar Lake silt sample.

Kanekar et al. (2002) explored alkaliphilic bacteria isolated from Lonar Lake for the production of enzyme protease. Among the 54 isolates screened, *Arthrobacter ramosus* and *Bacillus alcalophilus* exhibited high protease activity using soyacake as a sole source of carbon and nitrogen. Production of the enzyme was optimized for growth conditions. The enzyme was thermostable (up to 65 °C), stable at pH 12 and active in presence of commercial detergent. The enzyme removed blood stain from cotton fabrics indicating its potential use in detergent formulation. Nilegaonkar et al. (2002) studied extracellular protease from the alkaliphilic strain of *Arthrobacter ramosus* isolated from Lonar Lake. The enzyme production was optimized and the enzyme was purified. Characterization of the enzyme revealed that it was thermostable (up to 80 °C), alkalistable (pH 12), and detergent stable.

Bansode and Bejakal (2006) studied alkaline chitinase from microbial strains isolated from the alkaline soil of Lonar Lake. The enzyme from *Streptomyces* sp. and *Nocardia* sp. exhibited stability up to pH 10 with optimal at pH 9.0. Chitinase from *Bacillus* sp. showed a broad range of pH tolerance from 7 to 11 with maximum activity at pH 9.

Loni and Bajekal (2009) isolated chitinase-producing actinomycetes from littoral soil samples collected from Lonar Lake using chitin agar with pH 10. Six isolates exhibited chitinase enzyme activity in the range of 0.30–1.2 U/ml.

Loni and Bajekal (2011) have studied the alkaline chitinase enzyme from *Bacillus firmus* SBPL-05 isolated from Lonar Lake. The enzyme exhibited maximum activity at pH 10.

Thombre et al. (2013) have reported the production of cyclodextrin glycosyl transferase (CGTase) using alkaliphilic strain of *Paenibacillus* sp. L55 isolated from Lonar Lake. The enzyme CGTase converts starch into cyclodextrins (CDs) which are closed-ring structure having applications in food, pharmaceutical, and cosmetic industries. The production of CGTase by *Paenibacillus* sp. L55 was optimized with respect to three medium components namely starch, peptone and yeast extract and initial pH of the medium by Taguchi Design of Experiments. Starch was found to contribute the highest to the production of the enzyme.

3.15.2 Antimicrobial Compounds

Alkaliphilic microorganisms from Lonar Lake have been explored from antimicrobial activity.

Kharat et al. (2009) reported *Streptomyces* sp. having antimicrobial and cytotoxic activity from Lonar Lake. The organism exhibited inhibitory effect against *Bacillus subtilis*, *Staphyllococcus aureus*, *Escherichia coli*, *Proteus vulgaris*, *Pseudomonas aeruginosa*, *Salmonella typhi*, and *Enterococcus faecalis* (all strains from National Collection of Industrial Microorganisms, NCIM, National Chemical Laboratory,

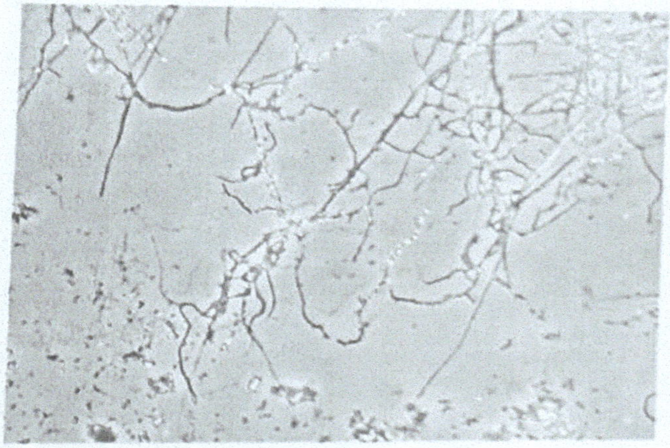

Fig. 3.5 Alkalithermophilic *Laceyella sacchari* isolated from Lonar Lake (Borgave et al. 2012)

Pune). The metabolites of the culture caused cell death of human lung carcinoma cells—A549 cell lines.

Borgave et al. (2012) screened alkaliphilic haloalkaliphilic bacteria and alkalithermophilic actinobacteria (Fig. 3.5) isolated from Lonar Lake for antimicrobial activity. The alkaliphilic bacteria exhibited inhibitory effect against clinical isolates of pathogenic bacteria and phytopathogenic fungi. Among the isolates, *Halomonas campisalis* was the most promising organism.

Tambekar et al. (2014a, b) isolated *Bacillus* species from Lonar Lake using Horikoshi medium, for the purpose of production of antimicrobial substances. Six isolates of *Bacillus* species namely *B. fusiformis* (3 isolates), *B. circulance*, *B. cereus*, and *B. krulwichiae* exhibited inhibitory activity against clinical isolates of pathogenic bacteria secondary metabolites of the isolates and were detected by gas chromatography–mass spectroscopy (GC-MS).

Shinde et al. (2017) have investigated the antimicrobial potential of phospholipid compound produced by halophilic and alkaliphilic *Bacillus subtilis* isolated from Lonar Lake.

3.15.3 Biodegradable Plastic, Polyhydroxyalkanoate (PHA)

Environmental pollution caused by discarded, non-degradable synthetic plastics is a serious problem. Researchers all over the globe are in search of biodegradable plastics.

Joshi (2007) and Joshi et al. (2007) reported the formation of PHA granules in moderately haloalkalitolerant *Halomonas campisalis* (Figs. 3.6 and 3.7) isolated from Lonar Lake. Extensive work has been carried out by Kulkarni (2010) on the

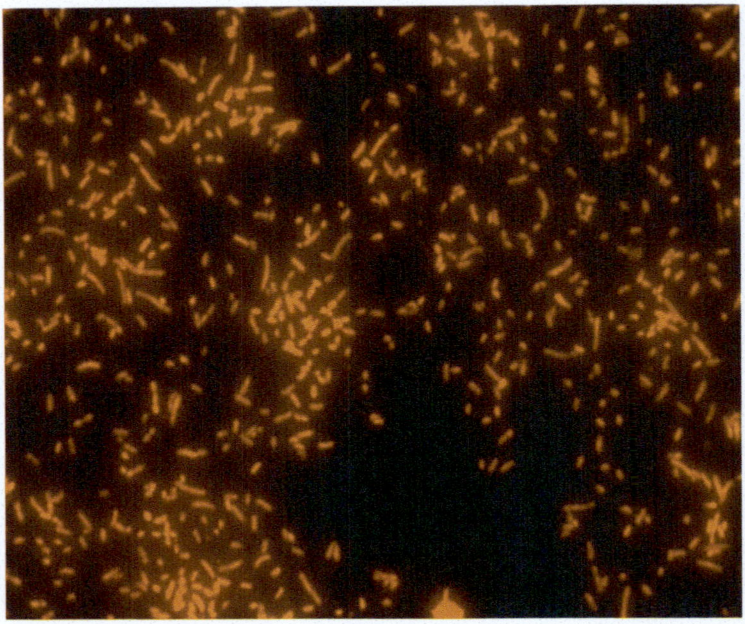

Fig. 3.6 Nile blue sulfate staining of cells of *Halomonas campisalis* showing orange fluorescence under UV indicating formation of PHA granules, observed under Nikon FX 35 Japan, microscope (Joshi 2007)

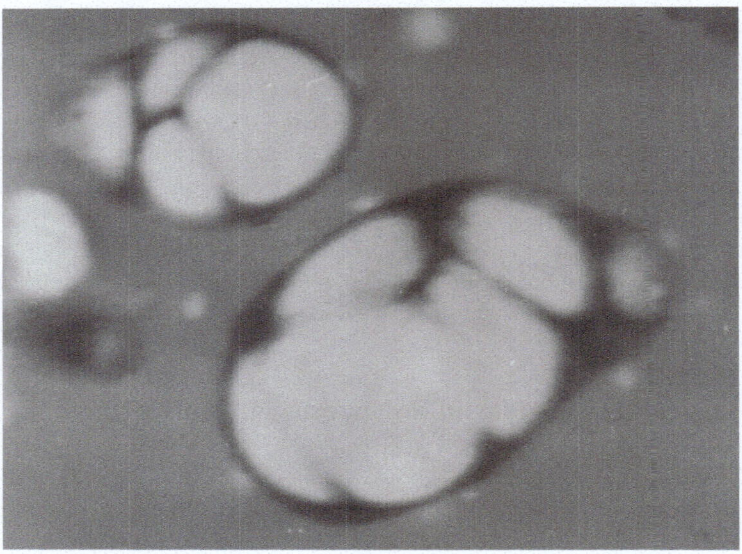

Fig. 3.7 Transmission electron micrograph (TEM) of cells of *Halomonas campisalis* showing granules (in white) of PHA (Joshi 2007)

production of biodegradable plastic, polyhydroxyalkanoate (PHA) using culture of moderately haloalkalitolerant *Halomonas campisalis*, MCM B-1027 isolated from Lonar Lake. Kulkarni et al. (2010) have described the production of PHA from *Halomonas campisalis* in the range of 45–81% on dry cell weight basis. The polymer was solvent extracted from cell mass of *Halomonas campisalis*, casted into thin films of 50–100 μm and polymer bags were prepared for packaging of food and grocery items (Fig. 3.8, Kulkarni 2010). The polymer was found to be a copolymer of poly(hydroxybutyrate-co-hydroxyvalerate) (PHB-Co-PHV) having molecular weight of 1.3×10^6 and formed using a simple carbon source as maltose. Kulkarni et al. (2011) further characterized the copolymer for melting temperature, X-ray diffraction, spherulitic morphology, microtensile strength, dynamic mechanical thermal analysis (DMTA), dielectric spectroscopic measurement, and biodegradable nature by soil burial method. The copolymer was found to have mechanical

Fig. 3.8 Polymer bags prepared using PHA films produced by *H. campisalis* and filled with food commodity items like wheat semolina, puffed rice, vegetable oil, cowpea grains, etc. (Kulkarni 2010)

strength of 18.8 MPa, melting temperature of 166.5 °C, composition of monomer units HB and HV as 96 and 4 mol %. The polymer was biodegradable in nature indicating its possible application as packaging material. Scale up production of the polymer by *Halomonas campisalis* was done at different aeration, agitation and controlled dissolved oxygen levels in fermenter (Kshirsagar et al. 2012). Maximum production of the polymer was at 1 vvm aeration and 100 rpm agitation. Controlled dissolved oxygen in the range of 1–5% during PHA accumulation phase enhanced PHA content. Production of PHA was scaled up to 120 L fermenter level. Kshirsagar et al. (2013) studied the recovery of PHA produced by *Halomonas campisalis* by several methods. The highest rate of PHA recovery obtained was 63 mg/h with solvent extraction using rotary film evaporator. The extraction process was further optimized for time, rotation and temperature. Maximum recovery of PHA was obtained at 25 rpm, 60 °C temperature within 3 h. Kulkarni et al. (2014) studied the effect of nutritional and process parameters on the production of PHA by *Halomonas campisalis* using Taguchi design of experiments. Statistical analysis of the experimental results indicated that the incubation period was the most influencing factor contributing 28.8% to PHA production while the substrate (maltose) concentration was the least significant (1.8%) in PHA production. Since the cost of production of PHA is high as compared to that of synthetic plastic, attempts were made to use cheap substrates. Kanekar et al. (2014) have reviewed environmental friendly microbial polymers, polyhydroxyalkanoates (PHAs) for packaging and biomedical applications. Kulkarni et al. (2015) studied the production of PHA by *Halomonas campisalis* using agrowastes such as bagasse, orange peel, and banana peel. Aqueous extracts of these agrowastes provided sugar for the production of PHA.

Kanekar et al. (2018) have reviewed material properties, biodegradability, and applications of polyhydroxyalkanoates (PHAs) in food packaging (Fig. 3.8, Kulkarni et al. 2010) agricultural mulch films and pharmaceutical industry in encapsulation of tablets and capsules for slow drug delivery.

3.15.4 Exopolysaccharide

Exopolysaccharides are important biomolecules having application in food industry, oil recovery, etc. Alkaliphilic strain of *Vagococcus carniphilus* isolated from Lonar lake was explored for the production of exopolysaccharide (EPS) by Joshi and Kanekar (2011). The organism produced 1.1 g/L EPS under optimized conditions. The gross chemical composition of the EPS produced indicated the presence of 20% protein and 75% neutral sugars while monosaccharide composition revealed the presence of galactose and mannose.

3.15.5 Other Products

Deshmukh and Puranik (2012a, b) have studied phycocyanin production using the culture of a cyanobacterium *Synechocystis* isolated from Lonar Lake. Optimization of eight media components was carried out using Plackett-Burman design of 12 experimental trials. The authors found that $CaCl_2$, $2H_2O$ and Na_2CO_3 were the most influencing media components at 95% significance. The optimum concentration of these two components were found by following Box Wilson Central Composite Design (CCD) with four star points and five replicates at the center points for each of two factors. The maximum yield of phycocyanin (12% dry wt) was obtained at 0.058 g/L of $CaCl_2$, $2H_2O$ and 0.115 g/L of Na_2CO_3.

3.16 Alkaliphilic Microorganisms Reported from Alkaline Environments in India Other Than Lonar Lake

Alkaline environments from India, other than Lonar Lake, also have been explored for isolation of alkaliphilic microorganisms and studied for their diversity and biotechnological potential (Table 3.2).

As listed in Table 3.2, many researchers from India have investigated alkaliphilic microorganisms from different parts of the country.

Kapoor et al. (2001) have reported application of an alkaline and thermostable polygalacturonase from *Bacillus* sp. MG-CP-2 in degumming of ramie (*Boehmeria nivea*) and sunnhemp (*Crotolaria juncea*) bast fibers.

Desai et al. (2004) isolated obligate alkaliphilic bacteria from various estuarine ecosystems of Goa. Most of the strains out of total 28 isolated were capable of aerobic growth up to pH 12 (optimum pH 10.5). The isolates exhibited high buffering capacity confirming their alkaliphilic nature. Significantly high difference was found in the buffering capacities of the alkaliphiles and neutrophiles indicating their ability to maintain their internal pH. One obligate alkaliphilic isolate A52 showed cytoplasmic buffering capacity of 8500 nanomoles OH^- ions/pH unit/mg protein. The isolates produced enzymes amylase, protease, and lipase and showed activity under alkaline conditions.

Kumar et al. (2004) have reported the compatibility of alkaline xylanase from an alkaliphilic *Bacillus* NCL (87-6-10) with commercial detergents and proteases.

Jagtap and Rao (2005) have described the purification and properties of a low molecular weight 1, 4-beta-D-glucan glucohydrolase from an alkalothermophilic *Thermomonospora* sp.

Dodia et al. (2006) have isolated halophilic alkaliphilic bacteria from saline habitat of coastal Gujarat, Western India using nutrient media containing 20% NaCl and pH 10.0. The isolates were screened for production of protease. The alkaline protease enzyme was salt tolerant.

Savergave et al. (2008) have described the production and single-step purification of cyclodextrin glycosyltransferase from alkalophilic *Bacillus firmus*.

Table 3.2 Reports of alkaliphilic microorganisms from other sites in India

Name/group of microorganisms	Source/place	Biotechnological potential	References
Corynebacterium sp.	Estuarine mangrove regions of Goa	Enzymes amylase, protease and lipase	Desai et al. 2004
Bacillus alcalophilus			
Bacillus sp.			
Actinomyces sp.			
Micrococcus lylae			
Bacillus schlegelii			
Bacillus stearothermophillus			
Bacillus brevis			
Flavobacterium sp.			
Bacillus sp.	Coastal Gujarat	Alkaline protease	Patel et al. 2005
Streptomyces sannanensis strain RJT-1 (salt tolerant and alkaliphilic)	Alkaline soil of Saurashtra University campus	Potent antibiotic producer	Vasavada et al. 2006
Alkaliphilic and halophilic bacterial 8 isolates	Saline habitat of coastal Gujarat	Alkaline protease	Dodia et al. 2006
Kocieria sp.	Red mud pond impoundment of National Aluminium Company Ltd (NALCO), Damanjodi, Orissa	Bioremediation of alkaline waste water	Krishna et al. 2008
Kurthia sp.	Activated sludge from ETP of a textile industry nearby Delhi	Treatment of wastewaters textile industry	Kumar et al. 2008
Stenotrophomonas maltophilia (Alkaliphilic psychrotolerant)	Gangotri glacier Western Himalaya soil samples	Extracellular cold active detergent stable alkaline protease	Kuddus and Ramteke 2009
Streptomyces clavuligerus strain Mit-1	Mithapur coastal region of Gujarat	Alkaline protease	Thumar and Singh 2009
Streptomyces tanashiensis strain A2D (salt tolerant and alkaliphilic)	Soil of phoomdi in Loktak lake, Manipur	Bioactive molecule producer	Singh et al. 2009
Streptomyces aburaviensis strain Kut-8	Saline desert of Kutch, Rajasthan	Antimicrobial agent	Thumar et al. 2010
Gracilibacillus sp. TSC pVG	Salt fields near Sambhar district of Rajasthan	Haloalkali thermostable xylanase	Giridhara and Chandra 2010
Paenibacillus tezpurensis sp. nov.	Soil sample Assam	Detergent stable alkaline protease	Rai et al. 2010

(continued)

Table 3.2 (continued)

Name/group of microorganisms	Source/place	Biotechnological potential	References
Alkalophilic *Bacillus subtilis* ash	Kurukshetra, Haryana	Production and purification of cyclodextrin glycosyltransferase	Sanghi et al. 2010
Alkalitolerant bacteria	Pulp and paper mill sludge, New Delhi	Decolorization and detoxification of pulp and paper mill effluent	Mishra and Thakur 2010
Enterobacter aerogenes	Soil near oil extraction unit	Alkaline lipase by fermentation of agro industrial wastes	Mitra et al. 2010
Paenibacillus montaniterrae RMV1	Red mud pond impoundment of National Aluminium Company Ltd (NALCO), Damanjodi, Orissa	Alkalistable xylanase	Arora et al. 2014
Agromyces indicus	Red mud pond impoundment of National Aluminium Company Ltd (NALCO), Damanjodi, Orissa	Bioremediation	Krishna et al. 2014
Bacillus litoralis			
Bacillus anthracis			
Chungangla korensis			
Kokuria flava			
Kokuria Polaris			
Microbacterium horainis			
Planococcus plakortidis			
Pseudomonas alkaliphile			
Salinicoccus roseus			
Bacillus, Virgibacillus, Micrococcus	Marine coast of Odisha	Extracellular alkaline protease	Maharana et al. 2016
Bacillus halodurans strain KG1	Rhizospheric soil of citrus plant (pH 8.2), Thaper Institute campus, Patiyala	Treatment of alkaline cement kiln dust	Kunal and Siddique 2016

Kuddus and Ramteke (2009) have described an extracellular, cold-active alkaline and detergent-stable protease from a novel psychro-tolerant alkaliphilic bacterium *Stenotrophomonas maltophilia* (MTCC7528) isolated from soil samples collected from Gangotri glacier, Western Himalaya, India. Production of the enzyme was optimized with respect to different culture conditions. The enzyme was characterized and indicated to be a metalloprotease. The enzyme was found to be stable and compatible with commercial detergent indicating potential application in cold washing and in environmental bioremediation in the cold regime.

Sayyed and Chincholkar (2010) have studied growth and siderophores production in *Alkaligenes faecalis* and found that it is regulated by metal ions.

Antony et al. (2012a, b) have reviewed microbial diversity of soda lakes including Lonar Lake of India. Haloalkaliphilic bacteria and archaea have been described from many soda lakes which are both saline and alkaline environments. They harbor diverse functional groups of microbes such as methanogens, methanotrophs, phototrophs, denitrifiers, sulfur oxidizers, sulfate reducers, and syntrophs. Some of these microorganisms produce biotechnologically important enzymes.

Krishna et al. (2014) have reported bacterial diversity of bauxite residue (red mud) a unique and extreme environment generated during the extraction of alumina from bauxite ore, NALCO, Orissa. The red mud has high pH, high concentration of soluble ions but no organic matter. The cultivable bacteria were identified (listed in Table 3.2). The isolates were alkalitolerant, enzyme producers, and organic acid producers. Non-cultivable diversity revealed the presence of proteobacteria and bacteroidetes.

Maharana et al. (2016) have isolated alkalitolerant bacteria from sediment samples of different coasts of Odisha. Among the 80 isolates obtained from 25 samples, 11 isolates were selected based on the protease activity exhibited. The isolates belonged to genera *Bacillus*, *Virgibacillus*, and *Micrococcus* and showed growth in pH range of 4–12 and 12% NaCl concentration. The authors have thus reported alkalitolerant marine bacteria from Odisha coastal areas.

Kunal and Siddique (2016) have isolated an acid-producing, alkaliphilic strain of *Bacillus halodurans* KG1 from rhizospheric soil of citrus plant from Thaper Institute Campus, Patiyala, Punjab, which reduced the alkalinity (by 83.64%) and chloride content (by 86.96%) of the cement kiln dust. The cement industry waste (cement kiln dust) generated during the manufacturing of cement clinker, is alkaline. The alkaliphilic bacterial strain grew well at high pH (~11 to 12) and thus could be used for the removal of alkalinity from solid alkaline waste like cement kiln dust.

Sreenivasulre et al. (2017) have characterized the fatty acid profile of alkaliphilic isolates *Bacillus subtilis* SVUNM4, *Bacillus licheniformis* SVUNM8, *Bacillus methylotrophicus* SVUNM9, and *Paenibacillus dendritiformis* SVUNM11 by GC-MS analysis. The authors suggest that fatty acid profiling will help in the classification of these organisms.

Thus, Indian researchers have contributed to the knowledge of alkaliphilic microorganisms from India.

3.17 Role of Alkaliphiles in Sustainability

Many of the industrial waste waters are alkaline and hence cannot be biologically treated using neutrophilic microorganisms. Alkaliphilic microorganisms play an important role in bioremediation of such waste waters and thus help in sustainability of the environment.

Kanekar et al. (1999) explored alkaliphilic bacteria isolated from Lonar Lake for bioremediation of alkaline waste water generated in the manufacture of a basic dye methyl violet and phenolic waste from an industry manufacturing cumin phenols. Both the waste waters were alkaline (pH ranging from 9.3 to 10.7) and their phenol content ranged from 258 to 875 mg/L. The alkaliphilic strains of *Citrobacter freundii*, *Pseudomonas putida*, and *Arthrobacter* sp. were able to remove phenol by 86–100% from the alkaline waste waters.

Kumar et al. (2008) have described the use of alkaliphilic bacterial strain isolated from 6 months old activated sludge from the effluent treatment plant (ETP) of a textile industry. The textile industry wastewater having high pH (12) was inoculated with the culture in bench scale experiment. Effect of amendment with glucose on neutralization of the wastewater was also studied. The isolate was found to grow at pH 12 and bring down pH to 7.1. Amendment with glucose further enhanced neutralization of the wastewater. The organism was identified as *Kurthia* sp.

Tambekar et al. (2012) have described biodegradation of phenol by alkaliphilic strain of *Pseudomonas stutzeri* isolated from Lonar Lake. The organism removed 87% phenol from peptone water–phenol medium.

Kulshreshtha et al. (2012) have described the neutralization of alkaline waste water using *Exiguobacterium* sp. strain 12/1 (DSM 21148). The organism is alkaliphilic and grows at high pH 12. The organism was found to produce organic acid (similar to formic acid) which helps in neutralization of alkaline waste water of pH 12 to pH 7.5.

Tambekar et al. (2014a, b) have studied bioremediation of chromium (VI) using haloalkaliphilic bacteria isolated from Lonar Lake by incorporating $K_2Cr_2O_7$ (100 μg/ml) in the nutrient medium. The organism capable of chromium remediation as estimated by spectrophotometric method was identified as *Proteus mirabilis*.

Tambekar and Rajgire (2015) have reported bioremediation of methanol using haloalkaliphilic methylotrophs namely *Pseudomonas aeruginosa* (DHT2) and *Enterobacter cloacae* (DHT8) isolated from Lonar Lake. The organism removed 75–78% methanol from a medium containing 2% methanol.

Zampieri et al. (2019) have reviewed work on alkaliphilic bacteria isolated from red mud of Bauxite processing plant. These bacteria produce organic acids that help in the treatment of alkaline residue.

3.18 Contribution to the Knowledge of Alkaliphilic Microbial Species

Researchers from India have isolated and identified some novel species of alkaliphilic microorganisms from Lonar Lake and other sites and thus have contributed to the knowledge of alkaliphilic microbial species over the globe. Three new genera and nine novel species were discovered and are added to the global collection of microorganisms. These novel species are listed in Table 3.3.

Table 3.3 Alkaliphilic microbial species, new to Science, from India

Sr. No.	Name of the microorganism	Source	Reference
1	*Nitritalea haloalkaliphilia* gen. nov. sp. nov.	Lonar Lake	Kumar et al. 2010a
2	*Indibacter alkaliphilus* gen. nov. sp. nov.	Lonar Lake	Kumar et al. 2010b
3	*Paenibacillus tezpurensis* sp. nov.	Soil sample Assam	Rai et al. 2010
4	*Cecembia lonarensis* gen. nov. sp. nov.	Lonar Lake	Kumar et al. 2008
5	*Methylophaga lonarensis* sp. nov.	Lonar Lake	Antony et al. 2012a
6	*Nitrincola alkalisediminis* sp. nov.	Lonar Lake	Joshi et al. 2016
7	*Cohaesibacter haloalkalitolerans* sp. nov.	Soda lake	Sultanpuram et al. 2013
8	*Nesterenkonia cremea* sp. nov.	Soda lake	Sultanpuram et al. 2017
9	*Nitrincola tapani* sp. nov.	Lonar Lake	Joshi et al. 2020

3.19 Conclusion

Alkaliphilic microorganisms have been extensively studied all over the world. They have been isolated from alkaline environments like soda lakes and explored for their biotechnological potential particularly industrially important enzymes. In India, alkaliphilic microorganisms have been studied mainly from alkaline soda lake of Lonar, the unique ecosystem on the Earth. Other alkaline environments like cement industry waste (cement kiln dust), bauxite residue (red mud) from NALCO, Orissa, estuarine mangrove regions of Goa, marine coast of Odisha, saline habitats of coastal Gujarat, saline desert of Kutch, salt fields near Sambhar District, Rajasthan, Gangotri glacier of Western Himalaya, soils from Assam, alkaline soils from Saurashtra, soils near Loktak lake, Manipur are explored for isolation of alkaliphilic microorganisms. Many other alkaline environments could be investigated for alkaliphilic diversity and their biotechnology. Alkaliphilic microorganisms isolated earlier and available as microbial resources could be further studied for their industrial applications. A moderately haloalkalitolerant, *Halomonas campisalis* isolated from Lonar Lake is found to have potential in the production of polyhydroxyalkanoate (PHA). Further work for commercial production of this environment-friendly, biodegradable plastic is needed. Nine novel species have been added to the global data collection of microorganisms.

References

Abd-el-Malek Y, Rizk SG (1963a) Bacterial sulphate reduction and the development of alkalinity. I. Experiments with synthetic media. J Appl Bacteriol 26:7–13

Abd-el-Malek Y, Rizk SG (1963b) Bacterial sulphate reduction and the development of alkalinity. II. Experiments with soils. J Appl Bacteriol 26:14–19

Abd-el-Malek Y, Rizk SG (1963c) Bacterial sulphate reduction and the development of alkalinity. III. Experiments under natural conditions. J Appl Bacteriol 26:20–26

Ahirrao SD, Pedge SS, Bhusare DU, Kulkarni SS, Mali RP (2011) Studies on Haloalkaliphilic Bacteria Isolated from Lonar Meteorite Crater in India. J Ecobiotechnol 3(6):5–8

Ahmed A, Othman M, Sarwade V, Kachru G (2012) Degradation of anthracene by alkaliphilic bacteria Bacillus badius. Environ Pollut 1(2):97–104

Akiba T, Horikoshi K (1980) A further separation and some properties of β-lactamase 1. Agric Biol Chem 44:2741–2742

Allen ON, Holding AJ (1974) Gram-negative aerobic rods and Cocci. Family III Rhizobiaceae. In: Buchanan RE, Gibbons NE (eds) Bergey's manual of derminative bacteriology, 8th edn. Williams and Wilkins, Baltimore, pp 264–267

Anderson AJ, Dawes EA (1990) Occurrence, metabolism, metabolic role and industrial use of bacterial polyhydroxyalkanoates. Microbiol Rev 54:450–472

Anoop A, Prasad S, Plessen B, Basavaiah N, Gaye B, Naumann R, Menzel P, Weise S, Brauer A (2013) Paleoenvironmental implications of evaporating gaylussite crystals from Lonar lake, Central India. J Quat Sci 4:349–359

Antony CP, Doronina NV, Boden R, Trotsenko YA, Shouche YS, Murrell JC (2012a) *Methylophaga lonarensis*, a novel moderately haloalkaliphilic methylotroph isolated from the soda lake sediments of meteorite impact crater. Int J Syst Evol Microbiol 62:1613–1618

Antony CP, Murrell JC, Shouche YS (2012b) Molecular diversity of methanogens and identification of *Methanolobus* sp. as active methylotrophic archaea in Lonar crater lake sediments. FEMS Microbiol Ecol 81:43–51

Aono R (2011) Adaptation processes in alakaliphiles when cell wall acidity is elevated. Extremophiles III:1–26

Aono R, Horikoshi K (1983) Chemical composition of cell walls of alkalophilic strains of Bacillus. J Gen Microbiol 129:1083–1087

Arias S, Moral AD, Ferrer MR, Tallon R, Quesada E, Bejar V (2003) Mauran, an exoploysaccharide produced by the halophilic bacterium *Halomonas maura*, with a novel composition and interesting properties for biotechnology. Extremophiles 7:319–326

Arora A, Krishna P, Malik V, Reddy MS (2014) Alkalistable xylanase production by alkalitolerant *Paenibacillus montaniterrae* RMV1 isolated from red mud. J Basic Microbiol 54(10): 1023–1029. https://doi.org/10.1002/jobm.201300357

Badve RM, Kumaran KPN, Rajshekhar C (1993) Eutrophication of Lonar lake, Maharashtra. Curr Sci 65:347–351

Baker BH (1958) Geology of the Magadi area. Geol Surv Kenya Rep 42:1–81

Bansode VV, Bejakal SS (2006) Characterization of chitinases from microorganisms isolated from Lonar lake. Indian J Biotechnol 5(1):357–363

Bath AH, Christofi N, Philip JC, Cave MR, McKinley IG, Berner U (1987) Trace element and microbiological studies of alkaline groundwaters in Oman, Arabian Gulf: a natural analogue for cement pore waters. Report Fluid Processes Res. Group, Brit. Geol. Surv, FLPU 87-2

Beadle LC (1932) Scientific results of the Cambridge expedition to the East African lakes 1930.4. The waters of some East African lakes in relation to their flora and fauna. J Linnean Soc Lond Zool 38:157–211

Boldareva EN, Bryantseva IA, Tsapin K, Nelson K, Sorokin DY, Tourova TP, Boichenko VA, Stadnichuk IN, Gorlenko V (2007) The new alkaliphilic bacteriochlorophyll a-containing bacterium *Roseinatronobacter monicus* sp. nov. from the Hypersaline Soda Mono Lake (California, United States). Microbiology 76(1):82–92

Borgave SB, Joshi AA, Kelkar AS, Kanekar PP (2012) Screening of alkaliphilic, haloalkaliphilic bacteria and alkalithermophilic actinomycetes isolated from alkaline soda lake of Lonar, India for antimicrobial activity. Int J Pharm Bio Sci 3(4):258–274

Borkar S (2015) Alkaliphilic bacteria: diversity, physiology and industrial applications. In: Borkar S (ed) Bioprospects of coastal eubacteria. Springer, Cham. https://doi.org/10.1007/978-3-319-12910-5_4

Bouchotroch S, Quesada E, Moral A, Umas I, Bejar V (2001) *Halomonas maura* sp. Nov., a novel moderately halophilic, exolpolysaccharide producing bacterium. Int J Syst Evol Microbiol 51: 1625–1632

Brewer PG, Goldman JC (1976) Alkalinity changes generated by phytoplankton growth. Limnol Oceanogr 21:108–117

Brock TD (1978) Thermophilic microorganims and life at high temperatures. Springer, Berlin

Brown MJ, Lester JN (1979) Metal removal in activated sludge: the role of bacterial extracellular polymer. Water Res 13:817–837

Carsaro MM, Grant WD, Grant S, Marciano CE, Parrilli M (1999) Structure determination of an exopolysaccharide from an alkaliphilic bacterium closely related to *Bacillus* spp. Eur J Biochem 264:554–561

Chaphalakar S, Dey S (1994) Some aspects of production of extracellular protease from *Streptomyces diastaticus*. J Microbiol Biotechnol 9:85–100

Chavdar MS, Bajekal SS (2009) Magnetotactic bacteria from Lonar lake. Curr Sci 8:957–959

Chinnathambi A (2015) Industrial important enzymes from alkaliphiles – an overview. Biosci Biotechnol Res Asia 12(3):2007–2016

Collins MD, Lund BM, Farrow JAE, Schleifer KH (1983) Chemotaxonomic study of an alkalophilic bacterium, *Exiguobacterium aurantiacum* gen nov., sp. nov. J Gen Microbiol 129:2037–2042

Desai RS, Krishnamurthy NK, Mavinkurve S, Bhosle S (2004) Alkaliphiles in estuarine mangrove regions of Goa (central west coast of India). Indian J Geo-Marine Sci 33(2):177–180

Deshmukh DV, Puranik PR (2012a) Statistical evaluation of nutritional compounds impacting phycocyanin production in *Synechocystis* sp. Braz J Microbiol 43(1):41

Deshmukh DV, Puranik PR (2012b) Statistical evaluation of nutritional components impacting phycocyanin production in *Synechocystis* sp. Braz J Microbiol 43(1):348–355. https://doi.org/10.1590/S1517-83822012000100041

Deshmukh D, Puranik P (2017) Alkaliphilism and carbonate transport in alkaliphilic cyanobacteria (Phormidium sp.) isolated from alkaline lake Lonar, India. Int J Curr Microbiol App Sci 6(2): 397–405

Deshmukh D, Puranik P (2020) Ecology and diversity of alkaliphilic cyanobacteria from Soda Lake, Lonar, India. Int J Recent Sci Res 11(7):39131–39137

Dhiraj P, Kumbhare SV, Mhatre SS, Chowdhury SP, Shetty SA, Marathe NP, Bhute S, Shouche YS (2016) Exploration of microbial diversity and community structure of Lonar lake: the only hypersaline meteorite crater lake within basalt rock. Front Microbiol 6:1553. https://doi.org/10.3389/fmicb.2015.01553

Dhundale VR (2021) Comparative analysis and characterization of lipase producing alkaliphilic bacteria isolated from Lonar soda lake, India. Indian J Ecol 48(1):123–127

Dodia MS, Joshi RH, Patel RK, Singh SP (2006) Characterization and stability of extracellular alkaline proteases from halophilic and alkaliphilic bacteria isolated from saline habitat of coastal Gujarat, India. Braz J Microbiol 37(3):276–282. https://doi.org/10.1590/S1517-83822006000300015

Duckworth AW, Grant WD, Jones BE, Van SR (1996) Phylogenetic diversity of soda lake alkaliphiles. FEMS Microbiol Ecol 19:181–191

Duckworth AW, Grant S, Grant WD, Jones BE, Meijer D (1998) *Dietzia natronolimna* sp. nov., a new member of the genus Dietzia isolated from an East African soda lake. Extremophiles 2: 359–366

Duckworth A, Grant W, Jones B, Meijer D, Marquez M, Ventosa A (2000) *Halomonas magadii* sp. nov., a new member of the genus *Halomonas*, isolated from a soda lake of the East African Rift Valley. Extremophiles 4:53–60

Eugster HP, Hardie LA (1978) Saline lakes and their deposits. In: Lerman A (ed) Lakes: chemistry and geology, physics. Springer, New York, pp 237–293

Fredriksson K, Dube A, Milton DJ, Balsundaram MS (1973) Lonar Lake, India: an impact crater in basalt. Science 180:862–864

Fritze D (1996) *Bacillus haloalkaliphilus* sp. nov. Int J Syst Bateriol 46(1):98–101

Ganeshkumar C, Joo H, Choi J, Koo Y, Chang C (2004) Purification and characterization of an extracellular polysaccharide from haloalkaliphilic *Bacillus* sp. I-450. Enzym Microb Technol 34:673–681

Gee JM, Lund BM, Metcalf G, Peel JL (1980) Properties of a new group of alkalophilic bacteria. J Gen Microbiol 117:9–17

Gessesse A, Gashe BA (1997) Production of alkaline protease by an alkaliphlic bacteria isolated from an alkaline soda lake. Biotechnol Lett 19:479–481

Giridhara PV, Chandra T (2010) Production of novel halo-alkali-thermo-stable xylanase by a newly isolated moderately halophilic and alkalitolerant *Gracilibacillus* sp. TSCpVG. Process Biochem 45(10):1730–1737

Grant WD (1992) Alkaline environments. In: Ledeberg J (ed) Encyclopedia of microbiology, vol 1. Academic, London, pp 73–80

Grant WD, Tindall BJ (1986) The alkaline, saline environment. In: Herbert RA, Codd GA (eds) Microbes in extreme environments. Academic Press, London, pp 22–54

Grant WD, Jones BE, Mwatha WE (1990) Alkaliphiles: ecology, diversity and applications. FEMS Microbiol Rev 75:255–270

Guffanti AA, Susman P, Blanco R, Krulwich TA (1978) The proton motive force and α aminoisobutyric acid transport in an obligately alkalophilic bacterium. J Biol Chem 253:708–715

Hashim SO, Delgado O, Hatti-Kaul R, Mulaa FJ, Mattiasson B (2004) Starch hydrolyzing *Bacillus halodurans* isolated from a Kenyan soda lake. Biotechnol Lett 26:823–828

Hensel R, Matussek K, Michlke K, Tacke L, Tindall BJ, Kohlhoff M, Siebers B, Dialenschnedder J (1997) *Sulfophobococcus zilligii* gen. nov., spe. nov. a novel hyperthermophilic archaeum isolated from hot alkaline spring of Iceland. Syst Appl Microbiol 20:102–110

Hoover RB, Pikuta EV, Bej AK, Marsic D, Whitman W, Tang B, Krader P (2003) *Spirochaeta americana* sp. nov., a new haloalkaliphilic, obligately anaerobic spirochaeta isolated from Mono Lake in California. Int J Syst Evol Microbiol 53:815–821

Horikoshi K (1991) Physiology. In: Horikoshi K (ed) Microorganisms in alkaline environment. VCH publishers Inc., New York, pp 25–80

Horikoshi K (1999) Alkaliphiles: some applications of their products for biotechnology. Microbiol Mol Biol Rev 63:735–750

Horikoshi K (2006) Alkaliphiles: genetic properties and applications of enzymes. Springer, Tokyo, p 256

Horikoshi K, Akiba T (1982) Alkalophilic microorganisms. Springer, Berlin

Issatchencko BL (1951) Chlorous, sulphate and soda lakes of the Kulunda Steppe and the biogenic process in them. In: Selected works, vol 2. Academia Naukova, Leningrad, pp 143–162

Jagtap S, Rao M (2005) Purification and properties of a low molecular weight 1, 4-beta-D-glucan glucohydrolase, having one active site, for carboxymethyl cellulose and xylan from an alkalithermophilic *Thermonospora* sp. Biochem Biophys Res Commun 329(1):111–116

Jenkin PM (1932) Reports on the Percy Sladen expedition to some Rift Valley lakes in Kenya in 1929. VII Summary of the ecological results with special reference to the alkaline lakes. Ann Mag Nat His Ser X 8:133–181

Jhingram AG, Rao KV (1954) Lonar lake and its salinity. Records Geol Survey India 85:313–334

Jones BE, Grant WD (1999) Microbial diversity and ecology of the soda lakes of East Africa. In: Bell CR, Brylinsky M, Johnson-Green P (eds) Microbial biosystems. Frontiers, Lausanne

Jones BE, Grant WD, Collins NC, Mwatha WE (1994) Alkaliphiles: diversity and identification. In: Priest FG (ed) Bacterial diversity and systematics. Plenum, New York, pp 195–230

Jones BE, Grant WD, Duckworth AW, Owenson GG (1998) Microbial diversity of soda lakes. Extremophiles 2:191–200

Joshi AA (2007) Studies on alkaliphilic bacteria from Lonar lake ecosystem. Ph.D. thesis, Pune University

Joshi AA, Kanekar PP (2011) Production of exopolysaccharide by *Vagococcus carniphilus* MCMB-1018 isolated from alkaline Lonar Lake, India. Ann Microbiol 61(4):733–740

Joshi AA, Kanekar PP, Kelkar AS, Sarnaik SS, Shouche YS, Wani AA (2007) Moderately halophilic, alkalitolerant *Halomonas campisalis* MCM B-365 from Lonar lake, India. J Basic Microbiol 47(3):213–221

Joshi AA, Kanekar PP, Kelkar AS, Shauche YS, Vani AA, Borgave SB, Sarnaik SS (2008) Cultivable bacterial diversity of Alkaline Lonar Lake, India. Microb Ecol 55:163–172

Joshi A, Thite S, Kulkarni G, Dhotre D, Joseph N, Ventakatramana V, Polkade A, Shouche Y (2016) *Nitrincola alkalisediminis* sp. nov., an alkaliphilic bacterium isolated from an alkaline lake. Int J Syst Evol Microbiol 66(3):1254–1259

Joshi A, Thite S, Dhotre D, Moorthy M, Joseph N, Venkata RV, Shouche Y (2020) *Nitrincola tapanii* sp. nov., a novel alkaliphilic bacterium from An Indian Soda Lake. Int J Syst Evol Microbiol 70(2):1106–1111. https://doi.org/10.1099/ijsem.0.003883

Jourdan F, Moymier F, Koeberl C, Eroglu S (2011) 40Ar/39Ar age of the Lonar crater and consequence for the geochronology of planetary impacts. Geology 39(7):671–674. https://doi.org/10.1130/931888.1

Kanekar PP, Sarnaik SS, Kelkar AS (1999) Bioremediation of phenol by alkaliphilic bacteria isolated from alkaline lake of Lonar, India. J Appl Microbiol Symp Suppl 85:128–133

Kanekar PP, Nilegaonkar SS, Sarnaik SS, Kelkar AS (2002) Optimization of protease activity of alkaliphilic bacteria isolated from an alkaline lake in India. Bioresour Technol 85:87–93

Kanekar PP, Joshi AA, Kelkar AS, Borgave SB, Sarnaik SS (2008) Alkaline Lonar lake, India – a treasure of alkaliphilic and halophilic bacteria. In: Proceedings of Taal, 2007: the 12th world lake conference, pp 1765–1774

Kanekar PP, Kulkarni SO, Nilegaonkar SS, Sarnaik SS, Kshirsagar PR, Ponraj M, Kanekar SP (2014) Environmental friendly microbial polymers, polyhydroxyalkanoates (PHAs) for packaging and biomedical applications. In: Alavi S, Sabu Thomas KP, Sandeep NK, Varghese J, Yaragalla S (eds) Polymers for packaging application. Apple Academic Press, New York, pp 197–226

Kanekar PP, Kulkarni SO, Kanekar SP (2018) Packaging: polyhydroxyalkanoates (PHAs) in encyclopedia of polymer applications. CRC Press, Boca Raton

Kapoor M, Beg QK, Bhushan B, Singh K, Dadhich KS, Hoondal GS (2001) Application of an alkaline and thermostable polygalactouranse from *Bacillus* sp. MG-CP-2 in degumming of ramie (*Boehemeria nivea*) and Sunhemp (*Crotolaria juncea*) bast fibres. Process Biochem 36(8–9):803–807

Karpeta WP (1989) Bedded cherts in the Riegat formation, Hartbeesfontain, South Africa: a late Archean to early Proteozoic magadiitic alkaline playa lake deposite? S Afr J Geol 92:29–36

Kharat KR, Kharat A, Hardikar BP (2009) Antimicrobial and cytotoxic activity of *Streptomyces* sp. from Lonar Lake. Afr J Biotechnol 8(23):6645–6648

Kitada M, Horikoshi K (1982) Energy coupling reactions of the active transport system for gamma amino isobutyric acid in alkalophilic Bacillus No. 8-1. Agric Biol Chem 46:1115–1120

Koga Y, Nishihara M, Mori H (1982) Lipids of alkalophilic bacteria: identification, composition and metabolism. J Occup Environ Health 4:227–240

Krishna P, Arora A, Reddy MS (2008) An alkaliphilic and xylanolytic strain of actinomycetes *Kocuria* sp. RM1 isolated from extremely alkaline bauxite residue sites. World J Microbiol Biotechnol 24:3079–3085

Krishna PA, Babu G, Reddy MS (2014) Bacterial diversity of extremely alkaline bauxite residue site of alumina industrial plant using culturable bacteria and residue 16S rRNA gene clones. Extremophiles 18:665–676

Krulwich TA (1982) The fine structure of obligately alkalophilic bacilli. FEMS Microbiol Lett 15: 299–301

Krulwich TA, Guffanti AA (1983) Physiology of acidophilic and alkalophilic bacteria. Adv Microb Physiol 24:173–214

Kshirsagar PR, Sutar R, Nilegaonkar S, Kulkarni S, Kanekar P (2012) Scale up production of polyhydroxyalkanoates (PHA) at different aeration, agitation and controlled dissolved oxygen levels in fermentor using *Halomonas campisalis* MCM B – 1027. J Biochem Technol 4(1): 512–517

Kshirsagar PR, Kulkarni SO, Nilegaonkar SS, Nivedita M, Kanekar PP (2013) Kinetics and model building for recovery of polyhydroxyalkanoate (PHA) from *Halomonas campisalis*. Sep Purif Technol 103:151–160

Kuddus M, Ramteke PW (2009) Cold-active extracellular alkaline protease from an alkaliphilic *Stenotrophomonas maltophilia*: production of enzyme and its industrial applications. Can J Microbiol 55(11):1294–1301. https://doi.org/10.1139/W09-089

Kulkarni SO (2010) Production of biodegradable polymer by Extremophilic bacteria from Lonar lake. Ph. D. thesis, Pune University

Kulkarni SO, Kanekar PP, Nilegaonkar SS, Sarnaik SS, Jog JP (2010) Production and characterization of a biodegradable poly (hydroxybutyrate-co-hydroxyvalerate) (PHB-co-PHV) copolymer by moderately haloalkalitolerant *Halomonas campisalis* MCMB-1027 isolated from Lonar Lake, India. Bioresour Technol 101(24):9765–9771

Kulkarni SO, Kanekar PP, Jog JP, Patil PA, Nilegaonkar SS, Sarnaik SS, Kshirsagar PR (2011) Characterization of co-polymer poly (hydroxybutyrate-co-hydroxyvalerate) (PHB-co-PHU) produced by *Halomonas campisalis* MCMB-1027, its biodegradability and potential application. Bioresour Technol 102(11):6625–6628. https://doi.org/10.1016/j.biortech.2011.03.054

Kulkarni SO, Kanekar PP, Rajwade JM (2014) Effect of nutritional and process parameters on production of copolymer, poly (hydroxybutyrate-co-hydroxyvalerate) by *Halomonas campisalis* MCM B-1027 as assessed using Taguchi design of experiments. Int J Pharma Biosci 5(1):1164–1172

Kulkarni SO, Kanekar PP, Jog JP, Sarnaik SS, Nilegaonkar SS (2015) Production of copolymer, poly(hydroxybutyrate-co-hydroxyvalerate) by *Halomonas campisalis* MCM B 1027 using agro-wastes. Int J Biol Macromol 72:784–789

Kulkarni S, Dhakar K, Joshi A (2019) Microbial diversity in the genomic era. In: Chapter 15 - alkaliphiles: diversity and bioprospection. Elsevier, London, pp 239–263

Kulshreshtha NM, Kumar A, Bisht G, Pasha S, Kumar R (2012) Usefulness of organic acid produced by *Exiguobacterium* sp. 12/1 on neutralization of alkaline wastewater. Sci World J 2012:345101. https://doi.org/10.1100/2012/345101

Kumar GG, Joo H, Choi J, Koo Y, Chang C (2004) Purification and characterization of an extra cellular polysaccharide from haloalkaliphilic *Bacillus* sp. I-450. Enzym Microb Technol 34: 673–681

Kumar A, Dhall P, Kumar R (2008) Application of an alkaliphilic bacterium, *Kurthia* sp. for treating waste waters. In: XVIth international conference on bioencapsulation, Dublin, Ireland, p 63

Kumar PA, Sriniwas TNR, Kumar PP, Madhu S, Shivaji S (2010a) *Nitritulea haloalkaliphila* gen. nov. sp. nov., an alkaliphilic bacterium of the family Cyanobacteriaceae, phylum bacteroidetes. Int J Syst Evol Microbiol 60:2320–2325

Kumar PA, Srinivas TNR, Madhu S, Manorama R, Shivaji S (2010b) *Indibacter alkaliphilus* gen. nov. sp. nov., an alkaliphilic bacterium isolated from a haloalkaline lake. Int J Syst Evol Microbiol 60:721–726

Kunal RA, Siddique R (2016) Bacterial treatment of alkaline cement Kiln dust using *Bacillus halodurans* strain KG1. Environ Microbiol 47(1):1–9

La Touche THD, Christie WAK (1912) The geology of the Lonar lake. Rec Geol Survey India 14: 266–289

Langworthy TA (1978) Microbial life in extreme pH values. In: Kushner DJ (ed) Microbial life in extreme environments. Academic, New York, pp 279–317

Lingojwar D, Jadhav R, Gawai K (2013) F_1 F_0 ATP synthase a-subunit cf *Stenotrophomonas* sp. DL18 from Indian Soda Lake, Lonar: a brief report. F1000Research 1:62

Loni PP, Bajekal SS (2009) Chitinase producing potential of actinomycetes from the extreme saline alkaline environment of Lonar Lake. J Environ Res Dev 4(1):161–164

Loni PP, Bajekal SS (2011) Alkaline chitinase from Bacillus firmus SBPL-05 isolated from alkaline-saline environment of Lonar Lake. Indian J Fundam Appl Life Sci 1(3):161–165

Loni PP, Patil JU, Phugare SS, Bajekal SS (2014) Purification and characterization of alkaline chitinase from *Paenibacillus pasadenesis* NCIM5434. J Basic Microbiol 54(10):1080–1089

Ma Y, Weizhou Z, Xue Y, Zhon P, Ventosa A, Grant WD (2004a) Bacterial diversity of the inner Mongolian Baer Soda lake as revealed by 16S rRNA gene sequence analysis. Extremophiles 8: 45–51

Ma Y, Xue Y, Grant WD, Collins NC, Duckworth AW, Steenbergn RPV, Jones BE (2004b) *Alkalimonas amylolytica* gen. nov., and *Alkalimonas delamerensis* gen. Nov., sp. nov., novel alkaliphilic bacteria from soda lakes in China and East Africa. Extremophiles 8:193–200

Maharana AK, Sinha I, Behera PC, Ray P (2016) Extracellular alkaline protease producing halo-alkalitolerant bacteria isolated from marine coasts of Odisha. Int J Pharm Pharm Sci 8(5): 379–386

Malu RA, Dhabhade DS, Kodarkar MS (2000) Diversity in Lonar lake. J Aquat Biol 15:16–18

Mandel KG, Guffanti AA, Kurlwich T (1980) Monovalent cation/proton antiporters in membrane vesicles from Bacillus alcalophilus. J Biol Chem 255:7391–7396

Martinez-Checa F, Toledo F, Vilchez R, Quesada E, Calvo C (2002) Yield, production, chemical composition and functional properties of emulsifier H28 synthesized by *Halomonas eurihalina* strain H-28 in media containing various hydrocarbons. Appl Microbiol Biotechnol 58:358–363

Martins RF, Davids W, Abu Al-Soud W, Levander F, Radstrom P, Hatti-Kaul R (2001) Starch hydrolyzing bacteria from Ethiopian soda lakes. Extremophiles 5:135–144

Mikami Y, Miyashita K, Arai T (1982) Diaminopimelic acid profiles of alkalophilic and alkaline-resistant strains of actinomycets. J Gen Microbiol 128:1709–1712

Milford AD, Achenbach LA, Jung DO, Madigan MT (2004) *Rhodobaca bogoriensis* gen. nov., an alkaliphilic purple nonsulfur bacterium from African Rift Valley soda lakes. Arch Microbiol 174:18–27

Mishra M, Thakur IS (2010) Isolation and characterization of alkalitolerant bacteria and optimization of process parameters and decolourization and detoxification of pulp and paper mill effluent by Taguchi experiment. Biodegradation. https://doi.org/10.1007/s10532-010-9356-x

Mitra S, Khare SK, Singh R (2010) Alkaline lipase production from *Enterobacter aerogenes* by solid state fermentation of agro-industrial wastes. Int J Environ Waste Manag 5(3-4):410–418

Mormile MR, Romine MF, Garcia MT, Ventosa A, Bailey TJ, Peyton BM (1999) *Halomonas campisalis* sp. nov., a denitrifying, moderately haloalkaliphilic bacterium. Syst Appl Microbiol 22(4):551–558

Nakamura T, Tokuda H, Unemoto T (1984) K+/H+ antiporter functions as a regulator of cytoplasmic pH in a marine bacterium, *Vibrio alginolyticus*. Biochin Biophys Acta 776:330–336

Nakanishi T (1975) Thesis on Osaka city Univ. Osaka

Nandy NC, Deo VB (1961) Origin of the Lonar lake and its alkalinity. Tata Iron Steel Company 8: 1–12

Nawani NN, Kapadnis BP (2003) Chitin degrading potential of bacteria from extreme and moderate environment. Indian J Exp Biol 41(3):248–254

Nicholaus B, Lama L, Esposito E, Manca MC, Importa R, Bellitti MR, Duckworth AW, Grant WD, Gambacorta A (1999a) *Haloarcula* spp. able to biosynthesise exo-endopolymers. J Ind Microbiol Biotechnol 23:489–496

Nicholaus B, Lama L, Manca MC, Gambacorta A (1999b) Extremophiles: polysaccharides and enzymes degrading polyascchrides. Recent Res Dev Biotechnol Bioeng 2:37–64

Nilegaonkar SS, Kanekar PP, Sarnaik SS (2002) Production, isolation and characterization of extracellular protease of an alkaliphilic strain of *Arthrobacter ramosus*, MCM B – 351 isolated from the alkaline lake of Lonar, India. World J Microbiol Biotechnol 18:785–789

Padan E, Zilberstein D, Rottenberg H (1976) The proton electrochemical gradient in Escherichia coli cells. Eur J Biochem 63:533–541

Patel R, Dodia M, Singh SP (2005) Extracellular alkaline protease from a newly isolated haloalkaliphilic *Bacillus* sp. Prod Optim Process Biochem 40(11):3569–3575

Pathak AP, Rathod MG (2019) Alkaliphilic microbes: distribution, diversity and applications. Int J Ind Biotechnol Biomater 5(2):32–36

Patil JV, Bejakal SS (2014) Characterization of xylanase and cellulase from extremely haloalkaliphilic archaeon *Natrinema* sp. SSBJUP-1 isolated from Lonar lake. Int J Pharma Biosci 5(1B):553–559

Patil J, Suryawanshi P, Bajekal SS (2016) Siderophores of haloalkaliphilic archaea from Lonar Lake, Maharashtra, India. Curr Sci 111(4):621–623. https://doi.org/10.18520/cs/v111/i4/621-623

Paul D, Kumbhare SV, Choudhary SP, Shetty SA, Marathe NP, Bhute S, Shouche YS (2016) Exploration of microbial diversity and community structure of Lonar Lake: the only hypersaline meteorite crate lake within Basalt rock. Front Microbiol 6(Article 1553):1–12

Pedersen K, Nilsson E, Arlinger J, Hallbeck L, O'Neill A (2004) Distribution, diversity and activity of microorganisms in the hyper-alkaline spring waters of Maqarin in Jordan. Extremophiles 8: 151–164

Pittarello LAP, Crosta C, Kazzuo-Vieira C, Koeberl S, Kenkmann T (2010) Geology and impact features of Vargeao Dome, Southern Brazil. Meteorit Planet Sci 47(1):51–71

Ponraj M (2007) Microbial production of biodegradable plastics – polyhydroxy alkanoic acids (PHA), Ph. D. thesis. Pune University

Quillaguman J, Hatti-Kaul R, Mattiasson B, Alvarez MT, Delgado O (2004) *Halomonas boliviensis* sp. nov., an alkalitolerant, moderate halophile isolated from soil around a Bolivian hypersaline lake. Int J Syst Evol Microbiol 54:721–725

Rai SK, Roy JK, Mukherjee AK (2010) Characterization of a detergent stable alkaline protease from a novel thermophilic strain *Paenibacillus tezpurensis* sp. nov. AS-524-II. Appl Microbiol Biotechnol 85(5):1437–1450

Raut AA, Bejakal SS (2009) Nitrogen fixing actinomycetes from the saline alkaline environment of Lonar lake: a meteorite impact crater. J Environ Res Dev 3(3):784–789

Raut AA, Bejakal SS (2010) Growth of non-symbiotic nitrogen fixing microaerophiles from Lonar crater on inorganic medium of Martian soil simulant components. J Pure Appl Microbiol 4(2): 819–822

Raut AA, Bejakal SS (2011) An agar degrading diazotrophic actinobacteria from hyperalkaline meteoric Lonar crater lake – A primary study. Microbiol Res 2(1):37–40

Raut AA, Phugare SS, Kalyani DC, Bajekal SS (2018) Biotreatment of paper mill effluent using alkaliphilic *Rhizobium* sp. NCIM 5590 isolated from meteoritic alkaline Lonar lake, Buldhana district, Maharashtra, India. Lakes Reserv 23(2):130–138

Rees H, Grant WD, Jones BE, Heaphy S (2003) Diversity of Kenyan soda lake alkaliphiles assessed by molecular methods. Extremophiles 8:63–71

Rich F (1933) Scientific results of the Cambridge Expedition to the East African lakes. 1930-1931, VII. The algae. Zoology 38:249–275

Rodriguez-Valera F, Ruiz BF, Ramos-Cormenzana A (1981) Characterization of the heterotrophic bacterial populations in hyper saline environments of differing salinities. Microb Ecol 7:235–243

Rupela OP, Tauro P (1973a) Isolation and characterization of *Thiobacillus* from alkali soils. Soil Biol Biochem 5:891–897

Rupela OP, Tauro P (1973b) Utilization of *Thiobacillus novellus* to reclaim alkali soils. Soil Biol Biochem 5:899–901

Sanghi A, Garg N, Gupta VK, Mittal A, Kuhad R (2010) One step purification and characterization of celluase free xylanase produced by alkalophilic *Bacillus subtilis* ash. Braz J Microbiol 41(2): 467–476

Sarethy IP, Saxena Y, Kapoor A, Sharma M, Sharma SK, Gupta V, Gupta S (2011) Alkaliphilic bacteria: applications in industrial biotechnology. J Ind Microbiol Biotechnol 38(7):769–790. https://doi.org/10.1007/s10295-011-0968-x

Savergave LS, Dhule SS, Jogdand VV, Nene SN, Gadre RV (2008) Production and single step purification of cyclodextrin glycosystransferase from alkalophilic *Bacillus firmus* by ion exchange chromatography. Biochem Eng J 39(3):510–515

Sayyed RZ, Chincholkar SB (2010) Growth and siderophores production in *Alcaligenes faecalis* is regulated by metal ions. Indian J Microbiol 50(2):179–182. https://doi.org/10.1007/s12088-010-0021-1

Schalachev VP (1978) Membrane-linked energy buffering as the biological function of Na+/K+ gradient. FEBS Lett 87:171–179

Sharma M, Gat Y, Arya S, Kumar V, Panghal A, Kumar A (2019) A review on microbial alkaline protease: an essential tool for various industrial approaches. Ind Biotechnol 15(2):69–78. https://doi.org/10.1089/ind.2018.0032

Shinde VA, Patil RB, Pawar VP (2017) Comparative study of antimicrobial potentials of phospholipid compound produced by halophilic and alkaliphilic *Bacillus subtilis* isolated from alkaline meteorite crater Lonar Lake. Int J Life Sci 5(3):420–424

Shivaji S, Ara S, Singh A, Pinnaka AK (2012) Draft genome sequence of *Cecembia lonarensis* strain LWg^T, isolated from Lonar Lake, a haloalkaline lake in India. J Bacteriol 194(23):6624–6625

Singh RN (1961) Role of blue green algae in nitrogen economy. ICAR, New Delhi

Singh LS, Mazumdar S, Bora TC (2009) Optimization of process parameters for growth and bioactive metabolites produced by a salt-tolerant and alkaliphilic actinomycete, *Streptomyces tanashiensis* strain A2D. J Mycol Med 19(4):225–233

Soledad A, Ana M, Ferrer MR, Tallon R, Quesada E, Bejar V (2003) An exopolysaccharide produced by the halophilic bacterium *Halomonas maura*, with a novel composition and interesting properties for biotechnology. Extremophiles 7(4):319–326

Sorokin DY, Tourova TP, Lysenko AM, Mityushina LL, Kuenen JG (2002) *Thioalkalivibrio thiocynoxidans* sp. nov. and *Thioalkalivibrio paradoxus* sp. nov., novel alkaliphilic, obligately autotrophic, sulfur oxidizing bacteria capable of growth on thiocynate from soda lakes. Int J Syst Evol Microbiol 52:657–664

Sorokin DY, Banciuh, van Loodrecht M, Kuenen JG (2003) Growth, physiology and competitive interaction of obligately chemolithoautotrophic, haloalkaliphilic, sulpher oxidizing bacteria from soda lakes. Extremophiles 7:195–203

Souza KA, Deal PH (1977) Characterization of a novel extremely alkalophilic bacterium. J Gen Microbiol 101:103–109

Souza KA, Deal PH, Mack HM, Turnbill EE (1974) Growth and reproduction of microorganisms under extremely alkaline conditions. Appl Microbiol 28:1066–1068

Sreenivasulre B, Paramageetham C, Sreenivasulu D, Suman B, Umamahesh K, Babu G (2017) Analysis of chemical signatures of alkaliphiles using fatty acid methyl ester analysis. J Pharm Bio-Allied Sci 9(2):106–114. https://doi.org/10.4103/jpbs.JPBS_286_16

Sultanpuram VR, Lodha TD, Chintalapati VR, Chintalapati S (2013) *Cohaesibacter haloalkalitolerans* sp. nov., isolated from a soda lake, and emended description of the genus *Cohaesibacter*. Int J Syst Evol Microbiol 63(11):4271–4276. https://doi.org/10.1099/ijs.0.050112-0

Sultanpuram VR, Mothe T, Chintalapati S, Chintalapati VR (2017) *Nesterenkonia cremea* sp. nov., a bacterium isolated from a soda lake. Int J Syst Evol Microbiol 67(6):1851–1866. https://doi.org/10.1099/ijsem.0.001876

Surakasi VP, Wani AA, Shouche YS, Ranade DR (2007) Phylogenetic analysis of methanogenic enrichment cultures obtained from Lonar Lake in India: isolation of *Methanocalculus* sp. and *Methanoculleus* sp. Microb Ecol 54:697–704

Surakasi VP, Antony CP, Sharma S, Patole MS, Shouche YS (2010) Temporal bacterial diversity and detection of putative methanotrophs in surface mats of Lonar crater lake. J Basic Microbiol 50:465–474

Surve RR, Shirke AV, Athalye RA, Sangare MM (2021) A review on chemical and ecological status of Lonar Lake. Curr World Environ 16(1):4929. https://doi.org/10.12944/CWE.16.1.07

Sutherland IW (1990) In biotechnology of microbial exolpolysaccharides. In: Baddiley J, Higgins NH, Potter WG (eds) Cambridge studies in biotechnology, vol 9. Cambridge University Press, Cambridge

Sutherland IW (1998) Novel and established applications of microbial polysaccharides. Trends Biotechnol 16:41–46

Syed S (2016) Haloalkaliphilic *Bacillus* species from solar salterns; an ideal prokaryote for bioprospecting studies. Ann Microbiol 66:1315–1327

Takahar Y, Tanabe O (1962) Studies on the reduction of indigo in industrial fermentation vat (XIX). Taxonomic characteristics of strain no. S-8. J Ferment Technol 40:77–80

Tambekar DH, Dhundale VR (2012a) Studies on the physiological and cultural diversity of *Bacilli* characterized from Lonar lake, (MS) India. Biosci Discov 3(1):34–39

Tambekar DH, Dhundale VR (2012b) Isolation and characterization of antibacterial substance produced from Lonar lake. Int J Res Rev Pharm Appl Sci 2:41–54

Tambekar DH, Rajgire AV (2015) Bioremediation of methanol by halo-alkaliphilic methylotrophs from Lonar Crater. Indian J Pharm Sci Res 5(4):290–294

Tambekar DH, Pawar AL, Dudhane MN (2010) Lonar Lake water: past and present. Nat Environ Pollut Technol 9(2):217–221

Tambekar DH, Bhorse PS, Gadakh PV (2012) Biodegradation of phenol by native microorganisms isolated from Lonar Lake in Maharashtra State (India). Int J Life Sci Pharm Res 2(4):26–30

Tambekar DH, Tiwari AA, Tambekar SD (2014a) Studies on production of antimicrobial substances from *Bacillus* species isolated from Lonar Lake. Indian J Appl Res 4(8):502–506

Tambekar DH, Waradkar PA, Tambekar SD (2014b) Bioremediation of chromium (vi) by haloalkaliphilic bacteria isolated from alkaline lonar lake. Int J Pharm Chem Biol Sci 4(3): 601–605

Thakker CD, Ranade DR (2002) An alkalophilic methanosarcina isolated from Lonar crater. Curr Sci 82:455–458

Thombre RS, Kanekar PP, Rajwade JM (2013) Production of cyclodextrin glycosyl transferase from alkaliphilic *Paenibacillus* Sp L55 Mcm B - 1034 isolated from alkaline Lonar Lake, India. Int J Pharm Bio Sci 4(1):515–523

Thumar JT, Singh SP (2009) Organic solvent treatment of an alkaline protease from salt-tolerant alkaliphilic *Streptomyces clavuligerus* strain Mit-1. J Ind Microbiol Biotechnol 36(2):211–218

Thumar JT, Dhulia K, Singh SP (2010) Isolation and partial purification of an antimicrobial agent from halotolerant alkaliphilic *Streptomyces aburaviensis* strain Kut-8. World J Microbial Biotechnol 26:2081–2087

Tiago I, Chung AP, Verissimo A (2004) Bacterial diversity in a nonsaline alkaline environment: heterotrophic aerobic populations. Appl Environ Microbiol 70:7378–7387

Tindall BJ (1980) Phototrophic bacteria from Kenyan soda lakes. Ph.D. Thesis, Leicester University

Tindall BJ (1988) Prokaryotic life in alkaline, saline athalassic environment. In: Rodriguez-Valera F (ed) Halophilic bacteria. CRC Press, Boca Raton, pp 31–67

Tindall BJ, Ross HNM, Grant WD (1984) *Natronobacterium* gen. nov. and *Natronococcus* gen. nov., two new genera of haloalkaliphilic archaebacteria. Syst Appl Microbiol 5:41–57

Tombs M, Harding SE (1998) An introduction to polysaccharide Biotechnology. Taylor and Francis, London

Ulukanli Z, Digrak M (2002) Alkaliphilic microorganisms and habitats. Turk J Biol 26:181–190

Vargas VA, Delgado OD, Hatti-Kaul R, Mattiasson B (2004) Lipase producing microorganisms from a Kenyan alkaline soda lake. Biotechnol Lett 26:81–86

Vasavada SH, Thumar JT, Singh SP (2006) Secretion of a potent antibiotic by salt tolerant and alkaliphilic actinomycete *Streptomyces sannanensis* strain RJT-1. Curr Sci 91(10):1393–1397

Wani AA, Surakasi VP, Siddharth J, Raghavan RG, Patole MS, Ranade DR, Shouche YS (2006) Molecular analyses of microbial diversity associated with the Lonar Soda lake in India: an impact crater in a basalt area. Res Microbiol 157(10):928–937

Watanabe N, Ota Y, Minoda Y, Yamada K (1977) Isolation and identification of alkaline lipase producing microorganisms, culture conditions and some properties of crude enzymes. Agric Biol Chem 41:1353–1358

Wormald RM, Rout SP, Mayes W (2020) Gomes H and Humphreys PN, Hydrogenotrophic Methanogenesis Under Alkaline Conditions. Front Microbiol 11:614227. https://doi.org/10.3389/fmicb.2020.614227

Yonezawa Y, Horikoshi K (1978) Polyamines in alkalophilic Bacillus Y-25. Agric Biol Chem 43: 271–278

Zampieri BDB, Nogueira EW, Quaglio OA, Brucha G (2019) Overview of known alkaliphilic bacteria from bauxite residue. J Mining Mech Eng 1(2):41–50. https://doi.org/10.32474/JOMME.2019.01.000107

Zavarzin GA (1993) Epicontinental soda lakes as probable relict biotopes of terrestrial biota formation. Microbiology 62:473–479

Zavarzin GA, Zhilina TN, Kevbrin VV (1999) The alkaliphilic microbial community and its functional diversity. Microbiology 68:503–521

Zhang W, Xue Y, Zhou P, Ventosa A, Grant WD (2002) *Salinicoccus alkaliphilius* sp. nov., a novel alkaliphilic and moderate halophile from Baer Soda Lake in Inner Mongolia Autonomous Region, China. Int J Syst Evol Microbiol 52:789–793

Zhilina TN, Zavarzin GA (1994) Alkaliphilic anaerobic community at pH 10. Curr Microbiol 29: 109–112

Thermophilic, Thermotolerant Microorganisms

4

Abstract

Thermophilic microorganisms are heat-loving microorganisms and grow optimally at temperatures above 50 °C and maximum 80 °C. Thermophilic microorganisms are considered as the "Universal ancestor" and thus bear importance in the evolutionary processes. They are found in thermal springs water, hydrothermal vents, ocean bottom, desert sand, compost, straw, etc. They develop resistance to high temperature by producing specialized proteins known as "Chaperonins" which are thermostable, saturated fatty acids in cell membrane, high melting temperature of DNA, heat shock proteins, etc. They are represented by spore-forming organisms like *Bacillus stearothermophilus*, *Bacillus thermoleovorans*, *Pyrococcus abyssi*, and *Thermococcus litoralis*. Thermophilic bacteria possess high metabolic rates, higher end product yield, and stable enzymes. Thermostable hydrolytic enzymes have applications in industrial processes.

In India, thermal springs are located in Himalayan ranges and many states. Thermophilic bacteria have been extensively studied from Manikaran hot spring in Himachal Pradesh. Thermophilic actinomycetes have been studied from hot springs in Western Maharashtra and explored for production of metalloproteases. Three novel species of thermophilic microorganisms have been reported from India.

Keywords

Thermophilic · Thermal springs · Hydrothermal vents · Evolutionary process · Origin of life · Heat shock proteins · Chaperonins · *Bacillus stearothermophilus* · Manikaran hot spring · Thermostable enzymes

P. P. Kanekar, S. P. Kanekar, *Diversity and Biotechnology of Extremophilic Microorganisms from India*, Microorganisms for Sustainability 41,
https://doi.org/10.1007/978-981-19-1573-4_4

The existence of life at high temperature is very fascinating. At elevated temperatures, only microorganisms are able to grow and survive. The term "thermophile" refers to any microbe capable of growth at 55 °C. Thermophilic bacteria were first described by Bergey in 1919. True thermophilic bacteria are those that grow at temperatures above the maximum temperature for the majority of the bacteria. The true thermophiles show no growth or only very poor growth, below 40–45 °C. Their development requires temperatures above 50 °C and some are able to develop at 80 °C. A group of facultative thermophilic bacteria were discovered which develop at about 20 °C and have their optimum temperature at about 50 °C and maximum about 60 °C. In this group belongs some of the well-known spore-forming soil organisms. The true thermophiles resist a temperature of 100 °C for two to 8 h indicating that the optimum temperature for growth is related to the heat-resisting power of the spores. The facultative thermophiles with their lower optimum temperature for growth show lower resistance to the boiling temperature (Bergey 1919).

Brock (1985) has described thermophiles from water environments with temperatures up to and above boiling and which are commonly found in association with geothermal activity. At temperatures above 60 °C, only bacteria are found and bacteria with temperature optima over the range of 65–105 °C have been obtained in pure culture. The upper temperature limit for life in liquid water has not yet been defined but is likely to be somewhere between 110 and 200 °C, since amino acids and nucleotides are destroyed at temperatures over 200 °C. Because bacteria capable of growth at high temperature are found in many phylogenetic groups, it is likely that the ability to grow at high temperature had a polyphyletic origin (Brock 1985).

Heat-loving microorganisms or thermophiles have attracted noticeable scientific interest nowadays, not only with the aim to elucidate the mystery of life at high temperature, but also due to huge field of biotechnological applications of the enzymes they produce. Satyanarayana et al. (2013) in their book Thermophilic Microbes in Environmental and Industrial Biotechnology have covered various aspects of thermophiles such as the diversity of thermophiles, their potential roles in pollution control and bioremediation, and composting and applications of their enzymes through articles written by researchers. Majumder (2007) and Majumder and Kanekar (2017) have reviewed work on thermophiles with reference to evolution, habitats, physiology, and enzymes proteases of thermophiles.

Maria Isabel Gonzalez-Siso (2019) has described the contents of book on Thermophiles and Thermozymes edited by her as a reprint of articles from the special issue published online in the open access journal Microorganisms (ISSN 2076-2607) from 2018 to 2019. In the Editorial of this special issue Thermophiles and Thermozymes, she has reviewed in brief the nine articles that study diverse aspects of thermophiles biology and their enzymes. Two research articles deal with genomics of thermophilic microorganisms. Four articles are related to different aspects of thermozymes. Three reviews deal with tRNA modification enzymes from thermophiles, application of thermophilic proteins as scaffolds for protein engineering proposed due to their robustness and evolvability and metagenomics approach in search of thermostable cellulases. The book thus focuses light on how the thermophilic microorganisms and their enzymes are stable and functional at high temperatures, and envisages new biotechnological applications and field for bioprospectation.

4.1 Origin of Life and Evolution of Thermophiles

"Big Bang" led to the generation of our solar system and Primitive Earth. 12.5 (±3) billion years ago (Bya) Big Bang occurred which led to rapid expansion of matter to ever-enlarging volume. Planet Earth was formed ~4.6 Bya (K_4O/Ar_4O radio dating). Primitive atmosphere on Primitive Earth surface was formed ~4 Bya. The "Biosphere" evolved deterministically as a self-organized system given the initial conditions of the sun-Earth system. Temperature, using a critical constraint, biogenesis, and the overall patterns of biotic evolution were the highly probable outcomes of this deterministic process. Emergence of life and its major evolutionary innovations occurred as soon as the temperature decreased to their upper temperature limits. These innovations include phototrophy and oxygenic. The stages of evolution were described by Kaula (1975).

Microorganisms appeared on the Earth's surface almost 3000 million years ago. The study of their origin and evolution is very complex. Primitive organisms capable of metabolism and reproduction first appeared ~3.6 billion years ago (Bya). They were most likely thermophilic anaerobes. Oxygenic photosynthesizers evolved ~1.2 Bya after the first organisms. After adequate oxygen concentration on Earth was available, aerobic thermophiles evolved around 0.5 Bya. Biogenesis and emergence of the universal ancestor common to all later forms of life known to currently exist on Earth was established. The evolutionary studies have shown that the last universal common ancestor was most likely a thermophilic organism living in a hydrothermal ecosystem (Reysenbach and Cady 2001; Lineweaver and Schwartzman 2004). Thus the importance of thermophiles in the evolutionary process is well established.

Plate tectonics plays an important role in Astrobiology. It is involved in the evolution of global planetary climate and life. The recent focus has been on the present-day Earth where plate tectonics is already occurring (Solomatob 2004). Natural geothermal habitats have a worldwide distribution and are primarily associated with tectonically active zones, where major movements of the Earth's crust occur (Brock 1985). There is strong molecular and physiological evidence from present-day microorganisms that the "Universal ancestor" was capable of growing at high temperatures. Thus, the importance of thermophiles in the evolutionary process is well established.

4.2 Physiology of Thermophiles and Mechanism of Thermophily

The most interesting problem posed by the thermophilic microorganisms is a physiological one. It is fascinating to know how these microbes are able to live and grow at high temperatures at which many proteins are coagulated and thus the existence of life appears as a biochemical anomaly. The answer to this anomaly would be that thermophilic bacteria, living at high temperature may gain more energy metabolizing available nutrients than life forms that live in more temperate environments (Allen 1953). The external environment of the thermophilic organisms, their structural features including structural modifications at molecular level and dynamic effects of metabolism in preventing or repairing thermal damage, leads to the thermal resistance of the organisms. Thermophiles are reported to contain specialized proteins known as "Chaperonins", which are thermostable and resist denaturation by refolding the proteins to their native forms and restoring their functions (Kumar and Nussinov 2001). Thus thermophiles can successfully resist proteolysis at high temperature. The cell membrane of thermophile is made up of saturated fatty acids, which provide a hydrophobic environment for the cell and keep the cell rigid enough to live at elevated temperatures. The DNA of thermophiles contains a reverse "gyrase" which produces positive supercoils in the DNA (Lopez 1999). This raises the melting point of the DNA to at least as high as the organisms' maximum temperature for growth. Thermophiles tolerate high temperature also by using electrostatic, disulfide bridge, and hydrophobic interactions.

4.3 Molecular Basis of Thermophily

In conventional organisms, macromolecules such as proteins and nucleic acids are inactivated irreversibly by heat, but in thermophiles these components are stable (Langworthy 1978). Supramolecular structures such as ribosomes and membranes are also thermostable in thermophiles although inactivated by heat in mesophiles. The most direct approach to determining the molecular basis of thermostability is to examine the amino acid sequence of a protein formed by a temperature-sensitive mutant and to compare this sequence to that of the wild-type parent. Grütter et al. (1979) experimentally investigated that differences in thermostability of proteins are, in general, due to subtle changes in hydrophobic interactions, hydrogen bonds, and so on and not to a single determinant such as metal binding or changes in secondary structure. Perutz (1978) analyzed the factors involved in the thermostability of proteins. Comparisons were made of amino acid sequences, and the three-dimensional structures of ferredoxins, hemoglobins, and glyceraldehyde phosphate dehydrogenases exhibiting different degrees of thermostability. The results showed that the greater heat stability of the thermostable proteins was due to extra salt bridges between portions of the folded molecules.

Matsumura et al. (1984) determined the nucleotide sequences of the gene encoded by the plasmid carrying a Kanamycin-inactivating enzyme in a thermophilic bacterium, *Bacillus stearothermophilus* as compared with the same enzyme in a mesophilic bacterium, *Staphylococcus aureus*. Despite the fact that the plasmids coding these two enzymes had completely separate origins, the nucleotide sequences were identical except for one base in the midst of the structural gene. This base change resulted in the substitution of a threonine residue with a lysine residue. The position of this lysine substituent was such that the protein surface could acquire increased electrostatic bridging without any significant change in a three-dimensional structure.

Solanki and Gupta (2013) studied the mechanism of adaptation of thermophiles to high temperature. Melting temperature of DNA from a thermophilic bacterium, *Thermonospora archaea* isolated from sand dunes of Bikaner region Kodemdeser (Rajasthan, India), and *E. coli* as the mesophilic organism were determined by incubating DNA samples at 30, 50, 70, and 90 °C for 5 min and observing absorbance at 260 nm. Melting temperature of *E. coli* was 60 °C while that of thermophilic bacterium 80 °C. The results suggest that one of the mechanisms of adaptation of thermophiles to high temperature could be the high melting temperature of their DNA. Since melting temperature of DNA is directly related to G + C mole %, thermophilic bacteria must have high G + C content which imparts stability to DNA at high temperature.

4.4 Habitats of Thermophiles

Hot springs or thermal springs are known since long as the natural sources of thermophiles.

4.4.1 Genesis of Thermal Springs

Mahala (2019) has discussed about origin of heat, water, gas, and mineral constituents related to thermal springs of Odisha, India and tried to correlate them with the genesis of thermal springs. The source of heat is considered as the geothermal gradient inside the Earth, from fault zones during tectonic movement in the crust, chemical changes (exothermic reactions), and disintegration of radioactive elements. The fluid phase may be derived from percolation of meteoric water inside the crust through structural openings which is subsequently heated up and manifest in the form of hot springs on the surface of the Earth. The gases such as methane and carbon dioxide are formed due to decomposition of organic materials. The rare gas helium, associated with the thermal springs, is considered to be originated by the disintegration of radioactive minerals. The mineral constituents present in the spring water are due to rock assemblages through which the hot meteoric water has been circulated. The chemical constituents are formed by the interaction between hot meteoric water and the mineralogical composition of the rocks.

The temperature conditions are favorable in hot springs for the growth of thermophiles. Therefore, the thermophilic bacteria, since the early nineteenth century was known as a normal flora of hot springs. Gradually, they became a subject of a lively field of investigation, and many of the fundamental facts of their existence were discovered.

Thermophilic microorganisms are also found in various hydrothermal vents.

4.4.2 Hydrothermal Vents

Hydrothermal vents are places where hot fluids (up to 400 °C) related to the volcanic activity are released from the ocean floor. Because of the high pressure exerted by the water at depth on the sea floor, hydrothermal fluids can exceed 100 °C without boiling. The most visible indications of ongoing volcanic activity are the plumes of hot fluids issuing from hydrothermal vents. These vents can occur as cracks in the top of cones of basalt (a dark, fine-grained rock that makes most of the Earth's crust), or the vents can issue from chimney-like structures that extend upward from the ocean floor. Often the fluids are black due to the presence of very fine sulfide mineral particles that precipitate out as the fluids cool. The sulfides present in these "black smokers" may contain amounts of iron, copper, zinc, and other metals that have been dissolved from underlying fresh basalt and concentrated in the hot solutions. These minerals can accumulate around the vents as sulfide deposits in mounds or chimney shapes up to 148 ft. high. Some vents have lower fluid temperatures and release light-colored precipitates of silica. These vents are called "white smokers". Hydrothermal vents usually occur along mid-ocean ridges where erupting basalt cools and creates new sea floor. The exact locations of the vents are controlled by cracks and faults in the basaltic rock (Encyclopedia.com).

Oceanographer Jack Corliss is credited with discovering the seafloor geysers in volcanic ridges in the Pacific Ocean in 1977. Hydrothermal vents have been then discovered in Atlantic Ocean, Indian Ocean, and even in the water under polar ice cap (Encyclopedia.com).

Reysenbach and Cady (2001) have discussed about the hydrothermal systems, microfossils, and autotrophs. Hydrothermal systems have prevailed throughout geological history on Earth, and ancient archaeon hydrothermal deposits could provide clues to understand Earth's earliest biosphere. Modern hydrothermal systems support a plethora of microorganisms and macroorganisms, and provide good comparisons for paleontological interpretation of ancient hydrothermal systems. However, all of the microfossils associated with ancient hydrothermal deposits reported are filamentous, and limited stable isotope analysis suggests that these microfossils were probably autotrophs. Therefore, the morphology and mode of carbon metabolism are attributes of microorganisms from modern hydrothermal systems that provide valuable information for interpreting the geological record using morphological and isotopic signatures.

Martin et al. (2008) have discussed about the hydrothermal vents which are geochemically reactive habitats that harbor rich microbial communities. There are two types of hydrothermal vents, (1) the hot (approximately 350 °C) black smoker type, the chemistry of which is driven by the magma-chamber that resides below ocean-floor spreading zones, and (2) the cooler (approximately 50–90 °C) Lost Type, the chemistry of which is driven by a process called serpentinization. Serpentinization is an H_2-producing geochemical reaction that has been in operation in hydrothermal systems for as long as there has been water on the Earth. Its reducing power is sufficient to generate substantial amount of abiogenic CH_4 and short hydrocarbons in the effluents of some modern hydrothermal vents. There are striking parallels between the chemistry of the H_2-CO_2 redox couple that is present in hydrothermal systems and the core energy metabolic reactions that initiated the chemistry of life.

4.4.2.1 Other Habitats

The thermophilic bacteria were isolated from soil, mud, water, air, desert sand, ocean bottom, sea water, marine solfatare, soda lakes, sewage, animal and human faces, cultivated soils, masses of decaying plant materials, moldy fodders and other vegetable matter, including cotton, straw, cereal grain, hay, composts, manures, mushroom compost, heated sugarcane bagasse, soil peat, sewage, dung and grass compost (Bergey 1919; Allen 1953; Goodfellow et al. 1984; Lacey 1989; Yallop et al. 1997; Rath and Subramanyam 1998; Haki and Rakshit 2003; Chitte and Dey 2002; Chitte et al. 1999; Patke and Dey 1998; Majumder and Kanekar 2017). Thus thermophilic microorganisms are found to be ubiquitous in nature.

Some of the examples of sources (habitats) of thermophilic microorganisms are listed below:

Hot springs—*Thermus* sp., *Bacillus* sp., *Bacillus thermoleovorans*, *Streptomyces megaspous* strain SD 5.

Deep sea hydrothermal vents—*Staphylothermus marinus*, *Pyrococcus abyssi*

Marine solfatare—*Thermococcus litoralis*

Compost—*Bacillus* sp., *Bacillus stearothermophilus*

Garbage—*Bacillus circulans*

Soil—*Bacillus thermoproteolyticus*

Soda lake of Lonar—*Streptomyces megasporus* strain SD5, *Streptomyces thermoviolaceus*

4.5 Diversity of Thermophilic Microorganisms

Thermophilic microorganisms have been isolated all over the globe from thermal springs, hydrothermal vents, compost, soil, sea water, soda lakes, decaying plant materials, etc.

Brock and Freeze (1969) described the isolation of a new thermophilic bacterium, *Thermus aquaticus* gen. n. and sp. n. The organism was isolated from a variety of thermal springs in Yellowstone National Park, and a thermal spring in California, USA. The organism has also been isolated from man-made thermal habitats such as hot tap water, in geographical locations quite distant from thermal springs. For successful enrichment, incubation at 70–75 °C and use of nutrient media relatively dilute with respect to the organic components is required. The isolates of this organism are Gram-negative nonsporulating nonmotile rods which frequently form long filaments at supraoptimal temperature or in the stationary phase. All isolates form a yellow cellular pigment, probably a carotenoid. They form a characteristic structure, a large sphere, larger than a spheroplast. The optimum temperature for growth is 70 °C, maximum 79 °C and minimum 40 °C. The organism is an obligate aerobe and has pH optimum of 7.5–7.8. The G+C mole % was found to be between

65.4 and 67.4. This is the organism which made revolution in enzyme biotechnology.

Uchino and Doi (1967) reported three strains of acido-thermophilic bacteria from hot spring waters of Tohoku district in Japan. They were aerobic spore-forming bacilli and identified to belong to genus *Bacillus*. They were acidophilic growing well in the pH range between 2.3 and 5.0. Optimum temperature for growth was 65 °C.

Thermophilic microorganisms are found to belong to different phyla. In the catalog of strains (German collection of microorganisms), Claus et al. (1983) have categorized thermophilic microorganisms in following groups:

Archaebacteria Aerobic acidophiles, autotrophs—*Sulfolobus acidocaldarius* type species, *Sulfolobus brierleyi*, *Sulfolobus solfataricus*. These organisms grow at 70–90 °C (optimum 75–85 °C), pH 1 to 4, use organic compounds and elemental sulfur as energy source and O_2 or Fe^{3+} as electron accepter.

Thermoproteales *Thermoproteus tenax*, *Desulfurococcus mobilis*, *Desulfurococcus mucosus*, *Thermophilum pendens*, *Thermococcus celer*. These organisms grow at 75–85 °C (optimum 85 °C), are anaerobic, acidophilic to neutrophilic, use organic compounds as energy sources, use elemental sulfur as electron acceptor.

Methanothermaceae *Methanothermus fervidus* which grows at 70–95 °C (optimum 85 °C), anaerobic, use H_2 and CO_2 exclusively.

Pyrodictium *Pyrodictium brockii*, *Pyrodictium occultum*. These organisms grow at 85–110 °C (optimum 105 °C, anaerobic, use H_2 and elemental sulfur, autotrophic

Other anaerobic species:

Clostridium thermohydrosulfuricum—optimum temperature 65 °C, produces ethanol

Clostridium thermosulfurogenes—optimum temperature 60 °C, ferments pectin, forms elemental sulfur from S_2O_3

Thermoanaerobacter ethanolicus—optimum temperature 65 °C, produces ethanol

Thermoanaerobium brockii—optimum temperature 65 °C, produces ethanol

Thermobacteroides acetoethylicus—optimum temperature 65 °C

Thermodesulfobacterium commune—optimum temperature 70 °C

Brock (1985) has described Thermoacidophiles as one group of thermophilic bacteria. They are acidophilic living only at low pH. These bacteria are remarkable because of their requirement for two extreme environmental factors at the same time. Most of these organisms are Archaebacteria capable of reducing elemental sulfur using H_2 as an energy source (sulfur respiration) or of oxidizing elemental sulfur using oxygen or ferric iron as an electron acceptor. Thermoacidophiles frequently have a sulfur-based energy metabolism, since the low pH environments are generally sulfur-rich. However, one thermoacidophile, *Thermoplasma acidophilum*, does not have a metabolism linked to sulfur. This heterotroph is found in self-heating coal refuse piles. It can live at moderately high temperature (55–60 °C) and low pH even though it lacks a cell wall.

Barns et al. (1994) carried out phylogenetic characterization of many archaeal small subunit r RNA gene sequences obtained by polymerase chain reaction amplification of mixed population DNA extracted directly from sediment of a hot spring in Yellowstone National Park. The analyses document the existence of crenarchaeal families, genera and species for which no close relatives were found. Large numbers of distinct archaeal sequence types were retrieved from this single hot spring.

Rysenbach et al. (1994) examined the phylogenetic diversity of a well-known pink filament community associated with the 84–88 °C outflow from Octopus Spring, Yellowstone National Park. Three phylotypes designated as EM3, EM17, and EM19 were identified and found to diverge deeply within the phylogenetic domain Bacteria. EM3 comprises a unique lineage with the Thermotogales group, and EM17 and EM19 are affiliated with the *Aquificales*. Majority of a single sequence type (EM17) were closely related to *Aquifex pyrophilus*.

Rysenbach et al. (2000) reported the occurrence of another member of the archaeal group "Korarchaeota" and a deeply rooted bacterial sequence from a thermal spring in Yellowstone National Park (USA). The DNA of a mixed community growing at 83 °C, pH 7.6 was extracted and 16S rRNA gene sequences were obtained using polymerase chain reaction. Five different phylotypes were identified, four archaeal phylotypes designated as PBA1, PBA2, PBA3, and PBA5 and only one bacterial phylotype, designated as PBB. Phylotype PBA5 is very closely related to the Korarchaeotal phylotype Pjp27 from Obsidian Pool in Yellowstone National Park. The PBB phylotype is a lineage within the *Aquificales*. The bacterial population was found to dominate the community and most likely contributed to biogeochemical cycling within the community.

Mohammad et al. (2017) have isolated thermophilic bacteria from hot springs in Jordan. Out of ten isolates, nine belonged to *Bacillus licheniformis* and one to *Thermomonas hydrothermalis*. The authors claim that this is the first report of *Thermomonas* species from Jordanian hot springs. The isolate produced thermostable enzymes, amylase, protease, cellulase, gelatinase, and lecithinase.

Thus a number of researchers all over the world have studied the diversity of thermophilic microorganisms.

4.6 Biotechnology of Thermophiles

4.6.1 Bioactive Molecules from Thermophiles and Their Advantages

As a consequence of growth at high temperature and unique macromolecular properties, thermophilic bacteria can possess high metabolic rates, physically and chemically stable enzymes and lower growth but higher end product yields than mesophilic species. Therefore, thermophilic processes are more stable, rapid and facilitate reactant activity and product recovery.

Chien et al. (1976) have reported the revolutionary enzyme deoxyribonucleic acid (DNA) polymerase from a thermophile. They purified a stable DNA polymerase (EC 2.7.7.7) with a temperature optimum of 80 °C from the extreme thermophile *Thermus aquaticus*. The pH optimum was 8.0 and molecular weight 63,000–68,000. The elevated temperature requirement, small size, and lack of nuclease activity distinguish this polymerase from the DNA polymerase of *Escherichia coli*.

The cellular proteins of thermophiles are of immense research interest in basic and applied areas like evolution, industrial, medical, biotechnology, and molecular biology (Zierenberg et al. 2000). A number of enzymes and bioactive molecules from several thermophilic microorganisms have been identified in recent years. Haki and Rakshit (2003) have described different enzymes of thermophiles and their biotechnological applications as detailed below:

Amylase—Bioconversion of starch to dextrose syrups in the temperature range of 90–100 °C; application in starch hydrolysis, brewing, baking, and detergents.

Xylanase—Bioconversion of craft pulp to xylan, lignin at 45–65 and 105 °C; application in pulp and paper industry.

Chitinase—Conversion of chitin into chitobiose and chitosan at 65–75 °C; application in food, cosmetics, and pharmaceuticals.

Cellulase—Conversion of cellulose to glucose at 45–55 and 95 °C; application in cellulose hydrolysis.

Protease—Hydrolysis of proteins to peptides and amino acids in the temperature range of 65–85 °C; application in baking, brewing, detergents and leather industry.,

Lipase—Removal of fat, hydrolysis of fat at 30–70 °C, interesterification, aminolysis alcoholysis; application in dairy, detergent, oleochemicals, pharmaceuticals, cosmetics, and leather industry.

DNA polymerase—DNA amplification at 90–95 °C; application in genetic engineering, polymerase chain reaction (PCR).

4.6.2 Proteases of Thermophilic Microorganisms

Among the enzymes studied, thermophilic proteases find major applications in research, biotechnology, and industrial processes carried out under harsh conditions. Several thermophilic microorganisms are reported for the production of thermostable proteases, such as *Bacillus stearothermophilus* (Matsubara 1970); *Thermus* sp., *Staphylothermus marimus*, *Thermococcus litoralis*, *Clostridium absonum*, *Bacillus thermoleovorans*, *Pyrococcus abyssi*, *Bacillus circulans*, *Bacillus thermoproteolyticus*, *Bacillus brevis*, *Bacillus licheniformis*, *Bacillus thermoruber*, Thermoacidophiles *Thermococcus celer*, *Thermococcus litoralis* and *Thermococcus maritema* (Haki and Rakshit 2003); *Thermoactinomyces* sp. 27a (Zabolotskaya et al. 2004); *Geobacillus caldoproteolyticus* (Chen et al. 2004); *Filobacillus* sp. RF2-5 (Hiraga et al. 2005); *Borrelia burgetoferi* (Bertin et al. 2005); *Geobacillus tepidamanas* GS5-977 (Kählig et al. 2005); *Ferridobacterium islandicum* (Gödde et al. 2005); *Thermoactinomyces* sp. 21E (Petrova et al. 2006); *Streptomyces* sp.594 (De Azeredo et al. 2006); *Pyrococcus furiosus* (Juhász et al. 2006); *Geobacillus collagenovorans* MO-1 (Itoi et al. 2006); *Sulfolobus solfactaricus* (De Felice et al. 2007); *Thermus aquaticus* YT 1 (Sakaguchi et al. 2008); *Thermus thermophiles* HB 27(Maehara et al. 2008); *Brevibacillus* sp. KH3 (Li et al. 2009); *Thermoplasma volcanium* GSS1(Kocabiyik and Demirok 2009); *Virgibacillus* sp.SK37 (Phrommao et al. 2011); *Thermococcus onnurieneus* NA1 (Yun et al. 2011); *Bacillus caldolyticus* (Bader et al. 2012); *Geobacillus thermodenitrificans* (Charbonneau et al. 2012); *Bacillus cereus* (Saleem et al. 2012); *Keratinibaculum paraultunense* (Huang et al. 2013); *Meiothermus ruber* H328 (Kataoka et al. 2014); *Coprothermobacter proteolyticus* (Toplak et al. 2013; Majeed et al. 2013); *Pyrococcus horikoshi* (Zhan et al. 2014) and many more.

Cost savings, associated with longer storage stability and higher activity at higher temperatures of thermophilic enzymes, provide good reason to select and develop thermophilic enzymes for industrial and biotechnological applications. Haki and Rakshit (2003) have described the following advantages of enzymes from thermophiles:

- Lesser stringency in the maintenance of sterile conditions as contamination is lower due to the increased temperature.
- Fermentation allows a higher operational temperature which has a significant influence on the bioavailability and solubility of organic compounds. Elevated process temperatures leads to higher reaction rates due to a decrease in viscosity and an increase in diffusion coefficient of substrates and higher process yield due to increased solubility of substrates and products and favorable equilibrium displacement in endothermic reactions.
- Volatile products can be easily recovered.
- Stability of bioactive molecules against temperature, denaturing agents like detergents and organic solvents, extreme pH conditions enhances their industrial and biotechnological applications.
- The fermentation is cost-effective as elaborate cooling conditions are not required.
- Thermostable enzymes can also be used as models for the understanding of thermostability and thermoactivity, which is useful for protein engineering.

Applications of thermophilic proteases include food processing aid, efficient aerobic and anaerobic sludge dissolution and digestion of organic particulates from high strength food waste water, fishery waste degradation, ecofriendly valorization of keratin containing waste by aerobic and anaerobic processes, protein engineering, proteolysis to develop products like hydrolysates, peptides and amino acids, and in dairy, pharmaceutical, detergent, leather and textile industries. They are also used in basic research where selective peptide bond cleavage is used in the elucidation of structure-function relationship, in the sequencing of proteins, etc.

4.6.3 Importance of Thermolysin Like Proteases (TLP)

Thermolysin is a thermophilic protease having immense commercial application as a synthetic enzyme in producing the artificial sweetener aspartame (Isowa et al. 1979). Besides, the commercial application, thermolysin has significant, degradative application in basic research like the role of thermolysin in protein structural analysis (Heinrikson 1977). Application of proteases in tissue culture got a new dimension with thermolysin. The number of cells after thermolysin treatment was tenfold higher than with trypsin treatment. Rat lungs were dissociated into individual viable cells after thermolysin perfusion through the vasculature and the trachea. This procedure was ideal for dispersing lung tissue into its cellular components (Price et al. 1985). Thermolysin is used to determine active α_2M (α-macroglobulin present in human plasma, ascites) in these fluids, as they contain no other known inhibitor of this enzyme (Salvesan and Nagase 1996). Thermolysin, an endopeptidase, is a neutral, zinc-containing metalloprotease. It is the most thoroughly characterized member of Clan MA that has the sequence HEXXH-E as the metal-binding site). Thermolysin hydrolyses peptide bonds involving the amino groups of hydrophobic amino acid residues. The unique specificity of thermolysin led to its broad and effective use in studies of protein sequences. The enzyme contains an essential Zn atom embedded in a cleft formed between twofold lobes of the protein and four calcium atoms (Veltman et al. 1997) which impart thermostability to the protein. The enzyme having a similar specificity, belonging to Clan MA and family M4 (thermolysin family), is found in various organisms including thermophilic namely *Thermoactinomyces* sp., *Bacillus thermoproteolyticus*, and *Bacillus stearothermophilus*.

4.7 Thermophilic Microorganisms from India

India is blessed with amazing natural habitats that support the growth of thermophiles. Hot springs, the primary habitats of thermophiles are located in Himalayan ranges, Uttarakhand, Himachal Pradesh, Arunachal Pradesh, Madhya Pradesh, Chhattisgarh, Bihar, Sikkim, Orissa, West Bengal, Jharkhand, Gujarat, Maharashtra, Karnataka, Andhra Pradesh, Tamil Nadu, Kerala, and Andaman and Nicobar Islands.

Sahu et al. (2020) have compiled information on the formation of hot springs, different types of hot springs, and occurrence of such hot springs in different parts of India. Hot springs which are natural habitats of thermophilic microorganisms are classified based on the type of elements present in them, e.g., sulfur spring containing sulfur. Salt spring, aluminum springs, mirabilite springs which contain sodium, melanterite springs containing iron and sulfur and highly acidic, iron spring containing Iron (>20 mg/kg of water), acidic iron springs, alkaline springs containing hydrogen carbonate and sodium. In India, hot springs are found to form due to organic activity and convectional circulation of the heat from one layer of the Earth crust to another layer which make it warm due to uneven temperature of both

Fig. 4.1 Sikkim hot spring (Google image)

layers of the Earth. Geologists have divided Indian hot water springs based on tectonic trends, continental rifts, and plate margins into (1) NW-SE Himalayan arc system with continuation to Andaman Nicobar Island, (2) Son-Narmada-Tapti, (3) West coast continental margins, (4) Parts of Gondwana-Grabens, and (5) Regions of Delhi folding. In India, hot springs have been located in Maharashtra, Himachal Pradesh, Karnataka, Madhya Pradesh, Chhattisgarh, Bihar, Sikkim (Fig. 4.1), West Bengal, Uttarakhand, Arunachal Pradesh, Orissa, Jharkhand, Andhra Pradesh, Tamil Nādu, Kerala, and Andaman and Nicobar.

Likewise, hydrothermal vents, the other important natural ecosystems where thermophiles exist are located in Indian Ocean. Indian researchers have reported different thermophilic microorganisms from such natural pristine habitats (Table 4.1).

4.7.1 Diversity

In Northern Himalayas, 21° 57′ to 37° S′N latitudes and 72° 40′ to 97° 25′ E longitudes with boiling hot water is famous among local people due to their thermophilic potential. Thermal hot springs represent extreme niches that have maintained some degree of pristine quality and their biotechnological potential has remained unrealized.

Sharma et al. (2013) isolated 101 thermophilic microbial strains from hot springs of Northern Himalayan Region of Himachal Pradesh. Water samples were collected from hot water springs of Tattapani (Dist. Mandi), (31° 13′ 50″ 04′, 30″ North and

Table 4.1 Thermophilic microorganisms reported from India

Sr. no.	Name of the microorganism	Source	Biotechnological potential	Reference
1	*Streptomyces diastaticus* strain SS1	Lonar Lake silt sample, Maharashtra	Thermostable alkaline metalloprotease	Chaphalkar and Dey 1994, 1998
2	*Streptomyces megaspores* strain SDP4	Lonar Lake Maharashtra	Proteolytic activity	Patke and Dey 1998
3	*Bacillus* sp.	Hot springs of Orissa	Thermotolerant enzyme	Rath and Subramanyam 1998, 2000
4	*Streptomyces thermoviolaceus*	Lonar Lake Maharashtra	Keratinolytic activity	Chitte et al. 1999
5	*Streptomyces megaspores* strain SD5	Thermal spring in Western Maharashtra	Fibrinolytic activity	Chitte and Dey 2000
6	*Streptomyces* sp.	Thermal spring in Western Maharashtra	Fibrinolytic activity	Chitte and Dey 2002
7	*Streptomyces megaspores* strain SD5	Thermal spring in Western Maharashtra	Process for isolation of novel fibrinolytic enzyme	Chitte and Dey 2003
8	*Shewanella* related thermophilic bacteria	Bakreswar hot spring, West Bengal	Diversity	Ghosh et al. 2006
9	Thermophilic bacteria	Hot springs of West Kamong District of Arunachal Pradesh	Extracellular enzymes protease, lipase, amylase, cellulase	Bora and Kalita 2006
10	Thermophilic bacteria growing at 54 °C	Hot spring, Tarabolo, Nayagarh Dist., Orissa	Thermophilic enzymes amylase, cellulase, protease, caseinase, lipase	Mohanta and Rath 2008
11	*Thiomonas bhubaneswarensis* sp. nov.	Hot spring in Orissa	Thiosulfate oxidizing	Panda et al. 2009
12	*Synechococcus bigranulatus*, *S. lividus*, *Glepcapsa gelatinosa*, *G. muralis*, *Phormidium laminosum*, *P. frigidum*, *Oscillatoria princes*, *O. fragilis*, *Lymgbya lutea*, *Pseudanabaena* sp., *Calothrix thermalis* *Fischerella thermalis*	Geothermal springs of Bakreswar, West Bengal	Pigment carotenoid chlorophyll nitrogen fixation capacity	Debnath et al. 2009

(continued)

Table 4.1 (continued)

Sr. no.	Name of the microorganism	Source	Biotechnological potential	Reference
13	*Geobacillus* sp.	Soldhar hot spring site at Garhwal, Himalaya	Diversity	Sharma et al. 2009
14	*Paenibacillus tezpurensis* sp. nov. AS-S23-	Tezpur, Assam	Alkaline protease	Rai et al. 2009
15	Thermophilic cyanobacteria	Tapoban geothermal field, Uttarakhand, Himalaya	Antibacterial compounds	Bhardwaj et al. 2010
16	*Bacillus amyloliquefaciens* TSWK-1	Tulsi Shyam hot spring reservoir, Gujarat	Thermostable calcium-independent amylase	Kikani and Singh 2011
17	*Bacillus subtilis* NM(G), *Bacillus megaterium* NM9, *Paenibacillus* N1	Hot springs of Northern Himalayan region of Himachal Pradesh	Amylase, cellulase, xylanase	Sharma et al. 2013
18	*Thermomonospora archaea*	Sand dunes of Bikaner region Kodomdeser soil temperature 50°C	Adaptation of thermophilic bacteria at molecular level	Solanki and Gupta 2013
19	Thermophilic *Bacillus* sp.	Hot springs of Tarabalo, Odisha	Protease	Panda et al. 2013
20	*Bacillus licheniformis*, *Paenibacillus ehimensis*, *Bacillus sonorensis*, *B*, *tequilensis* *Staphylococcus epidermidis*	Soldhar and Ringigad hot springs, Uttarakhand, Indian Himalayan region	Lipase, amylase, cellulase, protease, xylanase	Pandey et al. 2014
21	Thermophillic *Bacillus licheniformis*	Hot spring Gujarat	Thermostable and solvent tolerant serine protease	Dudhagara et al. 2014
22	*Thermodesulfobacterium commune* sp. Nov.	Hot water springs of Thane Maharashtra	Diversity	Patil and Unnithan 2014
23	Thermophilic bacilli	Hot spring reservoir Gir forest, Gujarat	Diversity	Kikani et al. 2015
24	*Bacillus*, *Aneurinibacillus* spp., *Brevibacillus* spp., *Deinococcus* spp., *Geobacillus* spp., *Lysinibacillus* spp., *Paenibacillus* spp., *Pseudomonas* spp.	Soldhar (high temperature) and Tapovan Kund (moderate temperature) hot springs from Chamali district, Uttarakhand, Himalaya	Lipase, amylase, protease enzymes producers	Ranawat and Rawat 2017

(continued)

Table 4.1 (continued)

Sr. no.	Name of the microorganism	Source	Biotechnological potential	Reference
25	*Geobacillus* sp. (K1C)	Manikaran hot spring Himachal Pradesh	Thermostable raw starch hydrolyzing α-amylase	Sudan et al. 2018
26	Thermophilic *Flavobacterium thermophilum* *Anoxybacillus* sp. *Bacillus licheniformis* *Bacillus tequilensis* *Bacillus paralicheniformis* *Bacillus licheniformis* *Staphylococcus sciuri*	Tatapani hot spring of Kishtwar District of Jammu and Kashmir Hot water ponds located in Yamunotri landscape of Garhwal Himalaya Thermal springs at Sativali, Nimbavali, Ganeshpuri (Vajreshwari group) Thane District of Maharashtra	Thermostable enzymes catalase, nitrate reductase, amylase, cellulase, glutaminase Diversity	Sharma et al. 2018 Kumar and Sharma 2020 Jaffer et al. 2019
27	Thermophillic *Bacillus* sp.	Thermal spring in Chumathang area of Leh district of area of Ladakh	Developing plant growth-promoting consortium	Verma et al. 2018
28	*Bacillus*, *Brevibacillus*, *Geobacillus*, *Anoxybacillus*, *Brevibacterium*	Tuva Timba, Hot springs, Gujrat	–	Patel et al. 2019a, b
29	*Brevibacillus boresteinsis*, *Paenibacillus dendritiformis*, *Streptococcus pyogenes*, *Geobacillus stearothermophilus*, *Thermoactinomyces candidus*, *Thermoactinomyces thalpophilum*, *Streptomyces albus*	Thermal springs Ringigad and Saldhar in Garhwal Himalaya Uttarakhand	–	Kumar and Sharma 2019
30	*Brevibacterium halotolerans*, *Tepidimonas taiwanese*, *Brevibacillus brostelensis*, *Bacillus subtilis*, *Bacillus sonorensis*	Dholera hot springs, Gujarat	Diversity studies	Patel et al. 2019a, b

(continued)

Table 4.1 (continued)

Sr. no.	Name of the microorganism	Source	Biotechnological potential	Reference
31	*Geobacillus thermoleovorens*	Thermal springs in Bandaru, Karnataka	Thermostable amylase and lipase producer	Kotresh et al. 2020
32	*Bacillus licheniformis*	Hot spring at Unkeshwar, Dist. Nanded, Maharashtra	Protease and amylase	Bhusare and Wakte 2011
33	*Thermoactinomyces thalpophilus*	Hot spring in Western Maharashtra	Thermolysin like protease	Majumdar and Kanekar 2012, Majumder et al. 2013
34	*Streptomyces thermovulgaris*, *Streptomyces thermoviolaceus*	Lonar Lake, Maharashtra	Protease, amylase, lipase at 55 °C temperature	Dey and Chaphalkar 1998
35	Thermophilic *Bacillus subtilis* RC1	Thermal spring in Western Maharashtra	Restriction endonuclease	Chitte et al. 2014

Fig. 4.2 Manikaran hot spring, Himachal Pradesh (Google image)

76° 37′ 20″ 77° 23′ 15″ East), Manikaran (Fig. 4.2) and Vashisht (Dist. Kullu) (31° 58′ N and 77° 06 4′ S) regions of Himachal Pradesh. Temperature of hot spring of Vashisht was 45 °C, of Manikaran 99 °C, and of Tattapani 52 °C while pH of all near neutral (7.67 and 6.59). The isolates were aerobic and sustaining at or above 50 °C. They produced enzymes amylase, cellulase, and xylanase. The isolates producing all the three enzymes were identified as *Bacillus subtilis* NM (G), *Bacillus megaterium*

Fig. 4.3 Tapovan thermal spring, Uttarakhand (Google image)

NM 9, *Paenibacillus* N1, and a fungal isolate as *Mycelophthora thermophila* SH 1 from Manikaran hot spring.

Pandey et al. (2014) have studied thermophilic bacteria from thermal springs in Uttarakhand. The Soldhar and Ringigad hot springs located in Uttarakhand State under the Indian Himalayan Region (IHR) were explored for colonization of thermophiles. The sampling was done from two springs located in the Tapovan area (Fig. 4.3) in the Dhauliganga river valley (District Chamoli in the State of Uttarakhand, India). The area is known as a promising geothermal field in the IHR. Several hot springs are found scattered in this region that maintain varied temperatures. Thermophilic bacteria were isolated from the sediment samples collected from two of these springs, Soldhar (latitude 39° 29′ 25″; longitude 79° 39′ 0.29; altitude 1900 m amsl) and Ringigad (latitude 30° 33′ 14″; longitude 79° 40′ 0.06″; altitude 1850 m amsl). Both the hot springs are situated in the same biogeographic area and the temperature gradient along the slope remains between 90 and 95 °C and 75 and 80 °C at Soldhar and Ringigad respectively. The peculiar feature of these hot springs is that their high temperatures are maintained throughout the year despite seasonal variations in the surroundings. 28 morphologically distinct bacterial isolates (16 from Soldhar and 12 from Ringigad) were obtained at a temperature range of 24–80 °C. The isolates exhibited tolerance to a wide range of temperature (20–80 °C). The optimum growth of the isolates was observed at 55 °C and pH 7.0. Identification of the isolates revealed the presence of *Bacillus licheniformis*, *Paenibacillus ehimensis*, *Bacillus sonorensis*, *Bacillus tequilensis*, and

Staphylococcus epidermidis. The isolates produced enzymes lipase, amylase, cellulase, protease, and xylanase at 55 °C and 65 °C.

Bhalla (2015) have reported microbial diversity of Himachal Pradesh (HP) of Indian Himalayan Region. HP has enormous climatic, geographic and sociocultural diversity. This state is divided into four agroclimatic zones namely Shivalik Hill, Mid Hill, High Hill, and cold dry zones which depict subtropical, subhumid, temperate wet, and temperate dry climatic conditions respectively. Thermophilic bacteria reported from thermal springs in HP included *Bacillus licheniformis*, *Bacillus megaterium*, *Bacillus sporothremodurans*, *Brevibacillus thermoruber*, *Geobacillus pallidus*, *Geobaciius subterraneus*, *Hydrogenobacter* sp., *Paenibacillus* sp., *Pyrobaculum aerophilum*, *P. caldifontis*, *Thermus brockianus*, *T. thermophiles*, and several thermophilic cyanobacteria.

Kotresh et al. (2020) have isolated a novel bacterium (strain I (5) from Bandaru hot spring Karnataka, India. The organism was found to be thermophilic growing optimally at temperature 55 °C, at pH 7.0 and tolerated 0–8% NaCl and exhibited good amylolytic and lipolytic activity. The isolate showed 99% similarity to *Geobacillus thermoleovorns* based on 16S rRNA gene sequencing. Maximum amylase activity was obtained at a temperature range of 50 °C to 60 °C and pH 7–8.

Ranawat and Rawat (2017) studied the bacterial diversity of Soldhar (90 °C) and Tapovan Kund (54 °C) hot springs from Chamoli district, Uttarakhand. *Bacillus* was found to be the dominant genus in both the hot springs and other genera were *Aneurinibacillus*, *Brevibacillus*, *Deinococcus*, *Geobacillus*, *Lysinibacillus*, *Paenibacillus*, and *Pseudomonas*. The isolates were found to produce enzymes lipase (73%), amylase and protease (50%) producers from Soldhar hot spring while 71% lipase, 35% protease, and 26% amylase producers from Tapovan Kund.

Ranawat and Rawat (2017) have studied thermophilic bacterial diversity of two hot springs, Soldhar and Suryakund of Uttarakhand Himalaya region. Identification of the isolates revealed the presence of *Bacillus*, *Paenibacillus*, *Geobacillus* spp., and *Brevibacillus* spp. in the hot springs. Among the bacterial population, 55% were amylase producers and 50% cellulase producers from suryakund while 34% amylolytic and 28% cellulolytic were from soldhar hot spring.

Kumar and Sharma (2020) assessed physico-chemical characteristics and microbial diversity of hot water samples collected from three important sacred hot water springs of the Uttarakhand Himalaya close to Yamunotri temple. A total of 15 physico-chemical parameters were studied for two consecutive years twice in a year. Twenty-two microbial strains were identified from Surya Kund having water temperature ranging from 86 to 89 °C, 22 microbial strains from Draupadi Kund having temperature in the range of 65–69 °C and 21 strains from Yamunotri Tapt Kund having temperatures from 45 to 48 °C. The organisms were identified as species of *Bacillus*, *Brevibacillus*, *Paenibacillus*, *Geobacillus*, *Actinobacillus*, *Streptomyces*, *Nocordiopsis*, *Thermobifida*, and some fungi.

Dutta et al. (2018) have explored microbial life within deep biosphere hosted by ~65 Myr old Deccan basalt and Archaean granitic basement by scientific deep drilling at Koyana, Western India. Quantitative PCR indicated a depth independent distribution of microorganisms predominated by bacteria. Abundance of dsrB and mcrA genes was found to be relatively higher (at least one order of magnitude/in basalt compared to granite. Bacterial communities were dominated by alpha-, beta-, Gammaproteobacteria, Actinobacteria, and Firmicutes while Euryarchaeota was the major archaeal group. Taxonomic affiliation at the genus level was possible only for 89% of total classified reads. *Acinetobacter*, *Corynebacterium*, *Pseudomonas*, and *Staphylococcus* were the most abundant genera. *Thermoplasma methanobacterium* and *methanosaeta* were the most abundant archaeal genera. Exploration of deep biosphere of granitic basaltic crustal system is highly imperative to answer the fundamental questions on microbial life within deep, dark, oligotrophic crust, their mode of interaction and role in Earth's biogeochemical processes, origin of life on Earth and even an extraplanetary locations (e.g., Mars). Compared to a reasonable level of understanding of deep life within marine subsurface, geomicrobiology of subterranean igneous rocks at greater depths remains almost elusive. It has been established that in contrast to sedimentary rocks crystalline igneous rocks are typically low in organic matter, porosity and interconnectivity and have been subjected to high temperature and/or pressure in the geologic past or at present. Scientific deep drilling in the Koyana/Warna region of Deccan traps, Western India provided a unique opportunity to investigate microbial life within deep terrestrial igneous rocks (basalts and granite-gneiss basement).

Bhusare and Wakte (2011) carried out microbiological and physicochemical analysis of water sample collected from hot sulfur spring located at Unkeshwar, Dist. Nanded, Maharashtra state. Four thermophilic isolates were obtained which were spore-forming and belonged to *Bacillus* sp. Optimum pH and temperature for the growth of one isolate BWU-1 were 9.0 and 50 °C. The isolate showed 97% similarity to *Bacillus licheniformis* and produced enzymes protease and amylase.

Patel et al. (2019a, b) have studied the bacterial diversity of Dholera hot springs of Gujarat, India during various seasons. Six well-isolated cultures out of total 18 thermophilic isolates obtained were found to belong to genera *Bacillus*, *Brevibacillus*, *Brevibacterium*, and *Tepidomonas* (species listed in table). The temperature of Dholera hot spring was 50 °C in summer, 45 °C in monsoon, and 40 °C in winter. Diversity of the isolates was studied in terms of Simpson, Shanon, Menhinick, Margalef, Berger-Parker, and Chao-1 based metabolic activity of thermophilic bacteria.

Solanki and Gupta (2013) studied thermophilic bacterium Thermonospora archaea isolated from sand dunes of Bikaner region Kodemdeser. Melting temperature of DNA from the thermophilic organism and *E. coli* as the mesophilic organism were determined by incubating DNA samples at 30 °C, 50 °C, 70 °C, 90 °C for 5 min and observing A260. Melting temperature on *E. coli* was 60 °C while that of thermophilic bacterium 80 °C. The results suggest one of the mechanisms of adaptation of bacteria to high temperature. As melting temperature is directly related to G+C%, thermophilic bacteria must have high G+C content which imparts stability to DNA at high temperature.

Sahoo et al. (2015) have investigated the bacterial diversity of hot springs of Odisha. Sediment samples from two hot springs (Atri and Tatapani, Fig. 4.4) of Odisha were subjected to 16S rRNA deep sequencing analysis, targeting V3 region

Fig. 4.4 Tattapani hot spring, Odisha (Google image)

using Illumina bar-coded sequencing. Atri and Tatapani metagenome were classified into 50 and 51 bacterial phyla. Proteobacteria (45.17%) dominated the Tatapani sample metagenome followed by Bacteriodetes (23.43%) and Cyanobacteria (10.48%). In Atri sample, Chloroflexi (52.39%), Nitrospirae (10.93%), and Proteobacteria (9.98%) were dominant. A large number of sequences remained taxonomically unresolved in both hot springs, indicating the presence of potentially novel microbes in these hot springs.

Dey and Chaphalkar (1998) investigated the thermophilic actinomycetes flora from a meteoritic crater of Lonar, Maharashtra, India. Five actinomycetes were selected on the basis of their bioactivity and antibiosis. The isolates were identified as *Streptomyces thermovulgaris* (SD1, SD3, and SDa) and *Streptomyces thermoviolaceus* (SD6 and SD7). All isolates produced protease, amylase, and lipase at 55 °C. The crude preparation was thermostable up to 85 °C with a half-life of 90 min and activity (Ca^{2+} dependent) at a pH range of 7.5–12. Culture broths from all isolates showed antimicrobial activity against *B. Subtilis*, *E. coli*, *M. luteus*, and *Aspergillus niger*.

Sharma et al. (2018) have explored more than 20 hot springs from the Himalayan regions of Jammu and Kashmir State (North-Western Himalayan State). Tatapani hot spring located in the Himalayan sub-range of Kishtwar district of Jammu and Kashmir, India (33° 19′0″ North, 75° 46′0″ East) was explored for isolation of thermophilic bacteria (Sharma et al. 2018). Two thermophilic bacteria (KSI and KSII) were isolated from the hot spring water sample. The isolates showed optimum growth at temperature 60 °C and pH 9.0. The isolate KSI showed 99% homology with *Flavobacterium thermophilum* and the isolate KSII showed 99% homology with *Anoxybacillus* sp. DR01. The isolates produced industrially important enzymes amylase, cellulase, and glutaminase.

Numerous hot springs dwell over a stretch of 300 km. along the West Coast of India grouped under West Coast geothermal province. Jaffer et al. (2019) explored thermal springs occurring in eight different localities in the Thane District of Maharashtra. Water samples were collected from Sativali spring (19° 37.85″ N and 72° 54.54″ E), Nimbavali spring (19° 30.434″ N and 73° 0.918″ E), and Ganeshpuri spring (19° 30.150″ N and 73° 0.841″ E) having temperature range 52–59 °C and pH 8.8 to 9. Thermophilic bacteria were obtained by enrichment in LB broth at 40 ± 2 °C and isolation on LB agar at 40 ± 2 °C. The isolate S1 (2014) from Sativali spring showed 100% similarity to *Bacillus licheniformis* strain UN1, the strain G (2014) from Ganeshpuri spring showed 100% similarity to *Bacillus licheniformis* strain UN1 and isolate N3 (2014) from Nimbavali showed 100% similarity to *Staphylococcus sciuri* strain 3–20.

Saxena et al. (2017) have performed a comprehensive metagenomic analysis of samples collected from seven different sites at three hot spring locations (Badi Anhoni, Chotti Anhoni, and Tattapani) in Central India. The hot springs namely Badi Anhoni and Chotti Anhoni are located (22.65° N latitude and 78.36° E longitude) approximately two km. apart from each other, near the hill station of Pachmarhi in the State of Madhya Pradesh, India. The Tattapani hot spring field is approximately 700 km. away from the Anhoni hot springs and is located in the Sarguja District of Chattisgarh State, India (23.41° N latitude and 83.39° E longitude). The temperature of Anhoni springs is mean 50.2 ± 5.98 °C and of Tattapani mean 73.88 ± 16.39 °C. Metagenomic analysis of the samples showed abundance of phyla *Proteobacteria* and *Thermi*. *Thermotogae* was abundant. The genus *Tepidimonas* was the most abundant genus, others were *Flavobacterium*, *Thermus*, *Acinatobacter*, *Fervidobacterium*, *Paracoccus*, *Acidovorax*, *Psudomonas*, *Stutzeria*, and *Thermus thermophilus*. An archaeal species *Pyrobaculum aerophilum* was also found. *Ramlibacter tataouinensis* was abundant in some samples. One of the interesting findings of this study is the presence of hydrocarbon degradation pathways in Anhoni samples. Complete metabolic pathways for degradation of toluene, xylene, benzoate, and benzene were found in various microbes from Anhoni samples. A high abundance of genes belonging to methane metabolism pathway was observed at Chhoti Anhoni hot spring with abundance of *Methylococcus capsulatus*.

Thakur et al. (2018) have investigated bacterial diversity of the Pangi-Chamba Himalaya (PCH) region, a very pristine, unique and virgin niche. A total of 454 pure bacterial isolates were obtained from diverse sites in PCH region and 111 isolates selected for 16S rDNA characterization. Identification of the bacteria revealed the presence of 28 genera representing phyla Firmicutes, Proteobacteria, Actinobacteria, and Bacteroidetes. In addition to abundant genera *Pseudomonas*, *Bacillus*, *Geobacillus*, *Arthrobacter*, *Paenibacillus*, and *Flavobacterium*, six putative novel bacteria and thermophiles from non-thermogenic sites were reported for the first time. Screening for cellulose degradation revealed that 70.92% of the bacteria were cellulolytic in broad pH and temperature range.

Patel et al. (2019a, b) studied the thermophilic bacterial diversity of Tuva Timba hot springs of Gujarat, India in different seasons. Diversity indices such as Simpson, Shanon, Menhinick, and Margalef index were also determined. The isolates were found to belong to genera *Bacillus*, *Brevibacillus*, *Geobacillus*, *Anoxybacillus*, and *Brevibacterium*. Many of the isolates showed potential in the production of enzymes catalase, caseinase, and amylase.

Kumar and Sharma (2019) studied physic-chemical properties (15 different parameters) and microbial diversity of water samples collected from two hot water springs (Ringigad and Soldhar) of the Garhwal Himalaya for a period of 2 years. The temperature of water ranged between 82 °C and 89 °C for Ringigad and between 87 °C and 92 °C for Saldhar. Seven bacterial strains, four archaeal and three fungal strains from Saldhar thermal spring, were identified and are listed in Table 4.1.

Pandey et al. (2015) have isolated 28 bacterial cultures from hot springs Soldhar (having 95 °C temperature) and Ringigad (having 80 °C temperature) of Uttarakhand, India. The isolates exhibited tolerance to a wide temperature range (20–80 °C) and a wide pH range (4–14). The isolates produced enzymes lipase, amylase, cellulose, protease, and xylanase. The thermophilic bacteria were identified as *Bacillus licheniformis*, *Paenibacillus chimensis*, *Bacillus sonorensis*, *Bacillus tequilensis*, and *Staphylococcus epidermidis*.

Debnath et al. (2009) have explored cyanobacterial flora of geothermal springs of Bakreswar, West Bengal. Geothermal springs in India, formed as a result of volcanic or techtonic activities, are characterized by high temperature and relatively abundant reduced compounds. These thermal springs are inhibited by characteristic thermophilic organisms including cyanobacteria. Cyanobacteria are among the few organisms that can occupy high-temperature aquatic environments including hot springs. In alkaline and neutral hot springs and streams flowing from them cyanobacteria can form thick colorful mats that exhibit bonding patterns. The present investigation involves the study of mat-forming cyanobacterial flora from hot springs located in Bakreswar, West Bengal, India (Fig. 4.5). The important species found are listed in Table 4.1. Their distribution pattern in relation to physico-chemical parameters of spring has also been studied. Three cyanobacterial strains were cultured and their pigment and nitrogen-fixing capacity were also studied. Nitrogen-fixing capacities of *Calothrix thermalis*, *Nostoc* sp., and *Fischerella thermalis* are 5.14, 0.29, and 2.60 nmole C_2H_4/μg of chl.a/h respectively. Carotenoid: chlorophyll-a ratio of four mat samples collected from Kharkunda, Suryakunda (Fig. 4.6), Dudhkunda and bathing pool are 2.45, 1.60, 1.48 and 1.34 respectively. Higher value of carotenoid: chlorophyll-a ratio coincided with higher temperature.

The studies reveal that India is rich in the diversity of thermophilic microorganisms.

4.7.2 Biotechnology of Thermophiles

Thermophilic microorganisms isolated from different ecological niches in India have been explored for their biotechnological potential and industrial applications.

Fig. 4.5 Bakreshwar hot spring, West Bengal (Google image)

Fig 4.6 Hot spring at Suryakund, Uttarakhand (Google image)

Verma et al. (2018) have investigated potential drought-tolerant plant growth-promoting bacterial consortium. The bacteria were isolated from the hot spring of the Chumthang area (Fig. 4.7) of Leh district of Ladakh region of India. *Bacillus* species (BHUJP-H, BHUJP-H_2, and BHUJP-H_3) were identified as *Bacillus subtilis* (KU312403), *Bacillus* sp. (KU312404), and *Bacillus licheniformis* (KU312405) based on 16S rRNA gene sequencing. Different treatment combinations of these three strains were evaluated for enhancing plant growth attributes of *Vigna radiata*. Such plant growth-promoting consortium can be further used as an effective microbial inoculant for enhancing the production of mungbean in field conditions.

Kiran et al. (2018) have isolated amylase-producing bacteria from the hot springs of Munger (Bihar, India). Water samples were collected from three different sites as Sita Kund, Rishikund, and Bhimbandh Kund with temperature 40–50 °C and pH 5.2–6.3.

Fig. 4.7 Hot spring in Chumthang area of Leh district of Ladakh region of India (Google image)

The isolate RK6 was selected for amylase characterization. The enzyme activity was highest at the optimum temperature of 80 °C. The isolate was identified as *Bacillus subtilis* RK6 (KX247637) based on 16S rRNA gene sequencing. The enzyme was thermoactive retaining its activity up to 90 °C after preheating for 30 min.

Rajakumari et al. (2018) have reviewed current existing technologies for the effective conversion of biomass, simultaneous saccharification and fermentation, and separate hydrolysis fermentation through consolidated bioprocessing techniques. Lignocellulosic biomass is the suitable raw material as a sustainable renewable resource in the production of biofuels. Thermophiles are robust, efficient, and can withstand very high range of temperatures during bioprocessing conditions. Thermophiles have high rates of conversion of biomass/lignocellulose to high rates of conversion of biomass/lignocellulose to biofuel due to the influence of thermostable activity at high temperatures. Thus, thermophiles would be a suitable candidate to develop bioconversion technologies.

Rai et al. (2009) have reported an alkaline protease-producing bacterial strain (AS-S24-II) from a soil sample in Assam. The organism was formed to grow in temperature ranging from 30 °C to 60 °C and salinity ranging from 0 to 7% (w/v) NaCl. The isolate represents novel species and identified as *Paenibacillus tezpurensis* sp. nov. (MTCC 8959). An alkaline serine protease, detergent stable, Ca^{2+}-independent with 43-kDA molecular weight was purified. The optimum activity was obtained at pH 9.5 and temperature 45–50 °C. The enzyme improved the wash performance of detergents.

Majumder et al. (2013) purified and characterized an extracellular thermolysin-like protease (TLP) from *Thermoactinomyces thalpophilus* MCMB-380 (Genebank accession no. EF397000) isolated from a thermal spring in Western Maharashtra, India. The molecular mass and isoelectric point of the enzyme were 34.4 kDa and 9.5 respectively. The proteolytic activity was inhibited by EDTA. The enzyme required Ca^{2+} for activity and thermostability. The enzyme was found to be a metalloprotease and showed potential in catalyzing the condensation reaction of two amino acids for the production of the dipeptide aspartame, an artificial sweetener. The TLP was stable at 80 °C for an hour.

Majumdar and Kanekar (2012) studied the production of thermolysin-like protease (TLP) from *Thermoactinomyces thalpophilus* PD1 isolated from thermal spring in Western Maharashtra, India at 55 °C and pH 7.2. The protease activity was studied using different nutrient media, carbon sources at 1% and 3% concentration, nitrogen sources, different ratios of carbohydrate to nitrogen source, tolerance to different concentrations of sodium, chloride, calcium carbonate, trace salts, and amino acids. Effect of inoculum size for seed culture from stock and for fermenter, agitation in flask culture production was also studied. The enzyme was capable of bringing about condensation/biotransformation of two amino acids, viz. Z-L-aspartic acid and L-phenylalanine methyl ester into aspartame (Fig. 4.8), which is an artificial, non-saccharide sweetener used as a sugar substitute in some foods and beverages. Optimization studies resulted in obtaining a maximum yield of 181 U/ml/min at 55 °C in glucose yeast extract medium of pH 7.2 containing $CaCl_2$ (0.2 % w/v) with an inoculum size of 0.5% v/v after 18 h of incubation under shake culture conditions (130 rpm).

Chitte and Dey (2002) purified a strong fibrin-specific fibrinolytic enzyme from the cell-free spent culture broth of a thermophilic organism, *Streptomyces megasporus* SD5. The enzyme was stable at a broad range of pH ranging from 5 to 9 and highly thermostable with 50% activity after storage at 60 °C for 6 months. The enzyme belonged to the serine endopeptidase group. In vitro clot lysis revealed that the enzyme was active at 37 °C.

Chitte et al. (2011) studied the production of fibrinolytic enzyme using 2.5 L glass fermenter. Culture of thermophilic *Streptomyces* sp. and glucose yeast extract peptone medium of pH 8.0. Five successive batches were carried out under controlled fermentation conditions viz. Agitation 140 rpm, aeration 0.5 vvm, 55 °C temperature and 18 h with yield of extracellularly produced protein of ~300–500 mg/L. The enzyme belongs to serine endopeptidase type and found to cleave fibrin filaments. The ex vivo activity of the enzyme was compared with 500 IU of urokinase and 350 IU of streptokinse. The ex vivo clot lysis was found to be faster as compared to the commercially available enzymes.

Chitte et al. (2014) studied endonuclease topoisomerase-like enzyme activity from thermophilic *Bacillus* sp. isolated from hot spring in Western Maharashtra. The molecular weight of the enzyme was ~ 33 kDa and optimum temperature for activity was 37 °C. Activity of the enzyme was enhanced in presence of ATP. The restriction pattern studies indicated that the enzyme cuts standard plasmid DNA.

Majumder and Kanekar (2017) have reviewed work on thermophiles with reference to the history of evolution of thermophiles, habitats of thermophiles, physiology of thermophiles, and proteases from thermophiles. Information on thermophilic microorganisms reported for the production of thermostable proteases is compiled.

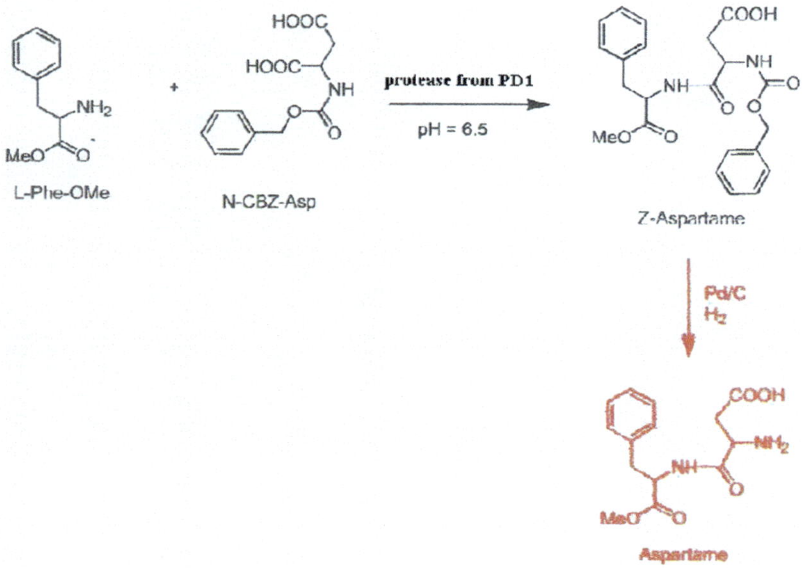

Fig. 4.8 Schematic presentation of production of Aspartame (Majumder 2007)

Over 50 thermophilic bacteria and archaea are enlisted. Commercially market-driven demand is to isolate novel wild-type thermophilic protease-producing microorganisms which will have good activity and stability under harsh conditions like the presence of organic solvents, detergents, surfactants, and oxidants. Thus, the thermophilic protease could be robust protease. Waste degradation by thermophilic proteases from aerobic and anaerobic organisms is a viable area.

Singh and Panda (2011) isolated and characterized thermophilic lipase-producing bacteria from an Indian hot spring. The isolate AK-P1 showed 100% homology with *Acinetobacter* sp.0180, AK-P2 showed 99.9% with *Brevibacillus borstelensis*, AR9 and isolate AK-p3 showed 99.2% with *Porphyrobacter cryptus* (T), ALC-2. Maximal lipase production by AK-P2 was observed at the end of the stationary phase (36h). The optimum temperature for this crude enzyme activity was at 60 °C and pH 10.0.

Mohanta and Rath (2008) isolated thermophilic bacteria from sulfur hot spring Tarabalo, Orissa, India. Out of 49 isolates, 77.75% isolates exhibited amylase, 97.95% isolates protease, 87.75% isolates caesinase, 83.67% isolates lipase, and 87.71% isolates cellulose activity.

Pathak and Mukundraj (2017) isolated a thermophilic bacterium from the Unkeshwar hot spring of Nanded district in Maharashtra. It was identified as *Bacillus globisporus* (APP11). The organism exhibited luxuriant growth and thermostable gelatinase production at 50 °C temperature. Maximum gelatinase production (675 ± 6.16 U/mg) from the organism was obtained using fish scale powder in the medium, after 24 h of time period. Potentially purified enzyme was remarkably active at pH range of 6–9 and temperature range of 35–65 °C.

Chitte and Dey (2000) studied a fibrinolytic enzyme from thermophilic *Streptomyces megasporus* strain SD5 produced at 55 °C temperature and pH 8. The enzyme

Fig. 4.9 Hot spring Vajreshwari, Maharashtra (Google image)

was purified and found to have a molecular weight of 35 kDa. The enzyme was resistant to a broad range of pH (6–9) and temperature (37–60 °C). The in vitro clot lysis by the enzyme at 37 °C was encouraging and isolated from a hot spring situated in the Western Maharashtra (Fig. 4.9), India. The enzyme being fibrinolytic has a potential as a therapeutic agent in thrombosis.

Venugopalan et al. (2008) have isolated six thermophilic bacterial strains from municipal solid waste, Delhi, India in 2006. All the isolates were screened for their antimicrobial activity. Four strains identified as *Brevibacillus borstelensis*, *Bacillus galactosidilyticus*, and *Bacillus licheniformis* RH101 and RH104 showed antimicrobial activity.

Chaphalkar and Dey (1998) have reported thermostable alkaline metalloprotease (EC3.4.24.4) from *Streptomyces diastaticus* strain SS1 isolated from Lonar Lake. The enzyme was characterized and found to have requirement of Ca^{2+} for both activity and thermal stability. The half-life of the enzyme at 55 °C was 24 h.

Desai and Dhala (1969) isolated proteolytic enzymes from two selected cultures of thermophilic actinomycetes, *Thermomonospora fusca* (A29) and *Thermoactinomyces vulgaris* (A60). Molecular weights of 21,500 and 23,800 were calculated for the active peaks of the enzymes. The enzymes were most active between the temperature range of 60 and 70 °C.

From the studies, major interest in biotechnological aspects appears to be in thermophilic enzymes.

4.7.3 Role of Thermophiles in Sustainability

Environmental pollution caused by toxic metals is a severe threat to biota and human life. Metal-microbe interactive processes have resulted in bioremediation of sites contaminated with metals. Sar et al. (2013) have reviewed the role of thermophilic microorganisms in metal bioremediation. Thermophilic microorganisms, owing to their natural ability to survive and flourish under elevated temperatures along with other stressful environmental conditions including high concentrations of heavy metals, have developed various adaptation strategies to cope up with harsh environments. Thermophilic microorganisms, being common in geological and anthropogenic thermal environments with high concentration of dissolved metal, possess unique cell wall structures and metabolic and enzymatic properties that may contribute in metals-thermophiles interaction. Biosorption or bioaccumulation of metals is the most effective and wildly used approach for bioremediation. Microbial transformation of metal through oxido-reduction reactions plays a critical role in metal speciation, distribution, and thus altered toxicity in the ecosystems implemented in metal remediation. Thermophilic bacteria with higher metal tolerance and metabolic characteristics at high temperature may exhibit enhanced metal solubilization through sulfur- or iron-oxidizing processes. Thermophilic microbial community can perform both degradative and productive functions through coupling of metal reduction with oxidation of a variety of organic and inorganic substrates. Thermophilic bacteria are also able to reduce a wide spectrum of metals and can be used for the immobilization of toxic metals/radionuclides during bioremediation of hot waste water of disposable sites of radioactive wastes having a temperature range favorable for thermophiles for a long period of time.

Chitte et al. (1999) have reported keratinolytic activity in the culture broth of a feather-degrading thermophilic *Streptomyces thermoviolaceus* SD8 isolated from Lonar Lake, Maharashtra. The optimum pH and temperature for the production of the enzymes were 8 and 55 °C respectively. The enzyme was stable at a pH range of 6.5–8.5 and 65 °C temperature. The enzyme could hydrolyze fibrin, collagen, nail and hair and could produce leucine, threonine and tyrosine from feather. It could be a useful enzyme treatment of fish waste, waste hair, waste feather (poultry waste), etc.

Chitte and Dey (2003) have filed a US patent. The present invention relates to novel microorganism *Streptomyces megasporus* SD5. The present invention also relates to a process for the isolation of said *Streptomyces megasporus* SD5. The invention also relates to a novel fibrinolytic enzyme actinokinase extracted from said microorganism and to a process for the extraction of said enzyme. In another aspect, the invention also pertains to a method for the treatment of thrombolytic disorders using said enzyme.

Patke and Dey (2002) have reported multiple proteases from a thermophilic actinomycete *Streptomyces megasporus* SD5 after 18 h growth at 55 °C temperature. The enzyme preparation exhibited activity over a broad pH and temperature range of 6–12 and 25 to 85 °C respectively. Thermostability was enhanced in presence of calcium and half-life was 30 min at 85 °C.

4.8 Contribution to the Knowledge of Thermophilic Microbial Species, New to Science, from India

Three novel species of thermophilic microorganisms have been added to the global collection of microorganisms and are listed in Table 4.2

Table 4.2 Thermophilic microbial species, new to Science, from India

Ref no.	Microorganism name	Location	Ref
1	*Thiomonas bhubaneswarensis* sp. Nov.	Hot spring in Orissa	Panda et al. 2009
2	*Paenibacillus tezpurensis* sp. nov. AS-S23-	Tezpur, Assam	Rai et al. 2009
3	*Thermodesulfobacterium commune* sp. Nov.	Hot water springs of Thane Maharashtra	Patil and Unnithan 2014

4.9 Future of Thermophiles

Thermophilic bacteria may be among the last life on Earth, perhaps surviving 2.8 Gyr into the future. According to Malley James et al. (2013), future biosphere on Earth (as with its past) will be made up of predominantly of unicellular microorganisms. Unicellular life was probably present for at least 2.5 Gyr before multicellular life appeared and will likely be the only form of life capable of surviving on the planet in the far future when the aging of the Sun causes environmental conditions to become more hostile to more complex forms of life. The end stage of habitability on Earth was the focus of this study. A simple, latitude-based climate model incorporating eccentricity and obliquity variations was used as a guide to the temperature evolution of the Earth over the next 3 Gyr. This allows inferences to be made about potential refuges for life, particularly in mountains and cold-trap (ice) caves and what forms of life could live in these environments. Results suggest that in high latitude regions, unicellular life could persist for up to 2.8 Gyr from present. In two billion years' time, life on Earth will be confined to pockets of liquid water deep underground. A combination of slow and rapid environmental changes will result in the extinction of all species on Earth, with the last inhabitants disappearing within 2.8 Gyr from present. The main driver for these changes will be the Sun. As it ages over the next few billion years, the sun will remain stable but become steadily more luminous, increasing the intensity of its heat felt on Earth and warming the planet to such an extent that the oceans evaporate. Within the next billion years, increased evaporation rates and chemical reactions with rain water will draw more and more carbon dioxide from the Earth's atmosphere. The falling levels of CO_2 will lead to the disappearance of plants and animals and our home planet will become a world of microbes. At the same time the Earth will be depleted of oxygen and will be drying out as the rising temperatures lead to the evaporation of the oceans. A billion years after that the oceans will have gone completely. The far-future Earth will be very hostile to life by this point. All living things require liquid water, so any remaining life will be restricted to pockets of liquid water, perhaps at cooler, higher altitudes or in caves or underground. This life will need to cope with many extremes like high temperature and intense ultraviolet radiation and only a few microbial species known on Earth today could cope with this.

Thus the thermophiles, very amazing microorganisms, are said to be near to origin of life and next the only future of life.

References

Allen MB (1953) The thermophilic aerobic spore-forming bacteria. Bacteriol Rev 17(2):125–173

Bader J, Skelac L, Wewetzer S, Senz M, Popović MK, Bajpai R (2012) Effect of partial pressure of CO_2 on the production of thermostable alpha-amylase and neutral protease by *Bacillus caldolyticus*. Prikl Biokhim Mikrobiol 48(2):206–211

Barns SM, Fundyga RE, Jeffries MW, Pace NR (1994) Remarkable archaeal diversity detected in a Yellowstone National Park hot spring environment. Proc Natl Acad Sci U S A 91(5): 1609–1613. https://doi.org/10.1073/pnas.91.5.1609

Bergey DH (1919) Thermophilic bacteria. J Bacteriol 4:301–306

Bertin PB, Lozzi SP, Howell JK, Restrepo-Cadavid G, Neves D, Teixeira AR, de Sousa MV, Norris SJ, Santana JM (2005) The thermophilic, homohexameric aminopeptidase of Borrelia burgdorferi is a member of the M29 family of metallopeptidases. Infect Immun 73(4): 2253–2261

Bhalla TC (2015) Microbial diversity of Indian Himalayan Region: Himachal Pradesh. ENVIS Newsl Himalayan Ecol 12(4):2015

Bhardwaj KN, Tiwari SC, Bahuguna YM (2010) Screening of thermophilic cyanobacteria isolated from Tapoban geothermal field, Uttarakhand, Himalaya for the production of antibacterial compounds. Asian J Exp Biol Sci 1(4):787–791

Bhusare DU, Wakte PS (2011) Microbiological and physiochemical attributes of hot water sulphur spring of Unkeshwar. J Exp Sci 2(4):4–6

Bora L, Kalita M (2006) Extracellular enzymatic activity profiles of bacterial strains isolated from hotsprings of West Kameng district of Arunachal Pradesh, India. J Microbiol 4(1):1095

Brock TD (1985) Life at high temperatures. Science 230:132–138

Brock TD (1986) Thermophiles. General, molecular and applied microbiology. Wiley, New York, pp 1–16

Brock TD, Freeze H (1969) *Thermus aquaticus* gen. n. and sp. n., a nonsporulating extreme thermophile. J Bacteriol 98(1):289–297. https://doi.org/10.1128/jb.98.1.289-297.1969

Chaphalkar SR, Dey S (1994) Some aspects of production of extracellular protease from *Streptomyces diastaticus*. J Microbiol Biotechnol 9(2):85–100

Chaphalkar SR, Dey S (1998) Thermostable alkaline metallooprotease from newly isolated alkalophilic *Streptomyces diastaticus* strain SSI. Indian J Biochem Biophys 35:34–40

Charbonneau DM, Meddeb-Mouelhi F, Boissinot M, Sirois M, Beauregard M (2012) Identification of thermophilic bacterial strains producing thermotolerant hydrolytic enzymes from manure compost. Indian J Microbiol 52(1):41–47

Chen XG, Stabnikova O, Tay JH, Wang JY, Tay ST (2004) Thermoactive extracellular proteases of *Geobacillus caldoproteolyticus* sp. nov., from sewage sludge. Extremophiles 8(6):489–498

Chien A, Edgar DB, Trela JM (1976) Deoxyribonucleic acid polymerase from the extreme thermophile *Thermus aquaticus*. J Bacteriol 127(3):1550–1557

Chitte RR, Dey S (2000) Potent fibrinolytic enzyme from a thermophilic *Streptomyces megasporus* strain SD5. Lett Appl Microbiol 31:405–410

Chitte RR, Dey S (2002) Production of a fibrinolytic enzyme by thermophilic *Streptomyces* species. World J Microbiol Biotechnol 18:289–284

Chitte RR, Dey S (2003) *Streptomyces megasporus* SD5, process for the isolation thereof, novel fibrnolytic enzyme prepared here from, process for the production of said enzyme and method of treatment of thrombolytic disorders using said enzyme. United States Patent No.: US 6,638,503 B2

Chitte RR, Nalawade VK, Dey S (1999) Keratinolytic activity from the broth of a feather degrading thermophilic *Streptomyces thermoviolaceus* strain. SD8. Lett Appl Microbiol 28:131–136

Chitte RR, Deshmukh SV, Kanekar PP (2011) Production, purification, and biochemical characterization of a fibrinolytic enzyme from thermophilic *Streptomyces* sp. MCMB-379. Appl Biochem Biotechnol 165(5-6):1406–1413. https://doi.org/10.1007/s12010-011-9356-2

Chitte RR, Deshmukh SV, Kanekar PP (2014) Detection and isolation of restriction endonuclease like topoisomerase from thermophilic *Bacillus subtilis* strain RC1. Int J Pharm Sci Res 5(12): 5250–5254

Claus D, Lack P, Neu B (1983) Catalog of strains (German Collection of Microorganisms, Gottigen, Fedearl Republic of Germany, 1983 and Supplement 1984)

De Azeredo LA, De Lima MB, Coelho RR, Freire DM (2006) Thermophilic protease production by Streptomyces sp. 594 in submerged and solid-state fermentations using feather meal. J Appl Microbiol 100(4):641–647

De Felice M, Medagli B, Esposito L, De Falco M, Pucci B, Rossi M, Grùz P, Nohmi T, Pisani FM (2007) Biochemical evidence of a physical interaction between Sulfolobus solfataricus B-family and Y-family DNA polymerases. Extremophiles 11(2):277–282

Debnath M, Mandal NC, Ray S (2009) The study of cyanobacterial flora from geothermal springs of Bakreswar, West Bengal, India. Algae 24(4):185–193. https://doi.org/10.4490/ALGAE.2009.24.4.185

Desai AJ, Dhala SA (1969) Purification and properties of proteolytic enzymes from thermophilic actinomycetes. J Bacteriol 100(1):149–155

Dey S, Chaphalkar SR (1998) Thermophilic streptomycete flora of a meteoritic crater. Actinomycetes 9(3):37–45

Dudhagara P, Bhavsar S, Ghelani A, Bhatt S (2014) Isolation, characterization and investing the industrial applications of thermostable and solvent tolerant serine protease from hot spring isolated Thermophillic *Bacillus licheniformis*U1. Int J Appl Sci Biotechnol 1:75–82

Dutta A, Gupta SD, Gupta A, Sarkar J, Roy S, Mukherjee A, Sar P (2018) Exploration of deep terrestrial subsurface microbiome in Late Cretaceous Deccan traps and underlying Archean basement, India. Sci Rep 8:17459. https://doi.org/10.1038/s41598-018-35940-0

Ghosh P, Mandal S, Pal SL, Bandyopadhyaya G, Chattopadhyay B (2006) Development of bioconcrete material using an enrichment culture of novel thermophilic anaerobic bacteria. Indian J Exp Biol 44(4):336–339

Gödde C, Sahm K, Brouns SJ, Kluskens LD, van der Oost J, de Vos WM, Antranikian G (2005) Cloning and expression of island is in, a new thermostable subtilisin from Fervidobacterium islandicum, in Escherichia coli. Appl Environ Microbiol 71(7):3951–3958

Gonzalez-Siso MI (2019) Thermophiles and thermozymes. Microorganisms 7:62

Goodfellow M, Mordarski M, Williams ST (1984) The biology of the actinomycetes. Academic Press Inc., London, pp 481–513

Grütter M, Hawkes R, Matthews B (1979) Molecular basis of thermostability in the lysozyme from bacteriophage T4. Nature 277:667–669

Haki GD, Rakshit SK (2003) Developments in industrially important thermostable enzymes: a review. Bioresour Technol 89:17–34

Heinrikson RL (1977) Applications of thermolysin in protein structural analysis. Methods Enzymol 47:175–189

Hiraga K, Nishikata Y, Namwong S, Tanasupawat S, Takada K, Oda K (2005) Purification and characterization of serine proteinase from a halophilic bacterium, *Filobacillus* sp. RF2–5. Biosci Biotechnol Biochem 69(1):38–44

Huang Y, Sun Y, Ma S, Chen; L., Zhang, H., Deng, Y. (2013) Isolation and characterization of *Keratinibaculum paraultunense* gen. nov., sp. nov., a novel thermophilic, anaerobic bacterium with keratinolytic activity. FEMS Microbiol Lett 345(1):56–63

Isowa Y, Ohmori M, Ichikawa T, Mori K, Nonaka Y, Kihra K, Oyama K, Satoh H, Nishimura S (1979) Thermolysin-catalyzed condensation reactions of N-substituted aspartic and glutamic acids with phenylalanine alkylesters. Tetrahedron Lett 28:2611–2612

Itoi Y, Horinaka M, Tsujimoto Y, Matsui H, Watanabe K (2006) Characteristic features in the structure and collagen-binding ability of a thermophilic collagenolytic protease from the thermophile Geobacillus collagenovorans MO–1. J Bacteriol 188(18):6572–6579

Jaffer Y, Ganie PA, Vinothkumar R, Bhat I, Ishfaq NM, Ganie A, Bhat A, Alagarsamy V (2019) Isolation and characterization of thermophilic bacteria from Maharashtra hot springs: *Bacillus* sp. and *Staphylococcus* sp. J Entomol Zool 7(1):691–695

Juhász T, Szeltner Z, Polgár L (2006) Properties of the prolyl oligopeptidase homologue from Pyrococcus furiosus. FEBS Lett 580(14):3493–3497

Kählig H, Kolarich D, Zayni S, Scheberl A, Kosma P, Schäffer C, Messner P (2005) N-acetylmuramic acid as capping element of alpha-D-fucose-containing S-layer glycoprotein glycans from *Geobacillus epidamans* GS5–97T. J Biol Chem 280(21):20292–20299

Kataoka M, Yamaoka A, Kawasaki K, Shigeri Y, Watanabe K (2014) Extraordinary denaturant tolerance of keratinolytic protease complex assemblies produced by Meiothermus ruber H328. Appl Microbiol Biotechnol 98(7):2973–2980

Kaula WH (1975) Seven ages of a planet. Icarus 26:1–15

Kikani BA, Singh SP (2011) Single step purification and characterization of a thermostable and calcium independent amylase from, *Bacillus amyloliquefaciens* TSWK-1 isolated from Tulsi Shyam hot spring reservoir, Gujarat (India). Int J Biol Macromol 48(4):676–681. https://doi.org/10.1016/j.ijbiomac.2011.02.010

Kikani BA, Sharma AK, Singh SP (2015) Culture dependent diversity and phylogeny of thermophilic bacilli from a natural hot spring reservoir in the Gir forest, Gujarat (India). Microbiology 84(5):687–700

Kiran S, Singh A, Prabha C, Kumari S, Kumari S (2018) Isolation and characterization of thermostable amylase producing bacteria from hot springs of Bihar, India. Int J Pharm Med Biol Sci 7(2):28–34

Kocabiyik S, Demirok B (2009) Cloning and overexpression of a thermostable signal peptide peptidase (SppA) from *Thermoplasma volcanium* GSS1 in *E. coli*. Biotechnol J 4(7):1055–1065

Kotresh KR, Neelagund YS, Gurumurthy DM (2020) Novel *Geobacillus thermoleovorns* KNG 112 thermophilic bacteria from Bandaru hot spring: a potential producer of thermostable enzymes. Asian J Pharm Clin Res 13(1):134–141

Kumar S, Nussinov R (2001) How do thermophilic proteins deal with heat? A review. Cell Mol Life Sci 58:1216–1233

Kumar R, Sharma RC (2019) Microbial diversity and physico-chemical attributes of two hot water springs in the Garhwal Himalaya, India. J Microbiol Biotechnol Food Sci 8(6):1249–1253

Kumar R, Sharma RC (2020) Microbial diversity in relation to physico-chemical properties of hot water ponds located in the Yamunotri landscape of Garhwal Himalaya. Heliyon 6:e04850

Lacey J (1989) Bergey's manual of systematic bacteriology, vol 4. Williams and Wilkins Publication, Philadelphia, pp 2573–2585

Langworthy TA (1978) Biochemistry of thermophily: membranes and lipids of extremely thermoacidophilic microorganisms. In: Langworthy TA, Shilo M, Hirsch P (eds) Strategies of microbial life in extreme environments: report of the Dahlem workshop on strategy of life in extreme environments. Wiley, Berlin, p 513

Li X, Ma H, Wang Q, Matsumoto S, Maeda T, Ogawa HI (2009) Isolation, identification of sludge-lysing strain and its utilization in thermophilic aerobic digestion for waste activated sludge. Bioresour Technol 100(9):2475–2481

Lineweaver CH, Schwartzman D (2004) Cosmic thermobiology. Thermal constraints on the origin and evolution of life in the Universe. In: Seekbach J (ed) Genesis, evolution and diversity of life. Kluwer, Dordrecht, pp 233–248

Lopez G (1999) DNA super coiling and temperature adaptation: a clue to early diversification of life. J Mol Evol 46:439–452

Maehara T, Hoshino T, Nakamura A (2008) Characterization of three putative Lon proteases of *Thermus thermophilus* HB27 and use of their defective mutants as hosts for production of heterologous proteins. Extremophiles 12(2):285–296

Mahala SC (2019) Genesis of thermal springs. In: Geology, chemistry and genesis of thermal springs of Odisha, India, SpringerBriefs in earth sciences. Springer, Cham. https://doi.org/10.1007/978-3-319-90002-5_7

Majeed T, Tabassum R, Orts WJ, Lee CC (2013) Expression and characterization of *Coprothermobacter proteolyticus* alkaline serine protease. Sci World J 25:396156

Majumdar DR, Kanekar PP (2012) Different aspects of production of thermolysin-like protease from *Thermoactinomyces thalpophilus*. Int J Pharm Bio-Sci 3(1):610–627

Majumder DR (2007) Production of thermolysin like bioactive molecules by thermophilic bacteria from pristine habitats. Ph.D. thesis, Pune University, Pune, India

Majumder DR, Kanekar PP (2017) Proteases from thermophiles and their industrial importance. In: Thangadurai D, Jayabalan S (eds) Industrial biotechnology – sustainable production and bioresource utilization. Apple Academic Press Inc., Burlington, pp 189–208

Majumder DR, Gaikwad SM, Kanekar PP (2013) Purification and characterization of a thermolysin like protease from Thermoactinomyces thalpophilus MCMB-38. Protein Pept Lett 20(8):918–925

Malley-James JTO, Greaves JS, Raven J, Cockell CS (2013) Swansong biospheres: refuges for life and novel microbial biospheres on terrestrial planets near the end of their habitable lifetimes. Int J Astrobiol 12(2):99–112. https://doi.org/10.1017/5147355041200047x

Martin W, Baross J, Kelley D, Russell MJ (2008) Hydrothermal vents and the origin of life. Nat Rev Microbiol 6:805–814. https://doi.org/10.1038/nrmicro1991

Matsubara H (1970) Purification and assay of thermolysin methods. Enzymology 19:642–651

Matsumura M, Katakura Y, Imanaka T, Aiba S (1984) Enzymatic and nucleotide sequence studies of a kanamycin-inactivating enzyme encoded by a plasmid from thermophilic bacilli in comparison with that encoded by plasmid pUB110. J Bacteriol 160(1):413–420. https://doi.org/10.1128/jb.160.1.413-420.1984

Mohammad BT, Al Daghistani HI, Jaouani A, Abdel-Latif S, Kennes C (2017) Isolation and characterization of thermophilic bacteria from Jordanian hot springs: *Bacillus licheniformis* and *Thermomonas hydrothermalis* isolates as potential producers of thermostable enzymes. Int J Microbiol 2017:6943952. https://doi.org/10.1155/2017/6943952

Mohanta H, Rath C (2008) Extracellular enzymatic activity of bacterial strains isolated from a local Hotspring Tarabalo, Nayagarh District, Orissa, India. Internet J Microbiol 7(2):11

Panda SK, Jyoti V, Bhandra B, Nayak KC, Shivaji S, Rainey FA, Das SK (2009) *Thiomonas bhubaneswarensis* sp. nov., an obligately mixotrophic moderately thermophilic Thiosulphate oxidizingbacterium. Int J Syst Evol Microbiol 59(9):2171–2175

Panda MK, Sahu MK, Tayung K (2013) Isolation and characterization of a thermophilic *Bacillus* sp. with protease activity isolated from hot spring of Tarabalo, Odisha, India. Iran J Microbiol 5(2):159–165

Pandey A, Dhakar K, Sharma A, Priti P, Sati P, Kumar B (2014) Thermophilic bacteria that tolerate a wide temperature and pH range colonize the Soldhar (95 °C) and Ringigad (80 °C) hot springs of Uttarakhand, India. Ann Microbiol 65(2):809–816. https://doi.org/10.1007/s13213-014-0921-0

Pandey A, Dhakar K, Sharma A, Priti P, Sati P, Kumar B (2015) Thermophilic bacteria that tolerate a wide temperature and pH range colonize the Soldhar (95°C) and Ringigad (80°C) hot springs of Uttarakhand, India. Ann Microbiol 65(2):809–816

Patel DN, Dave SR, Braganza VJ, Modi HA (2019a) Assessment of culturable microbial diversity of Dholera thermal springs of Gujarat, India. J Exp Biol Agric Sci 7(1):57–64

Patel DN, Dave SR, Braganza VJ, Modi HA (2019b) Seasonal variation in bacterial diversity of Tuva Timba thermal springs of Gujarat, India. Int Res J Biol Sci 8(2):6–14

Pathak RAP, Mukundraj G (2017) Production and characterization of thermostable gelatinase from *Bacillus globisporus* isolated from Unkeshwar hot spring of Maharashtra. IJMS 46(9): 1883–1888

Patil SZ, Unnithan AR (2014) Biosulphidogenesis: isolation and characterization of *Thermodesulfo* bacterium commune sp. nov. isolated from hot water springs of Thane, Maharashtra. IQSR J Pharm Biol Sci 9:9–13

Patke D, Dey S (1998) Proteolytic activity from a thermophilic *Streptomyces megasporus*, strain SDP4. Lett Appl Microbiol 26:171–174

Patke D, Dey S (2002) Proteolytic activity from a thermophilic *Streptomyces megasporus* strain SDP4. Lett Appl Microbiol 26(3):171–174. https://doi.org/10.1046/j.1472-765X.1998.00300.x

Perutz MF (1978) Electrostatic effects in proteins. Science 201(4362):1187–1191. https://doi.org/10.1126/science.694508

Petrova DH, Shishkov SA, Vlahov SS (2006) Novel thermostable serine collagenase from Thermoactinomyces sp. 21E: purification and some properties. J Basic Microbiol 46(4): 275–285

Phrommao E, Rodtong S, Yongsawatdigul J (2011) Identification of novel halotolerant bacillopeptidase F-like proteinases from a moderately halophilic bacterium, *Virgibacillus* sp. SK37. J Appl Microbiol 110(1):191–201

Price NC, Duncan D, McAlister JW (1985) Inactivation of rabbit muscle phosphoglycerate mutane by limited proteolysis with thermolysin. Biochem J 229:167–171

Rai S, Roy J, Mukherjee AK (2009) Characterization of a detergent-stable alkaline protease from a novel thermophilic strain *Paenibacillus tezpurensis* sp. nov. AS-S24-II. Appl Microbiol Biotechnol 85(5):1437–1450. https://doi.org/10.1007/s00253-009-2145-y

Rajakumari A, Thiruchelvi R, Venkataraghavan R, Biswal AR (2018) Action of thermophiles on plant biomass to bioenergy- a review. Res J Pharm Technol 11(9):4191–4199. https://doi.org/10.5958/0974-360X.2018.00769.2

Ranawat P, Rawat S (2017) Characterization of bacterial diversity of hot springs of Chamoli region of Garhwal Himalaya. ENVIS Bull Himalayan Ecol 25:120–128

Rath CC, Subramanyam VR (1998) Isolation of thermophilic bacteria from hot springs of Orissa, India. Geobiology 25:113–119

Rath CC, Subramanyam VR (2000) Enhanced protease and beta-lactamase activity by immobilization of a thermophilic *Bacillus* sp. isolated from a local hot springs in Orissa, India. In: Proceeding of national conference on recent trends in biotechnology

Reysenbach AL, Cady SL (2001) Microbiology of ancient and modern hydrothermal systems. Trends Microbiol 9(2):79–86

Rysenbach AL, Wickham GS, Pace NR (1994) Phylogenetic analysis of the hyperthermophilic pink filament community in Octopus spring, Yellowstone National Park. Appl Environ Microbiol 60(6):2113–2119

Rysenbach AL, Ehringer M, Hershberger K (2000) Microbial diversity of 83 degree Celsius in calcite springs, Yellowstone National Park: another environment where the Aquificales and Korarchaeota co-exist. Extremophiles 4(1):61–67

Sahoo RK, Subudhi E, Kumar M (2015) Investigation of bacterial diversity of hot springs of Odisha, India. Genomics Data 6:188–190

Sahu A, Kaur P, Bhattaacharjya A, Amat D, Sahu N, Bharati K, Tiwari A, Singh AI (2020) Indian hot water springs: potential source of thermophilic microbes. Soil Health Manage 3(2):15–18

Sakaguchi M, Niimiya K, Takezawa M, Toki T, Sugahara Y, Kawakita M (2008) Construction of an expression system for aqualysin I in Escherichia coli that gives a markedly improved yield of the enzyme protein. Biosci Biotechnol Biochem 72(8):2012–2018

Saleem M, Rehman A, Yasmin R, Munir B (2012) Biochemical analysis and investigation on the prospective applications of alkaline protease from a Bacillus cereus strain. Mol Biol Rep 39(6): 6399–6408

Salvesan G, Nagase H (1996) Inhibition of proteolytic enzymes. In: Beynon RJ, Bord JS (eds) Proteolytic enzymes: a practical approach. IRL Press, Oxford, pp 83–104

Sar P, Kazy SK, Paul D, Sarkar A (2013) Metal bioremediation by thermophilic microorganisms. In: Satyanarayana T, Littlechild J, Kawarabayasi Y (eds) Thermophilic microbes in environmental and industrial biotechnology. Springer, Dordrecht. https://doi.org/10.1007/978-94-007-5899-5_6

Satyanarayana T, Little-child J, Kawarabayasi Y (2013) Therrmophilic microbes in environmental and industrial biotechnology: biotechnology of thermophiles, 2nd edn. Springer, Cham, p 954

Saxena R, Dhakan DB, Mittal P, Waikar P, Chowdhury A, Ghatak A, Sharma VK (2017) Metagenomic analysis of hot springs in Central India reveals hydrocarbon degrading thermophiles and pathways essential for survival in extreme environments. Front Microbiol 7: 2123. https://doi.org/10.3389/fmicb.2016.02123

Sharma A, Pandey A, Shouche S, Kumar B, Kulkarni G (2009) Characterization and identification of *Geobacillus* spp. isolated from Soldhar hot spring site of Garhwal, Himalaya, India. J Basic Microbiol 48:187–194

Sharma N, Vyas G, Pathania S (2013) Culturable diversity of thermophilic microorganisms found in hot springs of Northern Himalayas and to explore their potential for production of industrially important enzymes. Sch Acad J Biosci 1(5):165–178

Sharma S, Sharma P, Sourirajan A, Baumler DJ, Dev K (2018) Identification of thermophilic *Flavobacterium* and *Anoxybacillus* in unexplored tatapani hot spring of Kishtwar District of Jammu and Kashmir: a North Western Himalayan State. Curr Trends Biotechnol Pharm 12(3): 245–256

Singh BS, Panda AK (2011) Isolation and identification of new lipolytic thermophilic bacteria from an Indian hot spring. Int J Pharm Bio Sci 2(2):229–235

Solanki A, Gupta D (2013) Studies on adaptations of thermophilic bacteria at molecular level. Res J Chem Environ Sci 1(2):29–31

Solomatob VS (2004) The role of plate tectonics astrobiology newsletter. J Geophys 108:1029

Sudan SK, Kumar N, Kaur I, Sahni G (2018) Production, purification and characterization of raw starch hydrolyzing thermostable acidic α-amylase from hot springs, India. Int J Biol Macromol 117:831–839. https://doi.org/10.1016/j.ijbiomac.2018.05.231

Thakur V, Kumar V, Kumar S, Singh D (2018) Diverse culturable bacterial communities with cellulolytic potential revealed from pristine habitat in Indian trans-Himalaya. Can J Microbiol. https://doi.org/10.1139/cjm-2017-0754

Toplak A, Wu B, Fusetti F, Quaedflieg PJ, Janssen DB (2013) Proteolysin, a novel highly thermostable and cosolvent-compatible protease from the thermophilic bacterium Coprothermobacter proteolyticus. Appl Environ Microbiol 79(18):5625–5632

Uchino F, Doi S (1967) Acido-thermophilic bacteria from thermal waters. Agric Biol Chem 31(7): 817–822. https://doi.org/10.1080/00021369.1967.10858890

Veltman OR, Vriend G, Hardy F, Mansfeld J, Van Den Burg B, Venema G, Eijsink VG (1997) Mutational analysis of a surface area that is critical for the thermal stability of thermolysin like protease. Eur J Biochem 248:433–440

Venugopalan V, Singh B, Verma N, Nahar P, Bora TC, Das RH, Gautam HK (2008) Screening of thermophiles from municipal solid waste and their selective antimicrobial profile. Curr Res Bacteriol 1:17–22

Verma JP, Jaiswal DK, Krishna R, Prakash S, Yadav J, Singh V (2018) Characterization and screening of thermophilic bacillus strains for developing plant growth promoting consortium from hot spring of Leh and Ladakh Region of India. Front Microbiol. https://doi.org/10.3389/fmicb.2018.01293

Yallop CA, Edwards C, Williams ST (1997) Isolation and growth physiology of novel thermoactinomycetes. J Appl Microbiol 83:685–692

Yun SH, Choi CW, Kwon SO, Lee YG, Chung YH, Jung HJ, Kim YJ, Lee JH, Choi JS, Kim S, Kim SI (2011) Enrichment and proteome analysis of a hyperthermostable protein set of archaeon Thermococcus onnurineus NA1. Extremophiles 15(4):451–461

Zabolotskaya MV, Demidyuk IV, Akimkina TV, Kostrov SV (2004) A novel neutral protease from *Thermoactinomyces* species 27a: sequencing of the gene, purification, and characterization of the enzyme. Protein J 23(7):483–492

Zhan D, Sun J, Feng Y, Han W (2014) Theoretical study on the allosteric regulation of an oligomeric protease from Pyrococcus horikoshii by Cl^- ion. Molecules 19(2):1828–1842

Zierenberg RA, Adams MWW, Arp AJ (2000) Life in extreme environments. Proc Natl Acad Sci U S A 97(24):12961–12962

Acidophilic Microorganisms

5

Abstract

Acidophilic microorganisms thrive in acidic environment (pH 3.0 and below). They form diverse group like chemolithotrophic, heterotrophic prokaryotes and phototrophic eukaryotes. They occur in natural acidic habitats like sulfur springs, sulfuric pools and man-made environments such as acid mine drainage, acid rock drainage, mining of coal and metal ores, and environments with abundance of sulfur and iron. Acidophilic chemolithotrophic and heterotrophic microorganisms solubilize metals from sulfide minerals in acidic environment. Chemolithotrophic *Acidithiobacillus ferrooxidans* and *A. thiooxidans* and heterotrophic *Acidiphilum* are mainly responsible for solubilization of metals. Acidophilic chemoautotrophs have the ability to use CO_2 as their sole source of carbon and derive energy from oxidation of inorganic chemicals. They resist low pH by different mechanisms such as cellular bioenergetics, impermeability of the cell membrane to influx of protons into cytoplasm, pumping out excess protons by transporters, etc. Acidophiles are biotechnologically important organisms that are applied in biomining operations to solubilize metals from ores.

In India, acidophilic microorganisms have been reported from mining industries, e.g., Malanjkhand copper mine and their exploration for bioleaching of metals. One novel species as *Acidiphilium symbioticum* sp. nov. is reported from India.

Keywords

Acidophilic · Sulfur springs · Acid mine drainage · Chemolithotrophic · Chemoautotrophic · Heterotrophic bacteria · *Acidithiobacillus* · *Leptospirillum* · *Sulfobacillus* · Bioleaching · Biomining

P. P. Kanekar, S. P. Kanekar, *Diversity and Biotechnology of Extremophilic Microorganisms from India*, Microorganisms for Sustainability 41, https://doi.org/10.1007/978-981-19-1573-4_5

Acidophiles are ecologically and economically important groups of microorganisms that thrive in acidic natural habitats, like solfataric fields, sulfuric pools, sulfur springs, marine volcanic vents, and man-made acidic environments such as acid mine drainage (AMD) or acid rock drainage (ARD), mining of coal and metal ores, having highly acidic conditions (pH 3.0 and below). They are distributed in all three domains of life, namely, Archaea, Bacteria, and Eukarya, and comprise of diverse groups like aerobic, facultatively anaerobic, chemolithotrophic, heterotrophic prokaryotes and photoautotrophic eukaryotes. Mirete et al. (2017) have summarized various investigations on acidophiles mainly bacteria and archaea inhabiting AMD and ARD environments generated as a result of mining and natural weathering of rocks containing sulfide minerals. Oxidation of metal sulfides like pyrite (an iron sulfide, FeS_2) results in generation of highly acidic effluent rich in toxic metals. The process is accelerated in the presence of chemolithotrophic bacterium *Acidithiobacillus ferrooxidans* which obtains energy from oxidation of sulfidic minerals. Acidophiles from AMD or ARD environments are exposed to both low pH and high concentration of toxic metals and metalloids.

Bhowal and Chakraborty (2015) have reviewed microbial diversity of acidophilic heterotrophic bacteria. Acidophilic heterotrophs are regular inhabitants of mineral sulfide environments such as AMD. Many characteristic features of these heterotrophs have been suggested to have beneficial effect on the bioleaching as well as for the sustenance of the iron-oxidizing organisms in the environment. A significant symbiosis has been observed to exist between heterotrophic and certain autotrophic species. A larger degree of metabolic plasticity of facultatively acidophilic heterotrophs over obligately acidophilic heterotrophs is found to be present. The tolerance towards high inorganic acidity and metal concentration has been a common phenotype of acidophilic autotrophs and heterotrophs.

5.1 Habitats of Acidophiles

Habitat gives the first conception of a living system which may represent a group or type, mainly with an idea of chemical, physical, and organic or biotic embodiment of a particular region on the Earth's surface. Environments with abundance of inorganic compounds of sulfur and iron, such as hydrogen sulfide containing sediments, rivers, canals, estuarine, tidal flats, acid sulfate soils near sulfur stock piles, acid (hot) springs, mine drainage effluents, and waste dispose effluents are the habitats of acidophiles.

Sharma et al. (2016) have described habitats of acidophiles as the natural and man-made environments with varying degrees of acidity present on the Earth. The majority of the areas with pH lower than 3.0 are those where comparatively huge amounts of sulfur or pyrite are exposed to oxygen. Both sulfur and pyrite are oxidized abiotically through an exothermic reaction where sulfur is oxidized to sulfuric acid and ferrous ion to ferric ion. Both the processes are accelerated by the activity of acidophiles. The acidic pyrite areas are generated by mining around sulfur and coal mines. These areas have very high concentration of sulfide, pH values as low as 1.0, have low organic content, and are toxic due to high concentration of heavy metals. The acidity is mainly due to sulfuric acid.

Acid mine drainage (AMD) is a low pH, high metal concentrated, iron and sulfate bearing water, usually formed when rocks containing sulfide minerals are exposed to the atmosphere or an oxidizing environment, and subsequently leached by water (Gonzalez-Toril et al. 2003; Okabayashi et al. 2005). AMD differs significantly from one location to the other, and covers a wide range of pH, temperature, and mineral content. Such variations indicate that a variety of acidophilic microorganisms with a diverse physiological property can be found in AMD. It is known that bacteria play an important role in the formation of AMD. A wide range of heterotrophic and chemolithotrophic bacteria can be detected in the acid-leaching environments. Acidic environments are especially fascinating, because, in general, the low pH of the habitat is the consequence of microbial metabolism (Bhowal and Chakraborty 2015).

Acidic environments are distributed all around the world (Mirete et al. 2017). The Tinto River (Huelva, Spain) present in the Iberian Pyrite Belt has highly acidic pH (between 1.5 and 3.0) and high concentration of dissolved heavy metals like iron, zinc, and copper. These extreme conditions in the river water column and sediments are caused by chemolithotrophic bacteria *Leptospirillum ferrooxidans* and *Acidiphilium* sp., all related to iron cycle (Gonzalez-Toril et al. 2003; García-Moyano et al. 2012). Santofimia et al. (2013) studied diverse pit lakes having low pH and high concentration of toxic metals, located in the Iberian Pyrite Belt. These acidic lakes harbor iron-oxidizing bacteria and archaea.

In cold environments like Antarctica, sulfide-rich rocks are more exposed due to increasing glacier melting caused by global warming, and pyrite oxidation caused by acidophilic microorganisms has been reported from such ARD environments (Dold et al. 2013). García-Moyano et al. (2015) investigated the presence of an AMD environment in a coal mining area located in the Arctic where iron-oxidizing bacteria, *Acidithiobacillus ferrivorans*, Acidobacteria, and Actinobacteria were dominant within the bacterial community.

Other habitats include (1) geothermal waters of the Rio Agrio, a volcanic river which flows through the Copahue-Caviahue system in Patagonia, Argentina, and which harbors sulfur-oxidizing bacteria *Acidithiobacillus* and iron-oxidizing bacteria *Leptospirillum* and *Ferrimicrobium*, and archaea *Ferroplasma* sp. (Urbieta et al., 2012); (2) low temperature (8–9 °C) and extremely acidic Cae Coch mine in North Wales with iron-rich waters, having presence of the iron-oxidizing acidophiles, *Acidithiobacillus ferrivorans*, *Ferrovum myxofaciens,* and *Leptospirillum ferrooxidans* (Kimura et al. 2011); (3) diverse mining areas such as tailings in Sweden, China and acidic streams in the South Island of New Zealand; (4) walls and ceilings of caves where microbial films known as "snottites" are found, in an abandoned pyrite mine and caves enriched with hydrogen sulfide.

In India, a different habitat of acidophilic, autotrophic, and heterotrophic strains in the northern part of West Bengal is described (Bhowal and Chakraborty 2015). The geological formation of Darjeeling district is comprised of unaltered sedimentary rocks, confined to the hills on the south. Records of the Geological Survey India

(1982, GSI records, vol. 109-Part 1) showed the occurrence of magnetite in Khanikhola, south of Siyok bir (27° 02′ 00” N latitude and 88° 30′ 45″ E longitude). Lead-zinc ore has been found to occur in the area around Gorubathan of Kalimpong subdivision (Shah et al. 1974–1975). An area of 0.247 km^2 on scale 1:2000 was covered by plane table mapping in the Kharkhola block and Mal Khola block in the east and west of Gorubathan. Geological sampling showed presence of Pb, Zn, Cu, and Ag in the ore. An occurrence of wolframite associated with base metal mineralization was observed at a place near Mansong. Mineralization was located near Pedong, Rishi, and Rorathang.

Other habitats in India mainly include acid mine waters and tailings from copper, manganese mines (Paknikar and Agate 1987), Chitradurga mine water (Dave and Natarajan 1979), Sukinda mine, Orissa (Sukla and Panchanadikar 1993), Ambamata complex ore mine (Dave and Pandhi 1995), Khetri mine water (Marhual et al. 2008), Hutti gold mine (Dave et al. 2008), Jaduguda mine water, Jharkhand (Pal et al. 2010), Rajpardi lignite mine, silica containing soil (Patel et al. 2011, 2012), thermal spring, Atri, Orissa (Singh et al. 2011), native mine water, Jamshedpur (Abhilash et al. 2012), Sanindipur manganese mine, Odisha (Ghosh et al. 2016a, b), bauxite and pyrite mine (Joshi et al. 2017).

5.2 Diversity of Acidophiles

A variety of chemolithotrophic and heterotrophic microorganisms are responsible for the solubilization of metals from sulfide minerals in acidic environments. Although *Acidithiobacillus ferrooxidans* and *A. thiooxidans* in the presence of heterotrophic *Acidiphilium* bacteria were commonly regarded as the principle biological catalysts, other acidophilic heterotrophs also play an important role (Bhattacharyya et al. 1991; Hallberg and Lindstrom 1994; van Niel 1995). Bacterial populations indigenous to acidic drainage streams are heterotrophs like *Acidiphilium* sp., *Flavobacterium acidurans*, and *Bacillus* sp. (Hallberg and Johnson 2001). *Acidithiobacillus ferrooxidans* and *Leptospirillum ferrooxidans* were recognized as the major chemolithotrophic bacteria responsible for acid production in acid mine drainage (AMD) (Johnson 1998; Hallberg and Johnson 2001). 16S rRNA-based analysis has revealed that there are other organisms that occur as dominant microflora of AMD, such as *Thermoplasmoles*, *Sulfobacillus*, *Acidimicrobium*, etc. Archaebacteria including a group of iron-oxidizers, such as *Sulfolobus*, *Acidianus*, *Metalosphaera* have been reported from acidic environments (Edwards et al. 2000; Fuchs et al. 1995) and *Verrucomicrobia* and *Chlorobi* in acid rock drainage (ARD) (Okabayashi et al. 2005). A wide range of heterotrophic and chemolithotrophic bacteria can be detected in the acid-leaching environment (Dopson et al. 2003). Different acidophilic microorganisms with diverse physiological properties are found in AMD (Hallberg 2010).

The sulfur-oxidizing microorganisms are primarily the gram-negative bacteria classified as species of *Thiobacillus*, *Thiomicrospira*, and *Thiosphaera*. Heterotrophs such as *Paracoccus*, *Xanthobacter*, *Alcaligenes*, and *Pseudomonas* can also be grown on inorganic sulfur compounds. Reduced inorganic sulfur compounds are oxidized by prokaryotes, and sulfate is the major oxidation product. Aerobic-oxidizing prokaryotes belong to genera *Acidianus*, *Acidithiobacillus*, *Aquaspirillum*, *Aquifer*, *Bacillus*, *Beggiatoa*, *Methylobacterium*, *Paracoccus*, *Pseudomonas*, *Starkeya*, *Sulfolobus*, *Thermithiobacillus*, *Thiobacillus*, and *Xanthobacter* (Lu and Kelly 1983; Ghosh and Roy 2007).

Kelly and Wood (2014) have described the family Acidithiobacillaceae which is one of the two families of the order Acidithiobacillales, containing a single genus *Acidithiobacillus*, with five currently known species. All are obligately chemolithotrophic acidophiles, growing in the pH range of 0.5–5.5. Three species are mesophiles, one is moderately psychrophilic, and one is a moderate thermophile. They oxidize elemental sulfur and its compounds to generate energy for carbon dioxide fixation and cell biosynthesis. Two species also oxidize ferrous iron and sulfide minerals such as pyrite for energy generation. The family is currently assigned to the Gammaproteobacteria but it is proposed to be a group separate from both the Beta- and Gammaproteobacteria which justifies its recognition as a distinct class of Proteobacteria. These organisms have relatively small genomes (ca. 3 Mb), which encode all the enzyme proteins required for total biosynthesis from carbon dioxide and energy generation from the inorganic oxidations. The genomes exhibit high levels of gene acquisition by horizontal gene transfer.

5.3 Physiology of Acidophiles

Acidophilic chemoautotrophs have in common the ability to use CO_2 as their sole source of carbon and to derive energy from the oxidation of inorganic chemicals. *Acidithiobacillus thiooxidans* use elemental or reduced compounds of sulfur as substrates. *Thiobacillus acidophilus* is a member of the group acidophilic thiobacilli capable of both chemoautotrophic (limited to the oxidation of elemental sulfur) and heterotrophic growth (supported by sugars, amino acids, and carboxylic acids). The bacterium *Caulobacter crecentus* is known for the distinctive ability to live in low-nutrient environments, a characteristic of most heavily metal-contaminated sites (Bhowal and Chakraborty 2015). Thiosulfate has been postulated as a key compound in the oxidation of the moiety of pyrite (Schippers and Sand 1999). Iron (III) ions are exclusively the oxidizing agents for the dissolution. Thiosulfate is degraded in a cyclic process to sulfate, with elemental sulfur being a side product. Lithotrophic oxidation is an ancient metabolic process. Ecologically and taxonomically diverged prokaryotes have different abilities to utilize different reduced compounds as lithotrophic substrates (Ghosh and Dam 2009). Reduced inorganic sulfur compounds are oxidized by prokaryotes, and sulfate is the major oxidation product. Sulfur is oxidized by aerobic lithotrophs or anaerobic phototrophs. Prokaryotes oxidize hydrogen sulfide, sulfite, thiosulfate, and various polythionates under acidic conditions (Friedrich et al. 2001).

5.4 Mechanisms of Resistance to Low pH in Acidophiles

Mirete et al. (2017) have discussed different mechanisms of resistance to low pH. Acidophiles are exposed to pH gradients several orders of magnitude greater than neutrophiles. This pH difference across the membrane is connected to cellular bioenergetics since it is the main contributor to the proton motive force (PMF) in acidophiles. PMF is the force that promotes movement of proton across membranes downhill the electrochemical potential and drives the conversion of ADP to ATP via membrane-bound ATPase. It is also necessary to transport substrates across membranes. Thus, in acidophiles with a near neutral cytoplasm, the net force across the membrane can drive energy-dependent processes.

5.4.1 Impermeability of Cell Membrane to Protons

One of the adaptations that contributes to maintain the intracellular pH of acidophiles is the impermeability of the cell membrane, which restricts the influx of protons into the cytoplasm. In archaea, there is a strong association between the presence of tetraether lipids in the cell membrane and tolerance to acid pH, as reported in *Thermoplasma acidophilum* (Shimada et al. 2002), *Ferroplasma acidiphilum* YT and Y-2 (Pivovarova et al. 2002; Batrakov et al. 2002) and *F. acidarmanus* (Macalady et al. 2004).

5.4.2 Inhibition of Proton Influx by a Chemiosmotic Gradient Created by a Donnan Potential

Acidophiles are able to generate a transmembrane electric potential that is positive inside the cell relative to outside, which is opposite to the inside negative pH difference of neutrophils (Baker-Austin et al. 2007). A Donnan potential is formed when the diffusion of ions through the membrane in one direction equals the flux of ions in the opposite direction, resulting in net zero mass and charge transport. However, the membrane is selectively permeable and does not allow the passage of all the ions. It is proposed that this positive pH difference in acidophiles is generated by a Donnan potential of positively charged molecules and inhibits the influx of protons using a chemiosmotic barrier against the proton gradient. It is suggested that this potential is produced by a greater influx of potassium ions than the outward flux of protons. Many potassium-transporting ATPases and other cation transporters have been observed in the genomes of some acidophiles such as *Leptospirillum* (*L. ferriphilum*) (Tyson et al. 2004), *P. torridus* (Fütterer et al. 2004).

5.4.3 Pumping out the Excess Protons by Transporters

Removal of excess protons from the cytoplasm is one important strategy to maintain pH homeostasis. Michels and Bakkar (1985) reported that *Bacillus acidocaldarius* and *T. acidiphilum* have the ability to actively pump protons out of their cytoplasm to maintain pH homeostasis, and this process is associated with the respiratory chain. The genes encoding putative proton efflux (H^+ ATPases, antiporters, and symporters) have been identified in all the sequenced genomes from acidophiles. A predominance of proton efflux systems was reported in the genomes of Ferroplasms type II and *Leptospirillum* group II (*L. ferriphilum*) identified in a whole-community sequencing of an AMD biofilm from Iron Mountain (Tyson et al. 2004).

5.4.4 Maintaining the Intracellular pH by Cytoplasmic Buffering

One of the strategies of acidophiles is the buffering capacity to sequester and release protons if and when the protons enter into the cytoplasm. The cells contain buffer molecules that have basic amino acids such as lysine, histidine, and arginine and these molecules help in the proton sequestration. In acidophilic bacterium *Thiobacillus acidophilus*, the amino acid side chains were responsible for the cytoplasmic buffering (Zychlinsky and Matin 1983).

5.4.5 Protection of DNA and Proteins from Damage Caused by Low pH, by Chaperones

The genomes of different extreme acidophiles like *P. torridus* (Crossman et al. 2004) contain a large number of DNA and protein repair genes. Chaperones involved in protein refolding were highly expressed in microorganisms from an environmental AMD biofilm (Ram et al. 2005). Chaperones were also highly expressed in *F. acidamarnus* in aerobic (Dopson et al. 2005) and anaerobic (Dopson et al. 2006) cultures.

5.5 Resistance/Tolerance to Metals and Its Mechanism in Acidophiles

Metal resistance was thought to be emerged during the early period of prokaryotic evolution in the metal-contaminated environment (Ji and Silver 1995). Metals may accumulate above normal physiological concentrations by the action of unspecific, constitutively expressed transport systems, whereby they become toxic. Intracellular metals can exert toxic effect by forming coordinate bonds with anions blocking functional groups of enzymes, inhibiting transport systems, displaying essential metals from their native binding sites, and disrupting cellular membrane integrity

(Nies 1999). *Acidithiobacillus ferrooxidans* strains adapted to increased levels of Cu (II) have been found to be tolerant to 800 mM (Dew et al. 1999). Other acidophiles shown to be resistant to copper include *Leptosprillum ferrooxidans*. Other acidophiles resistant to Ni (II) include *Acidiphilium multivorum*, *Acidicola aminolytica* (Sampson and Philips 2001).

As a result of the oxidation of minerals containing sulfides by chemolithoautotrophic bacteria, acid mine drainage (AMD)-impacted sites contain high concentrations of toxic metals and metalloids such as copper, nickel, cadmium, zinc, arsenic, etc. that are more soluble in acidic conditions. Thus, in addition to acid tolerance, microorganisms which thrive in AMD sites have developed strategies of adaptation to elevated concentration of metals (Dopson et al. 2003). The acidophilic, aerobic, chemolithoheterotrophic bacteria isolated from AMD of Gorubathan (West Bengal, India) showed tolerance to high concentration of heavy metal divalent cations (Bhowal and Chakraborty 2015). Many other metals including uranium, molybdenum, and chromium are found in acidic environment. *A. ferrooxidans*, *L. ferrooxidans*, *A. thiooxidans*, *Thiomicrospira cuprina*, *Acidiphilium* sp., and other heterotrophic iron-oxidizing acidophiles have been isolated from uranium-containing environments (Tuovinen and Bhatti 1999).

5.5.1 Genetic Basis of Metal Resistance

Banerjee (2004) has reviewed work on genetic basis of metal resistance in acidophilic prokaryotic microorganisms from mine environments. Metal resistance is considered as the most suitable phenotypic trait for cloning vectors of chemolithotrophic and heterotrophic bacteria of mine environments. Majority of these bacteria contain one or more plasmids – the genetic elements that usually carry metal resistance genes. However, in case of chemolithotrophic *Acidithiobacillus ferrooxidans*, none of the plasmids present in the bacterium has been definitely proved to harbor metal resistance genes which are mostly found in its chromosomes. Plasmids of acidophilic heterotrophic bacteria namely *Acidiphilium* and *Acidocella*, however, carry metal resistance genes. The genes encoding arsenic resistance in *Acidiphilium multivorum* are similar to those reported from other sources while Cd and Zn resistance conferring cloned plasmid DNA fragments from *Acidiphilium symbioticum* KM2 and *Acidocella* GC 19 h strains were found to have no sequence similarity with the reported Cd and Zn resistance genes. Such observations indicate some novel aspects of metal resistance in acidophilic bacteria.

5.6 Biotechnology of Acidophiles

5.6.1 Bioleaching and Biomining

Acidophiles are economically important microorganisms being applied in biomining to solubilize metals from ores (Rawlings and Johnson 2007), and some of their enzymes can be used under harsh industrial conditions (Dopson 2012; Sharma et al. 2012). The most direct application of acidophiles in the biotechnological processes involves the organisms themselves. The most established one is the bioleaching, which in addition to biomining recovers heavy metals from industrial solid wastes (Asghari et al. 2013) and spent refinery waste (Pradhan et al. 2010). The use of acidophilic, chemolithotrophic microorganisms capable of oxidizing iron and in industrial processes to recover metals from minerals containing copper, gold, and uranium is a well-established biotechnology. The insoluble metal sulfides are oxidized to soluble metal sulfates by the chemical action of ferric iron. The main role of the microorganisms is the re-oxidation of the generated ferrous iron to obtain additional ferric iron.

Bioleaching or biomining occurs in heaps of crushed ores. Biooxidation is carried out in stirred reactors and is principally used for recovery of gold (Rawlings and Johnson 2007). These processes involve iron- or sulfur-oxidizing acidophilic microorganisms adapted to different temperature ranges, from mesophiles (*Acidithiobacillus*, *Leptospirillum,* and *Ferroplasma*) to thermophiles (*Sulfolobus*, *Metallosphaera,* and *Acidianus*) (Johnson and Hallberg 2003; Hallberg 2010). The biotechnological leaching processes mediated by acidophiles cells may not seem competitive when compared to conventional technologies such as mechanical or chemical methods, since a significantly longer leaching time is required to achieve the same extraction efficiencies when using traditional processes. However, the biotechnological processes are more cost-efficient, simpler, and more environmentally friendly than the chemical processes (Johnson 1995; Beolchini et al. 2010; Dopson 2012). Biomining has distinctive advantages over the traditional mining procedures. It does not require the high amounts of energy used during roasting and smelting, and does not generate harmful gaseous emissions such as sulfur dioxide (Rawlings 2002). Biomining is also of great advantage since discarded low-grade ores from standard mining procedures can be leached in an economically feasible way. Industrial bioleaching processes use acidophilic bacteria including *Acidithiobacillus thiooxidans* and *A. ferrooxidans* (Bruins et al. 2000).

The microorganisms responsible for bioleaching are inherently adapted to extremely low pH (from 2 to 3) and elevated concentrations of metals such as iron, aluminum, manganese, copper, cadmium, etc.), and metalloids like arsenic, depending upon the mineralogy of the host rock. Thus, the physiology and genetic adaptability of biomining microorganisms dynamically and continuously adapt to the particular conditions of environment (Mirete et al. 2017). During the biooxidation process in gold-bearing arsenopyrite bioreactors, increased resistance to arsenic was reported. For commercial bioleaching operations, this natural mechanism can contribute to the selection of more robust and efficient strains with reduced sensitivity to metal toxicity (Rawlings 2005; Podar and Reysenbach 2006).

5.6.2 Industrial Applications of Enzymes from Acidophiles

Enzymes from acidophilic as well as acidothermophilic microorganisms, such as amylases, proteases, ligases, cellulases, xylanases, alpha-glucosidases, endoglucanases, and esterases were shown to be stable at low pH and high temperature and functionally active under harsh industrial conditions (Sharma et al. 2012). The possibility of improving extremozymes by genetic engineering and directed evolution will further enhance their industrial applications. Since cultivation of such extremophiles is associated with many potential difficulties, many researchers consider that genetic engineering of the desired extremophiles into mesophilic hosts will allow large-scale production of them (Mirete et al. 2017).

5.7 Acidophiles and Astrobiology

Mirete et al. (2017) have enlightened on a different aspect of acidophiles in relation to astrobiology. Observational and remote sensing data from different regions of Mars confirmed the presence of iron minerals that can only be formed in the presence of water (Moore 2004; Morris et al. 2004; Squyres et al. 2004).

Chemolithoautotrophic acidophiles provide a link to early Earth when oxygen was unavailable and ecosystems were sustained by inorganic reactions (Gonzalez-Toril 2005; Canfield et al. 2006). Environments which are inhabited by acidophilic microorganisms, such as Tinto River (Spain) and Eagle plains (Canada), are also considered to be important terrestrial analogs of Mars (Fairén et al. 2010). Thus, acid-stable enzymes might provide useful information about early Earth enzymology.

A promising device of biotechnology that would be used to detect organic molecules on Mars was developed based on two acidophile-chemolithotrophic bacterial strains *L. ferrooxidans* and *A. ferrooxidans* (Parro et al. 2005). Both are native and abundant microorganisms in the Tinto River ecosystem and also inhabit AMD environments and have demonstrated ways of life for which it is feasible to identify a past or present hypothetical niche on Mars (Lopez-Archilla et al. 2001; Gonzalez-Toril et al. 2003). By integrated approaches, an antibody microarray-based biosensor based on *L. ferrooxidans* and *A. ferrooxidans* was developed (Parro et al. 2005). Due to its novelty, this autonomous and remote operation instrument called SOLID (Signs of Life Detector) is proven for searching biosignatures on acid and metal-rich environments.

5.8 Acidophiles from India

Many mining industries are spread through different states of India, such as Madhya Pradesh, Karnataka, Goa, Maharashtra, Gujarat, Odisha, Chhattisgarh, and many more. Acid mine drainage (Fig. 5.1) waste low-grade ores from these mines host different groups of acidophiles. Chemolithoautotrophic and heterotrophic acidophilic microorganisms have been reported by Indian researchers and are presented in Table 5.1.

5.8.1 Diversity of Acidophiles from India

Many researchers from India have studied diversity of chemolithoautotrophic and heterotrophic acidophilic microorganisms from different habitats.

Paknikar and Agate (1987) developed a culture collection and data bank with respect to biometallurgical processing in India for microbial leaching of copper (Fig. 5.2) and bioconversion of manganese. Around 207 potentially useful microorganisms such as iron, sulfur, and manganese-oxidizing bacteria were isolated from 60 copper and manganese mines in the country. The bacteria were used for copper leaching and converting manganese to a better variety. These processes were scaled up to 300 kg level and the isolates were stored in an active, ready-to-use form by devising suitable preservation methods. The compilation of information on ores and bacterial cultures, with a facility to supply useful cultures, was designed as a "Culture collection and Data Bank."

Dave and Upadhyay (1993) have isolated thiosulfate-oxidizing microorganisms from a thermal spring. Menon and Dave (1995) studied eight different isolates of *Thiobacillus ferrooxidans* isolated from complex sulfidic ores, hot water springs, mines, and lignite. The substrates tested included ferrous, sulfur, thiosulfate, pyrite, complex sulfidic ore. All isolates grew on ferrous and sulfur but only four could utilize thiosulfate as an energy source. Growth adaptation of various substrates resulted in enhancement of growth activity.

Dave and Pandhi (1995) developed two isolates of *T. ferrooxidans* (SRco) cultures 1 and 2 to tolerate Cd^{2+}, Co^{2+}, Cu^{2+}, and Ni^{2+} in different conditions. Cadmium, in the form Cadmium sulfate, has been found to be the most tolerated by the organisms, followed by copper, nickel, and cobalt in the form of sulfate salts. Ni^{2+} and Co^{2+} are tolerated equally by the organisms, but organisms need more time to oxidize 95% Fe^{2+}-Fe^{3+} in presence of Co^{2+} as compared to Ni^{2+} shows inhibitory action at higher concentration. The isolates have been developed to tolerate as high as (and shows 95% iron oxidation) 1.75, 1.527, 1.254 and 1.254% corresponding to 16, 24, 21, 21 mM cadmium, copper, nickel, and cobalt, respectively.

Shah et al. (1999) have described bioremediation of copper using waste streptomyces biomass. Joshi et al. (2000) found spermidine as the predominant polyamine in acidophilic and neutrophilic strains of chemolithotrophic iron- and sulfur-oxidizing bacteria, and associated organotrophic cultures inhibiting a copper mining ecosystem spermidine content was found to increase in a neutrophilic *Thiobacillus*

Fig. 5.1 Acid mine drainage (Google image)

thioparus strain when adapted to pH 5.0 although the biomass decreased. However, exogenous supplementation of spermidine could restore the biomass yield. The studies indicated that spermidine plays an important role in adaptive behavior of microorganisms in natural ecosystems, particularly in acid environment.

Table 5.1 Reports of acidophilic microorganisms from India

Sr. No.	Name of the organism/ group	Source	Diversity/ biotechnology	Reference
1	*Thiobacillus ferrooxidans*	Chitradurga mine water, Karnataka, India	Biooxidation in presence of copper and zinc	Dave and Natarajan (1979)
2	Iron and sulfur oxidizers	Copper mines of India	Biometallurgy of copper in India	Agate et al. (1984)
3	*Thiobacillus ferrooxidans*		Multi-metal ore leaching	Dave and Mathur (1987)
4	207 strains of iron and sulfur oxidizers and manganese-oxidizing bacteria	60 copper and manganese mines in India	Culture collection and data bank Copper leaching and conversion of manganese scale-up studies at 300 kg level	Paknikar and Agate (1987)
5	*Thiobacillus ferrooxidans*	Copper mines in India	Tolerance of unusually high concentrations of metals	Paknikar and Agate (1988)
6	Iron and sulfur oxidizers	Copper mines in India	Scale-up leaching of Malanjkhand copper ores	Agate et al. (1989)
7	*Acidophilum symbioticum* sp. nov.	From *thiobacillus ferrooxidans* culture from Indian mines	New species	Bhattacharyya et al. (1991)
8	Indigenous flora of nickeliferous lateritic ore	Sukinda, Odisha	Nickel bioleaching	Sukla and Panchanadikar (1993)
9	*Thiobacillus ferrooxidans*	From varied habitats in India	Bioleaching of complex sulfide ores	Menon and Dave (1993)
10	Thiosulfate-oxidizing organisms	Thermal spring	Thiosulfate oxidation	Dave and Upadhyay (1993)
11	*Thiobacillus ferrooxidans* (silver tolerant strains)	Thermal spring ore isolate	Oxidation of ferrous	Menon and Dave (1994)
12	*Thiobacillus ferrooxidans*	Complex sulfidic ores, hot springs, mines, and lignite	Growth behavior	Menon and Dave (1995)
13	*Thiobacillus ferrooxidans*	Ambamata complex ore mine, Gujarat, India	Increased metal tolerance to Cd, Co, Cu, and Ni	Dave and Pandhi (1995)
14	*Thiobacillus ferrooxidans* consortium		Improved metal extraction from polymetallic concentrate	Tipre et al. (1998)

(continued)

Table 5.1 (continued)

Sr. No.	Name of the organism/ group	Source	Diversity/ biotechnology	Reference
15	*Thiobacillus ferrooxidans* consortium		Copper and zinc extraction from GMDC polymetallic concentrate in stirred tanks and large reactors	Tipre et al. (1999)
16	*Citrobacter* sp.	Polluted creek, Pune, Maharashtra	Bioleaching of manganese by reduction and solubilization	Mohod et al. (1999)
17	Iron- and sulfur-oxidizing bacteria	Copper mines in India	Requirement of spermidine for growth under acid conditions	Joshi et al. (2000)
18	*Acidithiobacillus ferrooxidans* TfPo and TfB	Indian mines	Ferrous biooxidation ability under pH, ferrous, ferric, sodium chloride and temperature stress	Dave (2002)
19	*Acidithiobacillus ferrooxidans*, *Acidithiobacillus thiooxidans*, *Leptospirillum ferrooxidans*		Bioleaching of Cu-Pb-Zn bulk concentrate at high pulp density	Tipre and Dave (2004)
20	*Acidithiobacillus ferrooxidans*		Recovery of nickel from Sukinda chromite low-grade ore	Mohapatra et al. (2007)
21	*Acidithiobacillus ferrooxidans*		Bioleaching of low-grade copper	Pradhan et al. (2008a, b)
22	Mesophilic acidophilic chemolithotrophic consortia (MACC), moderately (50 °C) thermophilic acidophilic chemolithotrophic consortia (TACC) *and Sulfobacillus* acidophilic consortia (SAC)	Mine water from copper mine, Khetri, Rajasthan, India	Higher copper extraction ability of thermophilic consortia	Marhual et al. (2008)
23	Four arsenic-resistant strains of *Acidithiobacillus ferrooxidans*	Hutti Gold Mine Ltd. (HGML), Karnataka, India	Ferrous oxidation in presence of arsenic, oxidation of arsenopyrite	Dave et al. (2008)
24	*Leptospirillum ferrooxidans*	Polymetallic concentrate leachate	Ferric sulfate generation during bioleaching process, ferric iron regeneration from the leachate	Dave et al. (2008)

(continued)

Table 5.1 (continued)

Sr. No.	Name of the organism/ group	Source	Diversity/ biotechnology	Reference
25	*Acidithiobacillus ferrooxidans*	Jaduguda mine water, Singhbhum district, Jharkhand	Bioleaching of Uranium with Turamdih ore sample from Uranium Corporation of India Ltd., Jaduguda	Pal et al. (2010)
26	*Acidithiobacillus ferrooxidans* strain SRDSM2	Silica containing soil sample at Rajpardi lignite mine, Gujarat, India	Bioleaching of polymetallic concentrate, extraction of Copper and Zinc	Patel et al. (2011)
27	*Bacillus stearothermophilus*	Thermal spring, Atri, Odisha	Extraction of copper from low-grade chalcopyrite ore from Malanjkhand	Singh et al. (2011)
28	*Acidithiobacillus ferrooxidans* dominated mixed culture of acidophiles		Heap bioleaching of low-grade chalcopyrite ($CuFeS_2$) in 1000 tons scale	Panda et al. (2012)
29	*Acidithiobacillus ferrooxidans*	Native mine water, Jamshedpur, Jharkhand, India	Bioleaching of chalcopyrite ore on bench scale	Abhilash et al. (2012)
30	*Acidithiobacillus thiooxidans*		Bioleaching of metals from hazardous spent hydroprocessing catalysts	Singh et al. (2012)
31	*Leptospirillum ferriphilum* dominated consortium SR-BH-L	Rajpardi lignite mine, Gujarat, India	Extraction of copper and zinc from polymetallic bulk concentrate	Patel et al. (2012)
32	*Acidithiobacillus ferrooxidans* dominated consortium		Bioleaching of low-grade chalcopyrite (ball mill spillage material) for recovery of copper	Panda et al. (2013a, b)
33	*Acidophilum* spp., *Leptospirillum* spp., *Thiobacillus ferrooxidans*, *Thiobacillus caldus*, *Sulfobacillus*		Recovery of metals from electronic waste, namely, printed circuit boards (PCBs)	Adhapure et al. (2014)
34	*Leptospirillum ferriphilum* dominated consortium		Two-step process for biorecovery of copper, zinc, and nickel from computer printed circuit boards (C-PCB)	Shah et al. (2015)

(continued)

Table 5.1 (continued)

Sr. No.	Name of the organism/ group	Source	Diversity/ biotechnology	Reference
35	Manganese solubilizing bacteria	Sanindipur manganese mine of Sundargh district, Odisha state	Four morphologically distinct strains	Ghosh et al. (2016a, b)
36	*Acidophilum* spp., *Leptospirillum* spp., *Acidithiobacillus ferrooxidans*, *Thiobacillus caldus*, *Sulfobacillus*	Bauxite and pyrite mine	Recovery of metals from printed circuit boards (PCBs) of cellphone charger	Joshi et al. (2017)
37	*Acidithiobacillus ferrooxidans*		Dissolution of copper and zinc present in a complex sulfide concentrate by bioleaching technique	Pradhan et al. (2019a, b)
38	Iron-oxidizing *Leptospirillum ferriphilum*	Coal sample	Extraction of metals from waste television printed circuit boards (TV-PCBs)	Sodha et al. (2019)
39	*Leptospirillum ferriphilum* dominated consortium		Recovery of copper from waste printed circuit board (PCB)	Sodha et al. (2020)
40	*Leptospirillum ferriphilum* dominated consortium		Bioleaching for recovery of metals from electronic waste, waste cell-phone printed circuit boards (WPCBs)	Thacker et al. (2021)
41	*Acidithiobacillus ferrooxidans*		Leaching of Ni, V, Mo, and Al present in a spent catalyst	Pradhan et al. (2021a, b)

Dave (2002) studied two *Acidithiobacillus ferrooxidans* Tfpo and TfB isolated from Indian mines for their ferrous biooxidation ability under pH, ferrous, ferric, sodium chloride and temperature stress. Adaptation and selection of the isolates resulted in enhancement of tolerance up to the pH of 0.8 ferrous ferric, and NaCl concentrations of 200, 50, and 12.5 g/L respectively. The isolate TfPo was found to be more tolerant and further studied for toxic metal and cation tolerance and the developed isolate was able to oxidize ferrous in presence of 16, 21, 21, 480, and 115 mM of cadmium, cobalt, nickel, copper, and zinc respectively. The developed culture was found to be at least 10 times more efficient for ferrous oxidation in presence of 480 mM of calcium, magnesium, and sodium sulfate. The developed culture could be very useful for metal extraction from the concentrates and ores.

Fig. 5.2 Malanjkhand copper mine (Google image)

Dave et al. (2008) isolated four arsenic-resistant ferrous oxidizers from Hutti Gold Mine Ltd. (HGML) samples and characterized them using conventional microbiological, biochemical, and molecular methods. The ferrous oxidation rates with these isolates were 16, 48, 34, and 34 mg/L/h and 15, 47, 34, and 32 mg/L/h in absence and presence of 20 mM of arsenite (As^{3+}), respectively. The isolate HGM8 was able to grow in the presence of 14.7 g/L of arsenite with 25.77 mg/L/h ferrous oxidation rate. All the four isolates were able to oxidize iron and arsenopyrite from 20 g/L and 40 g/L refractory gold ore and 20 g/L refractory gold concentrate. The organisms were identified as *Acidithiobacillus ferrooxidans.*

Marhual et al. (2008) developed three acidophilic enrichment consortium from mine water sample of copper mine site at Khetri, India. Out of these, one was mesophilic (35 °C) and two were moderately thermophilic (50 °C) and were named as mesophilic acidophilic chemolithotrophic consortia (MACC), thermophilic acidophilic chemolithotrophic consortia (TACC) and *Sulfobacillus* acidophilic consortia (SAC). Copper extraction ability of both the thermophilic consortia (77–78% extraction) was almost double to that of mesophilic consortia (40% extraction) at 10% pulp density after 55 days. Both the thermophilic consortia were equally effective in leaching of other metals such as Ni, Co, Zn, and Mn.

Dhakephalkar (2008) has extensively studied PCR-based finger printing of iron and sulfur oxidizers and associated bacteria from mining ecosystem.

Patel et al. (2011) have described *Acidithiobacillus ferrooxidans* strain SRDSM2 isolated from silica containing soil sample collected at a Rajpardi lignite mine. The strain responded to the addition of 0.5 g/L peptone and 1.0 g/L tryptone soya broth in the ferrous sulfate tryptone soya broth (ITSB) medium with 35.3% and 29.6% increase in iron oxidation rate (IOR) but decrease in IOR at higher peptone or tryptone soy broth levels. The presence of 4 mM of zinc as zinc sulfate in the medium increased the IOR by 24.4%. Forty percent of the inoculated cells survived even after exposure at 80 °C for 120 min and showed 30% ferrous iron oxidation. The isolate was able to oxidize ferrous iron even in presence of 4.06 M ionic strength of medium and leached >85% copper and zinc from the polymetallic concentrate.

Ghosh et al. (2016a, b) isolated and characterized manganese solubilizing bacteria from low-grade ores from Sanindipur manganese mine of Sundargh district in Odisha, India. Four morphologically distinct bacterial strains showing visible growth on Mn-supplemented plates were isolated. Mn solubilizing ability of the isolates was assessed by visualizing the lightening of the medium appearing around the growing colonies.

5.8.2 Biotechnology of Acidophiles from India

By virtue of their ability to oxidize iron and sulfur, and metal resistance/tolerance, the acidophilic microorganisms find biotechnological applications in biomining and bioleaching processes, recovery of metals from low-grade ores and industrial solid wastes.

Dave and Natarajan (1979) carried out biooxidation studies with *Thiobacillus ferrooxidans* in the presence of copper and zinc. Dave and Natarajan (1987) studied effects of physicochemical parameters on the bacterial leaching of sphalerite concentrates.

Extensive studies have been carried out on copper bioleaching (Paknikar 1983; Paknikar 1988; Agate et al., 1989). Paknikar and Agate (1988) have described occurrence of a *Thiobacillus ferrooxidans* strain tolerating unusually high concentration of metals. The organism was screened for the purpose of improving the extraction efficiency of copper from copper concentrates.

Dave and Mathur (1987) described factors affecting multi-metal ore leaching by *Thiobacillus ferrooxidans*. Menon and Dave (1993) carried out amenability studies of complex sulfide ores from various mines of India. Various leaching tests like shake flask, glass columns, and tank leaching were performed using *T. ferrooxidans*, isolated from varied habitats. Ore samples from Ambamata mines showed 17, 63, and 38% of Cu, Zn, and Pb extraction, while samples from the Rajpura-Dariba mines showed 30% Cu extraction and 75% Zn extraction. Bioleaching using tank method showed 28% of Cu and 80% Zn extraction. Typical behavior was exhibited by the ore samples in respect to pH, redox potential, and acid production during bacterial activity.

Sukla and Panchanadikar (1993) attempted leaching of nickel from nickeliferous lateritic ore of Sukinda, Odisha, India, using indigenous microflora, which were isolated from the same ore. The results showed 90% of nickel and 34% cobalt were extracted from the lateritic nickel ore. Menon and Dave (1994) investigated effect of ultraviolet exposure on silver tolerance and growth of *Thiobacillus ferrooxidans* (ore isolate). Ultraviolet exposure of 30s was found to influence the ferrous oxidation activity. Ultraviolet-induced silver tolerance as high as 0.414 mM Ag was developed with complete (100%) ferrous oxidation within 163 h. Ultraviolet reexposure of the above culture enhanced ferrous oxidation by about 1.5-fold at 0.276 mM Ag concentration.

Tipre et al. (1998) have attempted improved metal extraction by selected *Thiobacillus ferrooxidans* consortium from polymetallic concentrate. Tipre et al. (1999) studied copper and zinc extraction from GMDC polymetallic concentrate in shake flasks, stirred tanks, and large reactors. The concentrate contained chalcopyrite, sphalerite, galena, and pyrite as main constituents. *Thiobacillus ferrooxidans* consortium was adapted to high pulp density up to 20% of the concentrate. Scale-up experiments were at 0.1 dm^3, 3 dm^3, and 100 dm^3 levels at 30 °C, 40 °C, and more than 40 °C temperatures. Solubilized Fe^{2+} and oxidation reduction potential ranged between 0.13 to 1.89 g/L and 337 to 456 mV, respectively, on 15th day of experiment.

Mohod et al. (1999) have described bioleaching of manganese from low-grade ores using manganese-reducing microbial cultures. *Citrobacter* sp. isolated from sediment sample of a polluted creek was found to reduce and solubilize manganese from manganese dioxide. The culture could extract 90% manganese from a low-grade ore pyrolusite ore in a stirred tank reactor in 13 days at 1% pulp density.

Dave et al. studied (Dave et al. 2001) influence of high pulp density on copper and zinc bioextraction from polymetallic concentrate. Dave et al. (2001) have studied effect of high pulp density on Copper and Zn bioextraction from polymetallic concentrate. Tipre and Dave (2004) studied bioleaching of metals from a Cu-Pb-Zn sulfide bulk concentrate using a developed consortium consisting of *Thiobacillus ferrooxidans*, *Thiobacillus thiooxidans*, *Leptospirillum ferrooxidans*, and heterotrophic organisms. In a shake flask, copper and zinc extraction increased up to 2% (w/v) pulp density at an overall extraction rate of 0.61 and 4.2 g/L/day for copper and zinc, respectively, at 32 °C but at 25% pulp density, the extraction rate decreased to 0.41 and 4.0 g/L/day. A semicontinuous process resulted in 79% and 82% copper and zinc extraction at rates of 0.66 and 4.92 g/L/day. A 5-L laboratory Stirred Tank Reactor (LSTR) gave a 1.15- and 1.55-fold increase in copper and zinc extraction compared to a shake flask at 20% pulp density with a further increase when operated semicontinuously. This resulted in 85.3% and 80% copper and zinc extraction in 5 and 6 days.

Mohapatra et al. (2007) attempted recovery of nickel from a low-grade chromite overburden using a mixed culture of mesophilic acidophiles (predominantly *Acidithiobacillus ferrooxidans*), and two fungal strains *Aspergillus niger* and *Aspergillus fumigatus*. The culture of *Acidithiobacillus ferrooxidans* was found to solubilize nickel effectively at temperatures ranging from 30 to 37 °C but not at higher temperature such as 45 °C. Increased supplemental ferrous iron increased the leaching efficiency of *At. ferrooxidans* culture. The maximum nickel solubilization was 40% at 2% pulp density and 24% at 10% pulp density. Thus, nickel recovery decreased with increase in pulp density.

Dave et al. (2008) has discussed role of *Leptospirillum ferrooxidans* in ferric sulfate generation during bioleaching process. An attempt was made to select *L. ferrooxidans* from the polymetallic concentrate leachate and further developed it for enhanced ferric iron regeneration from the leachate in shake flask stirred tank and column reactor. When ferric to ferrous iron ratio in the shake flask reacted to 20: 1, *L. ferrooxidans* outcompeted *Acidothiobacillus ferrooxidans* and accounted for more than 99% of the total population.

Pradhan et al. (2008a, b) have reviewed microbiological and other important aspects of chalcopyrite heap bioleaching process. Bioleaching is an emerging technology with significant potentials to add value to the mining industries so as to deliver attractive environmental and social benefits to all the concerned. Chalcopyrite, $CuFeS_2$ is the most important copper-bearing mineral in the world and known to be recalcitrant to hydrometallurgical processing. The main limitation of the biohydrometallurgical processing of chalcopyrite is its slow rate of dissolution. The modest nutritional requirements of bioleaching organisms may be provided with the aeration of iron and/or sulfur containing mineral suspensions in water or the irrigation of a heap, while working in a large scale. This chemolithotrophic metabolism makes the organisms industrially important.

Pradhan et al. (2008a, b) have investigated iron oxidation rate of *Acidithiobacillus ferrooxidans*. It was observed that the regeneration time of bacteria reduced from 168 h to 16 h repeated subculturing, giving rise to oxidation rate and thus increasing the kinetics of the process. The active strain was used for bioleaching of low-grade copper ore by varying two parameters, namely, pH and pulp density. The results showed that around 30% of copper could be leached at initial pH 2 and pulp density of 20%.

Sukla et al. (2009) undertook heap bioleaching of the lean copper ore in 15 to 30 ton scales at IMMT, Bhubaneswar, Odisha. M/S Hindustan Copper Limited (HCL), India, generates large amounts of lean sulfide ores of copper. The current production of lean copper ores in India is 0.55 million tons with ~0.3% average copper content. The leaching study showed 0.09% dissolution of copper from the ore body per day.

Pal et al. (2010) carried out bioleaching of uranium with Turamdih ore sample from Uranium Corporation of India Ltd., Jaduguda. The bacterial strain used in the leaching experiments was isolated from Jaduguda mine water sample and identified as *Acidithiobacillus ferrooxidans*. Efficiency of bioleaching was studied by varying parameters like pulp density and initial ferrous concentration as a source of energy. The results showed that the bioleaching efficiency was 49% at 10% pulp density (w/v) and initial pH 2.0.

Singh et al. (2011) attempted use of thermophilic bacteria for extraction of copper from low-grade ore. The thermophilic organism *Bacillus stearothermophilus* was isolated from a thermal spring at Atri, Orrisa, India. The bacterium was adapted to low-grade chalcopyrite ore containing 0.27% Cu. The organism could solubilize copper up to 81.25% at pH 6.8, 60 °C in 30 days of incubation.

Patel et al. (2012) developed novel polystress resistant iron-oxidizing consortium SR-BH-L enriched from Rajpardi lignite mine soil sample, which gave iron oxidation rate 1954 mg/L/h. The consortium oxidized ferrous iron and 11 cP viscosity, 7.47 M ionic strength, 2.3 pH, and g/L of 0.50 cadmium, 3.75 copper, 0.20 lead, 92.0 zinc, 6.4 sodium, 5.5 chloride, 154 sulfate, and 393.8 TDS. The developed consortium showed 78% and 70% copper and zinc extraction from polymetallic bulk concentrate in monophasic bioleaching process. The biogenerated ferric in leachate showed 80.81% and 54.0% copper and zinc leaching in only 30 and 90 min. The consortium showed presence of *Leptospirillum ferriphilum* as analyzed by Denaturing Gradient Gel Electrophoresis (DGGE) followed by 16S rRNA gene sequencing.

Abhilash et al. (2012) attempted bioleaching of chalcopyrite ore on bench scale using *Acidithiobacillus ferrooxidans* isolated from native mine water. The enriched culture when adapted to the ore and employed for bioleaching at 5% (w/v) pulp density, pH 2.0 and 25 °C with three particle sizes, namely, 150–76 μm and <50 μm, resulted in recovery of 38.31%, 29.68%, and 47.5% Cu, respectively, with a maximum rise in redox potential from 530 to 654 mV in 35 days. Copper biorecovery increased with a rise in bacterial count from 1×10^7 cells/ml to 1.13×10^9 cells/ml in 35 days.

Panda et al. (2012) carried out heap bioleaching of low-grade chalcopyrite ($CuFeS_2$) in 1000 tons scale over a time period of 383 days. The microorganisms used in the pilot scale were a mixed culture of acidophilic bacteria predominantly of the *Acidithiobacillus ferrooxidans* strain. The bacterial solution was prepared and recirculated in specially designed BACFOX (Bacterial Film Oxidation) tank. The leaching studies showed a cumulative copper dissolution rate of 0.14% per day (2.37 kg/d). The overall recovery of the heap was 30%.

Panda et al. (2013a, b) tested bioleaching of a low-grade chalcopyrite (ball mill spillage material) for recovery of copper in shake flasks. The original samples were thermally activated (600 °C, 30 min) to see the change in physicochemical and mineralogical characteristics of the host rock and subsequent effect on copper recovery. A mixed culture of acidophilic chemolithotrophic bacterial consortium predominantly comprising of *Acidithiobacillus ferrooxidans* strain was used for bioleaching studies and optimization of process parameters of both original and thermally activated samples. Mineralogical characterization studies indicated presence of chalcopyrite, pyrite in the silicate matrix of the granite rock. With pH 2, pulp density 10% w/v, inoculum 10% v/v, temperature 30 °C, 150 rpm, 49% copper could be recovered in 30 days from the finest particle size of the original spillage sample. Under similar conditions, 95% copper could be recovered from the thermally activated sample with the same particle size in 10 days. The study revealed that thermal activation leads to volume expansion in the rock with the development of cracks, micro and macro pores on its surface, thereby enabling bacterial solution to penetrate more easily into the body, facilitating enhanced copper dissolution.

Patel et al. (2014a, b) employed a biphasic leaching operation for the extraction of copper and zinc from the GMDC polymetallic bulk concentrate (PBC). Biogenerated ferric was used in cyclic manner for fed batch metal extractions from the PCB for four cycles under the preoptimized conditions. Zn extraction were gradually decreased with each succeeding fed batch cycle and it resulted in 67.89, 62.46, 47.37, and 14.74% of Zn extraction in first, second, third and fourth cycle respectively. In case of copper, the extraction pattern was 86.63, 81.98, 75.58, and 46.51% Cu extraction generated in each cycle was added in 15.0 L downflow packed bed column reactor having biofilm of developed consortium SR-BH-L for the ferrous biogeneration and the obtained maximum IOR was 2650 mg/L/h.

Shah et al. (2015) investigated a two-step leaching process using ferric sulfate generated by *Leptospirillum ferriphilum* dominated consortium for extraction of copper, zinc, and nickel from computer printed circuit boards (C-PCB). With 10 g/

L pulp density, it was seen that pH 2.0 was optimum which resulted in 87.50–97.80% Cu-Zn-Ni extraction. Pretreatment of PCB powder with acidified distilled water and NaCl solution showed 3.80–7.98% increase in metal extraction corresponding to 94.08% Cu, 99.8% Zn, and 97.99% Ni extraction. Particle size of 75 μm for Cu and Zn while 1680 μm for Ni showed twofold increase in metal extraction giving 97.35–99.80% Cu-Zn-Ni extraction in 2–6 days of reaction time.

Joshi et al. (2017) have carried out bioleaching studies for recovery of metals from pulverized and non-pulverized printed circuit boards (PCBs) of cell phone charger. Chemical analysis of PCBs showed presence of 10.8, 68.0 and 710.9 mg/L of Zn, Pb, and Cu respectively. For bioleaching, both pulverized and non-pulverized PCBs samples were used. A natural consortium of acidophilic microorganisms obtained from bauxite and pyrite mine was used. The consortium was found to compose of *Acidiphylum* sp., *Leptospirillum* sp., *Acidithiobacillus ferrooxidans*, *Thiobacillus caldus*, and *Sulfobacillus*. Pulverized PCB showed more copper leaching efficiency as compared to non-pulverized one. Copper was recovered using a simple setup of electrodeposition and 92.85% recovery was attained.

Qureshi et al. (2018) carried out optimization studies for bioleaching of potassium from feldspar by five fungal and two bacterial isolates for carbon, nitrogen, agitation, pH, and incubation time. Bacterial isolate SDS1 released 28 ppm K in sucrose-amended medium. The pH of the medium was found to decrease from 7.0 to 2.05 after growth of the organisms. The acids produced could be responsible for the solubilization of potassium.

Sukla et al. (2019) have reviewed work on beneficiation of low-grade bauxite. Bauxite is the major alumina (Al_2O_3) bearing ore used in aluminum manufacturing industries. The bauxite containing less than 50% Al_2O_3 is called low-grade bauxite ore which is commonly used for the alumina-based abrasives and refractories production. The alumina-silica and alumina-ferrite complexes are the major impurities present in low-grade bauxite, which affect its commercial utilities due to development of poor binding property in the alumina grains that creates mechano-physical problems. Different conventional methods are used to remove such impurities but they have limitations. Bio-beneficiation is a potential solution to the problems associated with conventional methods. In the bio-beneficiation process, microorganisms and their metabolic products can mobilize or polarize different impurities present in the low-grade bauxite, by creating active redox environment in the indigenous atmosphere. Many reports have suggested that *Paenibacillus polymyxa* efficiently removes calcium from the low-grade bauxite. Iron-oxidizing and silicate bacteria remove iron and solubilize silica, respectively, from the low-grade bauxite ore. However, pilot scale operation of the bauxite bio-beneficiation process has not been reported. Hence, there is a need to undertake biotechnology research to develop a comprehensive and efficient process for the bio-beneficiation of low-grade bauxite ore.

Pradhan et al. (2019a, b) examined dissolution of copper and zinc present in a complex sulfide concentrate by applying bioleaching with the aid of pure *Acidithiobacillus ferrooxidans*. The leaching rate of both copper and zinc depended on three different parameters such as pulp density, initial pH, and ferrous iron concentrate. The leaching rate increased with increase in ferrous iron addition, due to iron precipitation. The leaching efficiencies of both metals decreased below and above pH 2.0 indicating requirement of particular pH for the leaching reactions. The efficiency also decreased with increase in pulp density. The bacterial activity remained unaltered during the bioleaching experiments using the sulfide minerals.

Sodha et al. (2020) carried out extraction of copper from printed circuit board (PCB), an essential part of all the electronic waste which is a potential source of various metals thus known as "urban mine." A novel approach to enhance the extraction of copper from non-pulverized waste printed circuit board (WPCBs) was followed. WPCBs pretreatment with NaOH for removal of the epoxy coating was optimized before the bioleaching process. The pretreatment resulted in the exposure of the metal present on WPCBs. Copper was extracted by the two-step bioleaching process. In the first step, ferric sulfate was biogenerated by *Leptospirillum ferriphilum* dominating iron-oxidizing consortium with 1196.0 mg/L/h of average ferrous sulfate oxidation rate.

Thacker et al. (2021) attempted bioleaching for recovery of metals from electronic e-waste. Generally, e-waste recycling is done by pyrometallurgy and hydrometallurgy methods. *Leptospirillum*-dominated consortium was used to generate ferric iron lixiviant by oxidizing ferrous iron. The resulting lixiviant was used for bioleaching of metals from waste cell phone printed circuit boards (WPCBs). The pretreatment of pulverized WPCB (P-WPCB) and unpulverized WPCB (UP-WPCB) was done using 2 M NaOH to see the effect of pretreatment on metal extraction.

Pradhan et al. (2021a, b) evaluated the improvement in the leaching rate of Ni, V, Mo, and Al present in a spent catalyst sample by the adopted and unadapted bacterial cultures. The adaptation technique was found to be fruitful as the leaching equilibrium reduced from 240 to 40 h, the leaching rate of Ni and V was improved by 10% (w/w). The leaching rates of Mo and Al were low due to different factors such as presence of impervious elemental sulfur, refractory nature, and low solubility and refractoriness of the alumina matrix present in the spent catalyst sample. The bacterial culture comprised of mainly *Acidithiobacillus ferrooxidans.*

Pradhan et al. (2021a, b) have reviewed different aspects of development of bioleaching. Bioleaching is a useful tool to recover different metals from low-grade ores and wastes. It has benefits in terms of cost and environment. For processing large volume of ores, heap leaching is applied. Genetic improvement is proposed to understand the cellular metabolism of microorganisms for bioleaching purpose which helps the industries to open up renewed possibilities for mineral processing technology through improved kinetics.

5.8.3 Role of Acidophiles in Sustainability

The acidic environments are contaminated with many metals some of which are toxic, particularly the heavy metals. Acidophiles oxidize sulfur compounds and reduce their toxicity. Likewise, electronic wastes such as printed circuit boards (PCBs) of computers, cell phones, mobile chargers, etc. generate recalcitrant solid wastes. Acidophiles can play an important role in treating such electronic wastes and recovery of metals from them. Some work carried out by Indian researchers depict the role of acidophiles in remediation of metal-contaminated sites and solid wastes, and thus their importance in sustainability.

Singh et al. (2012) attempted bioleaching of metals from hazardous spent hydroprocessing catalysts in two stages. In the first stage, transformation of elemental sulfur particles to sulfuric acid was achieved through oxidation process by acidophilic bacteria, *Acidithiobacillus ferrooxidans.* In the second stage, the acid produced was used for the leaching process. Nickel, vanadium, and molybdenum contained within spent catalysts were leached from solid material to liquid by the action of sulfuric acid produced by acidophilic bacteria. At 50 g/L spent catalyst concentration and 20 g/L elemental sulfur, 94.3% Ni was extracted after 7 days.

Adhapure et al. (2014) have described method for recovery of metals from electronic waste, namely, printed circuit boards (PCBs) by bioleaching. Instead of using pulverized PCB, the authors have used large pieces of PCBs for removal of metals. Chemical treatment was used to solve problem of chemical coating on PCBs. Consortium of acidophilic microorganisms was developed from ore samples of bauxite and pyrite by enrichment using 9 k medium. Two enrichments were mixed to have mixed microbial consortium (mmc) which was used for bioleaching. The mixed microbial consortium (mmc) revealed presence of *Acidiphilum* spp., *Leptospirillum* spp., *Thiobacillus ferrooxidans*, *Thiobacillus caldus*, and *Sulfobacillus.* The results showed that 940 mg/L, 239 mg/L, and 72.7 mg/L of Cu, Zn, and Ni, respectively, were solubilized in 240 h from PCB pieces of size 12 cm × 6 cm. Lead recovery was comparatively less (3.53 mg/L after 48 h). From PCB pieces of size 4 cm × 2.5 cm, 536 mg/L, 152 mg/L, 35.64 mg/L, and 0.31 mg/L of Cu, Zn, Ni, and Pb, respectively, were solubilized in 240 h. It was observed that complete removal of metals takes place in 10 days except the solder traces indicating poor leaching of Pb. Better clearance of PCB pieces was observed for 4 cm × 2.5 cm as compared to 12 cm × 6 cm pieces of PCBs. The final metal-free PCBs can be easily recycled.

Sodha et al. (2019) analyzed the ability of *Leptospirillum* isolated from coal sample, to extract the metals from television printed circuit boards (TV-PCBs). Activation of iron oxidizer enriched from coal sample resulted in 8.5-fold enhancement in iron oxidation rate. The TV-PCBs' adapted culture showed 1.2, 1.3, and 1.6-fold higher Cu, Zn, and Ni leaching, respectively, from 10 g/L TV-PCBs waste pulp density compared to unadapted culture. It also showed 24-h reduction in leaching time. The two-step process showed 1.3-fold higher metal extractions compared to one-step process with both adapted and unadapted culture. Additionally, there was a 1.3-fold reduction in reaction time in the two-step leaching process as compared to the one step process.

5.9 Contribution to the Knowledge of Acidophilic Microbial Species, New to Science, from India

One novel species of acidophilic microorganism has been added to the global collection of microorganisms and is described in Table 5.2.

Table 5.2 Acidophilic microbial species, new to science, from India

Sr. No.	Name of the microorganism	Source	Reference
1	*Acidophilum symbioticum* sp. nov.	From *thiobacillus ferrooxidans* culture from Indian mines	Bhattacharyya et al. (1991)

5.10 Future Perspective of Acidophiles

For application of enzymes of acidophiles in industrial processes, genetic engineering and directed evolution to express them in mesophiles to carry out large-scale production could be a line of research in future. Likewise, improvement in strains of acidophiles to reduce leaching time will enhance the cost-benefit of the biomining/ bioleaching process. Astrobiological studies of acidophiles will be fascinating.

References

Abhilash, Mehta KD, Pandey BD (2012) Efficacy of bacterial adaptation on copper biodissolution from a low-grade chalcopyrite ore by *A. ferrooxidans*. Int J Nonferr Metall 1(1):18955. https://doi.org/10.4236/ijnm.2012.11001

Adhapure NN, Dhakephalkar PK, Dhakephalkar AP, Tembhurkar VR, Rajgure AV, Deshmukh AM (2014) Use of large pieces of printed circuit boards for bioleaching to avoid precipitate contamination problem and to simplify overall metal recovery. Methods X 1:e181–e186

Agate AD, Paknikar KM, Khinvasara NJ, Gupta SG (1984) Microorganisms associated with biometallurgy of copper in India. 7th International Biotechnology Symposium, pp 483–484

Agate AD, Paknikar KM, Khinvasara NJ (1989) Scale-up of leaching of Malanjkhand copper ores - a case study. In: Salley J, McCready RGL, Wichlacz PL (eds) Biohydrometallurgy. Canada Centre for Mineral and Energy Technology, p 577-58

Asghari I, Mousavi SM, Amiri F, Tavassoli S (2013) Bioleaching of spent refinery catalysts: a review. J Ind Eng Chem 19:1069–1081

Baker-Austin C, Dopson M, Wexler M, Sawers RG, Stemmler A, Rosen BP, Bond PL (2007) Extreme arsenic resistance by the acidophilic archaeon '*Ferroplasma acidarmanus*' Fer1. Extremophiles 11:425–434

Banerjee PC (2004) Genetics of metal resistance in acidophilic prokaryotes of acidic mine environments. IJEB 42(01):9–25. http://hdl.handle.net/123456789/23315

Batrakov SG, Pivovarova TA, Esipov SE, Sheichenko VI, Karavaiko GI (2002) β-D-Glucopyranosyl caldarchaetidylglycerol is the main lipid of the acidophilic, mesophilic, ferrous iron-oxidising archaeon *Ferroplasma acidiphilum*. Biochim Biophys Acta 1581:29–35

Beolchini F, Fonti V, Ferella F, Vegliò F (2010) Metal recovery from spent refinery catalysts by means of biotechnological strategies. J Hazard Mater 178:529–534

Bhattacharyya S, Chakrabarty BK, Das A, Kundu PN, Banerjee PC (1991) *Acidiphilium symbioticum* sp. nov., an acidophilic heterotrophic bacterium from *Thiobacillus ferrooxidans* cultures isolated from Indian mines. Can J Microbiol 37:78–85

Bhowal SS, Chakraborty R (2015) Microbial diversity of acidophilic heterotrophic bacteria: an overview, Chapter 17. In: Jha P (ed) Biodiversity, conservation and sustainable development, vol 1: issues and approaches. New Academic Publishers, New Delhi. ISBN: 9788186772751

Bruins MR, Kapil S, Oehme FW (2000) Microbial resistance to metals in the environment. Ecotoxicol Environ Saf 45:198–207

Canfield DE, Rosing MT, Bjerrum C (2006) Early anaerobic metabolisms. Philos Trans R Soc Lond Ser B Biol Sci 361:1819–1836

Crossman L, Holden M, Pain A, Parkhill J (2004) Genomes beyond compare. Nat Rev Microbiol 2:616–617

Dave SR, Natarajan KA (1979) Bio-oxidation studies with *Thiobacillus ferrooxidans* in the presence of copper and zinc. Inst Min Metall, Trans, Sect C, Miner Processing Extractive Metall 88(12):234–237

Dave SR (2002) Stress tolerance development in indigenous isolates of *Acidithiobacillus ferrooxidans*. In: Shukla LB, Mishra VN (eds) Mineral biotechnology, proceedings of the national seminar, vol 37. Allied Publ Pvt Ltd, New Delhi

Dave SR (2008a) Microbial interactions with inorganic pollutants: acid mine drainage, microbial accumulation of heavy metals and radionuclides. Appl Microbiol:1–25. http://nsdl.niscair.res.in/jspui/handle/123456789/646

Dave SR (2008b) Selection of *Leptospirillum ferrooxidans* SRPCBL and development for enhanced ferric regeneration in stirred tank and airlift column reactor. Bioresour Technol 99(16):7803–7806

Dave SR, Mathur P (1987) Factors affecting multi-metal ore leaching by *Thiobacillus ferrooxidans*. Indian J Microbiol 27(1–4):51–54

Dave SR, Natarajan KA (1987) Microbial ecology of some Indian sulfidic mines. Trans Indian Inst Metals 40(4):315–327

Dave SR, Pandhi ND (1995) Increased metal tolerance by *Thiobacillus ferrooxidans* (SRco) isolated from Ambamata complex ore mine. IJEMS 02(3):139–141. http://hdl.handle.net/123456789/29844

Dave SR, Upadhyay NM (1993) Thiosulfate oxidizing organism from thermal springs. Indian J Microbiol 33:241–241

Dave SR, Tipre DR, Vora SB (2001) Influence of high pulp density and its mode of addition on copper and zinc bioextraction from polymetallic concentrate: LSTR study. Proc Metall 11:561–572

Dave SR, Gupta KH, Tipre DR (2008) Characterization of arsenic resistant and arsenopyrite oxidizing *Acidithiobacillus ferrooxidans* from Hutti gold leachate and effluents. Bioresour Technol 99(16):7514–7520

Dew DW, Muhlbauer R, van Buuren C (1999) Bioleaching of copper sulphide concentrates with mesophiles and thermophiles. Presented in: Alta 1999 Copper Hydrometallrrgy Forum, Brisbane

Dhakephalkar AP (2008) PCR based fingerprinting of Iron-and sulphur-oxidizers and associated bacteria from mining ecosystem, Ph. D. Thesis. Pune University

Dold B, González-Toril E, Aguilera A, López-Pamo E, Cisternas ME, Bucchi F, Amils R (2013) Acid rock drainage and rock weathering in Antarctica: important sources for iron cycling in the Southern Ocean. Environ Sci Technol 47:6129–6136

Dopson M (2012) Physiological adaptations and biotechnological applications of acidophiles. In: Anitori RP (ed) Extremophiles: microbiology and biotechnology. Horizon Scientific Press, Norwich

Dopson M, Baker-Austin C, Koppineedi PR, Bond PL (2003) Growth in sulfidic mineral environments: metal resistance mechanisms in acidophilic micro-organisms. Microbiology 149:1959–1970

Dopson M, Baker-Austin C, Bond PL (2005) Analysis of differential protein expression during growth states of *Ferroplasma* strains and insights into electron transport for iron oxidation. Microbiology 151:4127–4137

Dopson M, Baker-Austin C, Bond P (2006) Towards determining details of anaerobic growth coupled to ferric iron reduction by the acidophilic archaeon '*Ferroplasma acidarmanus*' Fer1. Extremophiles 11:159–168

Edwards KL, Bond PL, Gihring TM, Banfield JF (2000) An archaeal iron-oxidizing extreme acidophile important in acid mine drainage. Science 279:1796–1799

Fairén AG, Davila AF, Lim D, Bramall N, Bonaccorsi R, Zavaleta J, Uceda ER, Stoker C, Wierzchos J, Dohm JM, Amils R, Andersen D, McKay CP (2010) Astrobiology through the ages of Mars: the study of terrestrial analogues to understand the habitability of Mars. Astrobiology 10:821–843

Friedrich C, Rother G, Bradischewsky DF, Quentmeier A, Fischer J (2001) Oxidation of reduced inorganic sulfur compounds by bacteria: emergence of a common mechanism? Appl Environ Microbiol 67:2873–2882

Fuchs T, Huber H, Teiner K, Burggraf S, Stetter KO (1995) *Metallosphaera prunae*, sp. nov., a novel metal-mobilizing, thermoacidophilic archaeon, isolated from a uranium mine in Germany. Syst Appl Microbiol 18:560–566

Fütterer O, Angelov A, Liesegang H, Gottschalk G, Schleper C, Schepers B, Dock C, Antranikian G, Liebl W (2004) Genome sequence of *Picrophilus torridus* and its implications for life around pH 0. Proc Natl Acad Sci U S A 101:9091–9096

García-Moyano A, González-Toril E, Aguilera A, Amils R (2012) Comparative microbial ecology study of the sediments and the water column of the Rio Tinto, an extreme acidic environment. FEMS Microbiol Ecol 81:303–314

García-Moyano A, Austnes AE, Lanzén A, González-Toril E, Aguilera Á, Øvreås L (2015) Novel and unexpected microbial diversity in acid mine drainage in Svalbard (78° N), revealed by culture-independent approaches. Microorganisms 3:667–694

Geological Survey India (1982) GSI records, vol. 109-Part 1

Ghosh W, Dam B (2009) Biochemistry and molecular biology of lithotrophic sulfur oxidation by taxonomically and ecologically diverse bacteria and archaea. FEMS Microbiol Rev 33:999–1043

Ghosh W, Roy P (2007) Chemolithoautotrophic oxidation of thiosulfate, tetrathionate and thiocyanate by a novel rhizobacterium belonging to the genus *Paracoccus*. FEMS Microbiol Lett 270:124–131

Ghosh S, Mohanty S, Akcil A, Sukla LB, Das AP (2016a) A greener approach for resource recycling: manganese bioleaching. Chemosphere 154:628–639

Ghosh S, Mohanty S, Nayak S, Sukla LB, Das AP (2016b) Molecular identification of indigenous manganese solubilizing bacterial biodiversity from manganese mining deposits. J Basic Microbiol 56(3):254–262

Gonzalez-Toril E (2005) Iron meteorites can support the growth of acidophilic chemolithoautotrophic microorganisms. Astrobiology 5:406–414

Gonzalez-Toril E, Llobet-Brossa E, Casamayor EO, Amann R, Amils R (2003) Microbial ecology of an extreme acidic environment, the Tinto River. Appl Environ Microbiol 69:4853–4865

Hallberg KB (2010) New perspectives in acid mine drainage microbiology. Hydrometallurgy 104:448–453

Hallberg KB, Johnson DB (2001) Biodiversity of acidophilic prokaryotes. Adv Appl Microbiol 49:37–84

Hallberg KB, Lindstrom EB (1994) Characterization of *Thiobacillus caldus* sp. nov., a moderately thermophilic acidophile. Microbiology 58:85–92

Ji G, Silver S (1995) Bacterial resistance mechanisms for heavy metals of environmental concern. J Ind Microbiol 14:61–75

Johnson DB (1995) Acidophilic microbial communities: candidates for bioremediation of acidic mine effluents. Int Biodeterior Biodegradation 35:41–58

Johnson DB (1998) Biodiversity and ecology of acidophilic microorganisms. FEMS Microbiol Ecol 27:307–317

Johnson DB, Hallberg KB (2003) The microbiology of acidic mine waters. Res Microbiol 154:466–473

Joshi NR, Agate AD, Paknikar KM (2000) Polyamine patterns in iron- and Sulphur-oxidizing bacteria isolated from an Indian copper mine indicate requirement of spermidine for growth under acid conditions. World J Microbiol Biotechnol 16(7):631–634. https://doi.org/10.1023/A:1008963829520

Joshi V, Shah N, Wakte P, Dhakephalkar P, Dhakephalkar A, Khobragade R, Naphade B, Shaikh S, Deshmukh A, Adhapure N (2017) Comparative bioleaching of metals from pulverized and non-pulverized PCBs of cell phone charger: advantages of non-pulverized PCBs. Environ Sci Pollut Res 24(36):28277–28286. https://doi.org/10.1007/s11356-017-0780-x

Kelly DP, Wood AP (2014) The family acidithiobacillaceae. In: Rosenberg E, DeLong EF, Lory S, Stackebrandt E, Thompson F (eds) The prokaryotes. Springer, Berlin, Heidelberg. https://doi.org/10.1007/978-3-642-38922-1_250

Kimura S, Bryan CG, Hallberg KB, Johnson DB (2011) Biodiversity and geochemistry of an extremely acidic, low-temperature subterranean environment sustained by chemolithotrophy. Environ Microbiol 13:2092–2104

Lopez-Archilla AI, Marin I, Amils R (2001) Microbial community composition and ecology of an acidic aquatic environment: the Tinto River, Spain. Microb Ecol 41:20–35

Lu WP, Kelly DP (1983) Thiosulphate oxidation, electron transport and phosphorylation in cell-free systems from *Thiobacillus* A2. J Gen Microbiol 129:1661–1671

Macalady JL, Vestling MM, Baumler D, Boekelheide N, Kaspar CW, Banfield JF (2004) Tetraetherlinked membrane monolayers in *Ferroplasma* spp: a key to survival in acid. Extremophiles 8:411–419

Marhual NP, Pradhan N, Kar RN, Sukla LB, Mishra BK (2008) Differential bioleaching of copper by mesophilic and moderately thermophilic acidophilic consortium enriched from same copper mine water sample. Bioresour Technol 99(17):8331–8336. https://doi.org/10.1016/j.biortech.2008.03.003

Menon AG, Dave SR (1993) Amenability studies of complex sulphide ores from various Indian mines. Biohydrometall Technol 1:137–146

Menon AG, Dave SR (1994) Biooxidation of ferrous by ultraviolet-exposed silver tolerant *thiobacillus ferrooxidans* (ore-isolate). J Biotechnol 38(1):7–9

Menon AG, Dave SR (1995) Growth behavior of various *Thiobacillus ferrooxidans* isolates on different substrates. Trans Indian Inst Met (India) 48(2):135–138

Michels M, Bakker EP (1985) Generation of a large, protonophore-sensitive proton motive force and pH difference in the acidophilic bacteria *Thermoplasma acidophilum* and *Bacillus acidocaldarius*. J Bacteriol 161:231–237

Mirete S, Morgante V, González-Pastor JE (2017) Acidophiles: diversity and mechanisms of adaptation to acidic environments. In: Stan-Lotter H, Fendrihan S (eds) Adaption of microbial life to environmental extremes. Springer. https://doi.org/10.1007/978-3-319-48327-6_9

Mohapatra S, Bohidar S, Pradhan N, Kar RN, Sukla LB (2007) Microbial extraction of nickel from Sukinda chromite overburden by *Acidithiobacillus ferrooxidans* and *Aspergillus* strains. Hydrometallurgy 85(1):1–8

Mohod AM, Salunkhe PB, Paknikar KM (1999) Bioleaching of manganese from low-grade ores using manganese- reducing microbial cultures. J Indian Inst Sci 79(4):275–285

Moore JM (2004) Mars: blueberry fields forever. Nature 428:711–712

Morris RV, Klingelhöfer G, Bernhardt B, Schröder C, Rodionov DS, de Souza PA, Yen A, Gellert R, Evlanov EN, Foh J, Kankeleit E, Gütlich P, Ming DW, Renz F, Wdowiak T, Squyres SW, Arvidson RE (2004) Mineralogy at Gusev Crater from the Mössbauer spectrometer on the Spirit Rover. Science 305:833–836

Nies DH (1999) Microbial heavy-metal resistance. Appl Microbiol Biotechnol 51:730–750

Okabayashi A, Wakai S, Kanao T, Sugio T, Kamimura K (2005) Diversity of 16S ribosomal DNA-defined bacterial population in acid rock drainage from Japanese pyrite mine. J Biosci Bioeng 100:644–652

Paknikar KM (1983) Studies on physiology of microorganisms associated with leaching of copper, Ph. D. thesis. Poona University, Pune

Paknikar KM (1988) Production of chelating compounds by yeasts in a microbial copper leaching system and its practical implications. Bull Mater Sci 10(5):477–478

Paknikar JM, Agate AD (1987) Development of a culture collection and data bank for biometallurgical processes in India. Mircen J 3:169–178. https://doi.org/10.1007/BF00933617

Paknikar KM, Agate AD (1988) Occurrence of a *Thiobacillus ferrooxidans* strain tolerating unusually high concentrations of metals and an associated metal-tolerant acidophilic heterotrophic bacterium. In: Norris PR, Kelly DP (eds) Biohydrometallurgy. Kew, Science & Technology Letters, pp 558–560

Pal S, Pradhan D, Das T, Sukla LB, Chaudhury GR (2010) Bioleaching of low-grade uranium ore using *Acidithiobacillus ferrooxidans*. Indian J Microbiol 50(1):70–75

Panda S, Sanjay K, Sukla LB, Pradhan N, Subbaiah T, Mishra BK, Prasad MSR, Ray SK (2012) Insights into heap bioleaching of low-grade chalcopyrite ores—a pilot scale study. Hydrometallurgy 125:157–165

Panda S, Parhi PK, Nayak BD, Pradhan N, Mohapatra UB, Sukla LB (2013a) Two step meso-acidophilic bioleaching of chalcopyrite containing ball mill spillage and removal of the surface passivation layer. Bioresour Technol 130:332–338

Panda S, Pradhan N, Mohapatra U, Panda SK, Rath SS, Rao DS, Nayak BD, Sukla LB, Mishra BK (2013b) Bioleaching of copper from pre and post thermally activated low grade chalcopyrite contained ball mill spillage. Front Environ Sci Eng 7:281–293. https://doi.org/10.1007/s11783-013-0484-5

Parro V, Rodríguez-Manfredi JA, Briones C, Compostizo C, Herrero PL, Vez E, Sebastián E, Moreno-Paz M, García-Villadangos M, Fernández-Calvo P, González-Toril E, Pérez-Mercader J, Fernández-Remolar D, Gómez-Elvira J (2005) Instrument development to search for biomarkers on mars: terrestrial acidophile, iron-powered chemolithoautotrophic communities as model systems. Planet Space Sci 53:729–737

Patel MJ, Tipre DR, Dave SR (2011) Isolation, identification, characterization and polymetallic concentrate leaching studies of tryptic soy-and peptone-resistant thermotolerant *Acidithiobacillus ferrooxidans* SRDSM2. Bioresour Technol 102(2):1602–1607

Patel BC, Tipre DR, Dave SR (2012) Development of *Leptospirillum ferriphilum* dominated consortium for ferric iron regeneration and metal bioleaching under extreme stresses. Bioresour Technol 118:483–489

Patel BC, Sinha MK, Tipre DR, Pillai A, Dave SR (2014a) A novel biphasic leaching approach for the recovery of Cu and Zn from polymetallic bulk concentrate. Bioresour Technol 157:310–315

Patel BC, Tipre DR, Dave SR (2014b) Biphasic leaching operation for the GMDC polymetallic bulk concentrate. In: Advanced materials research conference, vol 828. Trans Tech Publications Ltd Publisher, pp 117–121

Pivovarova TA, Kondrat'eva TF, Batrakov SG, Esipov SE, Sheichenko VI, Bykova SA, Lysenko AM, Karavaiko GI (2002) Phenotypic features of *Ferroplasma acidiphilum* strains YT and Y-2. Microbiology 71:698–706

Podar M, Reysenbach A-L (2006) New opportunities revealed by biotechnological explorations of extremophiles. Curr Opin Biotechnol 17:250–255

Pradhan D, Pal S, Sukla LB, Chaudhury GR, Das T (2008a) Bioleaching of low-grade copper ore using indigenous microorganisms. Indian J Chem Technol 15:588–592

Pradhan N, Nathsarma KC, Rao KS, Sukla LB, Mishra BK (2008b) Heap bioleaching of chalcopyrite: a review. Miner Eng 21(5):355–365

Pradhan D, Mishra D, Kim DJ, Ahn JG, Chaudhury GR, Lee SW (2010) Bioleaching kinetics and multivariate analysis of spent petroleum catalyst dissolution using two acidophiles. J Hazard Mater 175:267–273

Pradhan D, Kim DJ, Sukla LB, Pattanaik A, Samal DPK (2019a) Bacterial leaching of chalcopyrite concentrates using *Acidithiobacillus ferrooxidans*. Inglomay Sect C 16:1–9

Pradhan D, Sukla LB, Mishra BB, Devi N (2019b) Biosorption for removal of hexavalent chromium using microalgae *Scenedesmus* sp. J Clean Prod 209:617–629

Pradhan D, Sukla LB, Behera PK, Dash S (2021a) Development of bioleaching for extraction of metal values from low-grade ores and wastes. Aspects Min Miner Sci 7(1):000654. https://doi.org/10.31031/AMMS.2021.07.000654

Pradhan D, Sukla LB, Pattanaik A, Samal DPK, Biswal T, Jena SKB (2021b) Improvement in metal dissolution from spent catalyst by adapted *Acidithiobacillus ferrooxidans*. Biointerface Res Appl Chem 11:7794–7803

Ram RJ, VerBerkmoes NC, Thelen MP, Tyson GW, Baker BJ, Blake RC, Shah M, Hettich RL, Banfield JF (2005) Community proteomics of a natural microbial biofilm. Science 308:1915–1920

Rawlings DE (2002) Heavy metal mining using microbes. Annu Rev Microbiol 56:65–91

Rawlings DE (2005) Characteristics and adaptability of iron- and sulfur-oxidizing microorganisms used for the recovery of metals from minerals and their concentrates. Microb Cell Factories 4:1–15

Rawlings DE, Johnson DB (2007) The microbiology of biomining: development and optimization of mineral-oxidizing microbial consortia. Microbiology 153:315–324

Qureshi SA, Qureshi RA, Sodha AB, Tipre DR, Dave SR (2018) Bioextraction dynamics of potassium from feldspar by heterotrophic microorganisms isolated from ceramic and rhizospheric soil. Geomicrobiol J 35(2):127–131

Sampson MI, Phillips CV (2001) Influence of base metals on the oxidizing ability of acidophilic bacteria during the oxidation of ferrous sulfate and mineral sulfide concentrates, using mesophiles and moderate thermophiles. Miner Eng 14:317–340

Santofimia E, Gonzalez-Toril E, Lopez-Pamo E, Gomariz M, Amils R, Aguilera A (2013) Microbial diversity and its relationship to physicochemical characteristics of the water in two extreme acidic pit lakes from the Iberian Pyrite Belt (SW Spain). PLoS One 8:e66746

Schippers A, Sand W (1999) Bacterial leaching of metal sulfides proceeds by two indirect mechanisms via thiosulfate or via polysulfides and sulfur. Appl Environ Microbiol 65:319–321

Shah AB, Chakraborty S, Bandhopadhyaya S (1974–1975) Records of the geological survey of India, vol 109, p 94

Shah MP, Vora SB, Dave SR (1999) Copper remediation by waste Streptomyces biomass. Indian J Microbiol 39(2):109–112

Shah MB, Tipre DR, Purohit MS, Dave SR (2015) Development of two-step process for enhanced biorecovery of Cu–Zn–Ni from computer printed circuit boards. J Biosci Bioeng 120(2): 167–173

Sharma A, Kawarabayasi Y, Satyanarayana T (2012) Acidophilic bacteria and archaea: acid stable biocatalysts and their potential applications. Extremophiles 16(1):1–19. https://doi.org/10.1007/s00792-011-0402-3

Sharma A, Parashar D, Satyanarayana T (2016) Acidophilic microbes: biology and applications, Chapter 7. In: Rampelotto PH (ed) Biotechnology of extremophiles, grand challenges in biology and biotechnology 1, vol 215. Springer. https://doi.org/10.1007/978-3-319-13521-2_7

Shimada H, Nemoto N, Shida Y, Oshima T, Yamagishi A (2002) Complete polar lipid composition of *Thermoplasma acidophilum* HO-62 determined by high-performance liquid chromatography with evaporative light-scattering detection. J Bacteriol 184:556–563

Singh S, Sukla LB, Mishra BK (2011) Extraction of copper from Malanjkhand low-grade ore by *Bacillus stearothermophilus*. Indian J Microbiol 51(4):477–481. https://doi.org/10.1007/s12088-011-0073-x

Singh B, Lal A, Lal B (2012) Bioleaching of metal using acidophilic bacteria, AIChE spring meeting and global congress on process safety conference. In: Proceeding of 2012 spring meeting & 8th global congress on process safety

Sodha AB, Qureshi SA, Khatri BR, Tipre DR, Dave SR (2019) Enhancement in iron oxidation and multi-metal extraction from waste television printed circuit boards by iron oxidizing *Leptospirillum feriphillum* isolated from coal sample. Waste Biomass Valorizat 10(3):671–680

Sodha AB, Tipre DR, Dave SR (2020) Optimization of biohydrometallurgical batch reactor process for copper extraction and recovery from non-pulverized waste printed circuit boards. Hydrometallurgy 191:105170

Squyres SW, Grotzinger JP, Arvidson RE, Bell JF III, Calvin W, Christensen PR, Clark BC, Crisp JA, Farrand WH, Herkenhoff KE, Johnson JR, Klingelhöfer G, Knolls AH, Mclennan SM, Mcsween HY Jr, Morris RV, Rice JW Jr, Rieder R, Soderblom LA (2004) In situ evidence for an ancient aqueous environment at Meridiani Planum, Mars. Science 306:1709–1714

Sukla LB, Panchanadikar V (1993) Bioleaching of lateritic nickel ore using a heterotrophic microorganism. Hydrometallurgy 32(3):373–379

Sukla LB, Nathsarma KC, Mahanta JR, Singh S, Behera S, Rao KS, Subbaiah T, Mishra BK (2009) Recovery of copper values from bio-heap leaching of low grade Malanjkhand chalcopyrite ore. Korean J Chem Eng 266:1668–1674

Sukla LB, Pattanaik A, Pradhan D (2019) Advances in beneficiation of low-grade bauxite. In: Chesonis C (ed) Light metals, the minerals, metals and materials series. Springer, Cham. https://doi.org/10.1007/978-3-030-05864-7_1

Thacker SC, Nayak NS, Tipre DR, Dave SR (2021) Multi-metal mining from waste cell phone printed circuit boards using lixiviant produced by a consortium of acidophilic iron oxidizers. In: Environmental engineering science. Mary Ann Liebert, Inc

Tipre DR, Dave SR (2004) Bioleaching process for Cu–Pb–Zn bulk concentrate at high pulp density. Hydrometallurgy 75(1–4):37–43

Tipre DR, Vora SB, Dave SR (1998) Improved metal extraction by selected *thiobacillus ferrooxidans* consortium from polymetallic concentrate. J Scient Indust Res 57(10–11):805–808

Tipre DR, Vora SB, Dave SR (1999) Comparative copper and zinc bioextraction at various stages of scale up using *T. ferrooxidans* consortium. Process Metall 9:219–227

Tuovinen OH, Bhatti TM (1999) Microbiological leaching of uranium ores. Miner Metall Process 16:51–60

Tyson GW, Chapman J, Hugenholtz P, Allen EE, Ram RJ, Richardson PM, Solovyev VV, Rubin EM, Rokhsar DS, Banfield JF (2004) Community structure and metabolism through reconstruction of microbial genomes from the environment. Nature 428:37–43

Urbieta MS, Gonzalez TE, Aguilera A, Giaveno MA, Donati E (2012) First prokaryotic biodiversity assessment using molecular techniques of an acidic river in Neuquen, Argentina. Microb Ecol 64:91–104

van Niel CB (1995) Natural selection in the microbial world. J Gen Microbiol 13:201–217

Zychlinsky E, Matin A (1983) Cytoplasmic pH homeostasis in an acidophilic bacterium, *Thiobacillus acidophilus*. J Bacteriol 156:1352–1355

Metallophilic, Metal-Resistant, and Metal-Tolerant Microorganisms

6

Abstract

Metals are found in high concentration around metal processing industries or in dumping sites. The microorganisms present in such metal-rich environments resist high concentration of metals and are termed as metallophiles and metal-resistant microorganisms. Metal-tolerant microorganisms can survive at high concentrations of metals. Metals essential for bacterial life include copper, cobalt, nickel, chromium, molybdenum, and zinc. while cadmium, silver, mercury, and lead ions are toxic to living things. The natural habitats for metallophiles are deep-sea or terrestrial hydrothermal sources but they are found mostly in man-made metal-rich environments. Metal-resistant bacteria include *Thiobacillus ferrooxidans*, *Bacillus*, *Pseudomonas*, *Staphylococcus*, *Micrococcus*, and Coryneforms. In many microorganisms, the genes controlling metal resistance are carried on plasmids. The heavy metals cause environmental pollution and metallophiles work as their bioremediators by removing the metals through different mechanisms such as biosorption, bioreduction, and biooxidation.

In India, metal-rich environments exist mainly in mining areas, groundwater and aquifers, and metallophilic, metal-resistant microorganisms have been reported from such areas. These include *Acinetobacter*, *Pseudomonas*, *Citrobacter*, *Microbacterium*, *Thiobacillus*, and *Bacillus*. Applications of metallophiles have been described for recovery of metals. Five novel species of metallophilic microorganisms, new to Science, have been reported from India.

Keywords

Metallophilic · Metal resistant · Metal tolerant · Metal processing industries · Metal resistance genes on plasmids · Biosorption · Chromate reduction · Arsenic contaminated groundwater

P. P. Kanekar, S. P. Kanekar, *Diversity and Biotechnology of Extremophilic Microorganisms from India*, Microorganisms for Sustainability 41,
https://doi.org/10.1007/978-981-19-1573-4_6

Metallophiles are microorganisms that grow optimally at high concentrations of metals. These organisms are specifically adapted to metal-rich biotopes and can therefore be considered as a kind of extremophile. Metal-rich biotopes may be divided up to some extent into natural and anthropogenic biotopes. Natural biotopes correspond to a variety of geological events. Some deep-sea or terrestrial hydrothermal sources may also be considered as natural metal-rich biotopes (Mergeay 2007).

In their extreme form, anthropogenic metal-rich biotopes are mainly industrial and may be found around metal-processing factories or in dumping sites. They are also characteristic of the industrial revolution of the last 200–300 years. Their multiplication on the Earth's surface has promoted the large-scale redistribution of the bacterial population. The harsh selection pressure, linked to the chemical mixtures they contain, has likely forced some key genetic rearrangements in these populations, and is therefore significant from the point of view of evolution (Mergeay 2007).

6.1 Microorganisms and Elements

Life in prokaryotes (Bacteria and Archaea), as well as in eukaryotes, relies on various elements, the basic ones being carbon, nitrogen, oxygen, hydrogen, sulfur, and phosphorus. Metals are also required for life—calcium, magnesium, potassium, and sodium are required in substantial concentrations. The biological requirements for iron and manganese are satisfied at lower concentrations. Metallic trace elements essential for bacterial life include copper, cobalt, nickel, chromium, molybdenum, vanadium, tungsten, and zinc. But concentrations even slightly exceeding those satisfying the requirements for the trace elements are generally toxic. Tungsten (which can also substitute for molybdenum in various molybdoenzymes) deserves a special mention as far as extremophiles are concerned. Tungsten (W) is essential for the growth of the hyperthermophiles *Pyrococcus furiosus* and *Thermococcus littoralis*. Other than the basic metallic elements that are essential for life, all are generally toxic when bioavailable. Cadmium, silver, mercury, thallium, and lead ions are the most toxic for living beings (Mergeay 2007).

Among the other elements, chloride is required for ionic equilibrium and may be tolerated through a broad range of concentration.

The nonmetallic elements boron, fluoride, arsenic, selenium, and iodine may be added to the list of biological trace elements in prokaryotes or in eukaryotes. But outside the basic metallic elements that are essential for life, all the other metals and elements are generally toxic when they are bioavailable (bioavailable means soluble at biological pH, and/or having access to the various transport pathways that cross the biological membranes). Cadmium, silver, mercury, thallium, and lead ions are the most toxic for living beings and constitute major sources of concern in environmental toxicology.

For soil microbes, there is a special challenge as they may come in contact with metallic ores or with locally high concentrations of metals. This occurs in a lot of natural situations created by the geological events, for example, deep hydrothermal sources and volcanic areas. The metal-rich ultramafic soils of New Caledonia are a few examples of such natural situations.

Anthropogenic situations include the dumping sites of metallurgical waste from processed ores. These biotopes may contain very high amounts of metals of which a part becomes bioavailable. Metal release in agricultural practice may also affect bacterial communities from soil. Copper-based fungicides allowed the selection of plasmid-born resistance to copper in bacteria pathogenic for vegetable crops *Pseudomonas syringae* var *tomato*, *Xanthomonas campestris*. In the metal-rich biotopes that are reservoirs of extremophiles, most of the attention has been focused on metal-resistant heterotrophic bacteria. Nevertheless, these bacteria most likely form only a small part of the biodiversity of these biotopes (Mergeay 2007).

Geothermal environments are also a source of interesting microorganisms adapted to heavy metals. Bacteria were reported to rapidly oxidize inorganic As (III) to As (V) from arsenic-rich terrestrial geothermal environments or from hot springs in the Yellowstone National Park (USA). Some of these bacteria belong to the genus *Thermus*: One of these *Thermus* strains are able not only to oxidize As (III) to As (V) but also respire on As (V) in anaerobic conditions. Deep sea hydrothermal sources are also environments that are studied for the adaptation of bacteria of heavy metals.

The industrial revolution that has lasted for the last 200 years has generated plenty of new biotopes in soils, rivers, lakes, and sediments, mainly via the dumping of metal-rich wastes that were often mixed with recalcitrant synthetic chemical compounds (chloroaromatic compounds). The extreme variety of these industrial sites and their geographical dispersion across the surface of the Earth has provoked a huge and yet almost unnoticed redistribution of bacterial populations probably including those that were already adapted to heavy metals, and has surely exerted revolutionary pressures on their genotypes (Mergeay 2007).

6.2 Metal-Resistant Bacteria

Diverse group of microorganisms have been reported for resistance to one or more metals. Trevors et al. (1985) have reviewed different metal-resistant bacteria.

Thiobacillus ferrooxidans was found to tolerate Zn, Ni, Cu, Co, Mn, and Al at concentrations of 10 g/L (Tuovinen et al. 1971). Austin et al. (1977) studied metal-tolerant bacteria isolated from an estuary in Chesapeake Bay, USA, for metals Co, Pb, Mo at 100 mg/L and Hg at 10 mg/L concentration. *Bacillus*, *Erwinia*, *Mycobacterium*, *Pseudomonas*, and Coryneforms were predominant metal-resistant bacteria.

Cadmium-Resistant Bacteria Cadmium resistance was studied in *Staphylococcus aureus* having a plasmid harboring gene for cadmium resistance (Chopra 1971). The cadmium (Cd^{2+}) uptake occurred via the energy-dependent manganese transport system. Removal of the plasmid from the resistant strain caused cadmium sensitivity. Tynecka et al. (1981a, b) concluded that the resistance of *S. aureus* 17810R to Cd^{2+} appears to be caused by a plasmid-encoded Cd^{2+} efflux system. This efflux was not observed in a sensitive strain of the organism. Other cadmium-resistant bacteria studied include *Pseudomonas aeruginosa* and *Klebsiella aerogenes*.

Cobalt- and Nickel-Resistant Bacteria Soil organisms were found to be resistant to 0.62 and 1.09 mM $CoCl_2$ (Duxbury and Bicknell 1983). Sjogren and Port (1981) found coliforms to be resistant to 10^{-3}M $CoCl_2$. Bacterial isolates resistant to Zn, Cd, Cu, Hg, and Co were also found to be resistant to 1.24 mM $NiCl_2$ and 2.14 mM $NiCl_2$ (Duxbury and Bicknell 1983).

Lead-Resistant Bacteria Environmental contamination with lead is caused by combustion by products of leaded petrol, pesticides containing lead arsenate, phosphate fertilizers, and metal-smelting wastes (John 1971). *Micrococcus luteus* and *Azotobacter* sp. immobilized 4.9 and 3.1×10^2 mg of lead per gram dry cell weight, respectively (Tornabene and Edwards 1972). Some strains of *Staphylococcus aureus* contain plasmids that confer resistance to several metals including lead (Novick and Roth 1968). *Pseudomonas aeruginosa* showed resistance to lead acetate, 3200 μg/ml (Marques et al. 1979).

Mercury-Resistant Bacteria Mercury-resistant bacteria have been the most extensively studied of all the metal-tolerant organisms. In majority of these organisms, plasmids carry the genes coding for Hg resistance. The bacteria reported for Hg resistance are *Bacillus* sp., *Pseudomonas fluorescens*, *Acinetobacter* sp., *Pseudomonas* sp. (Olson et al. 1982); *Staphylococcus aureus* (Kondo et al. 1974); *Pseudomonas aeruginosa* (Nakahara et al. 1977); *Arthrobacter* sp., *Citrobacter* sp., *Enterobacter* sp., *Vibrio* sp., and *Flavobacterium* sp. (Nelson et al. 1973).

Zinc-Resistant Bacteria Duxbury and Bicknell (1983) have reported soil organisms resistant to 1.10–1.90 mM $ZnSO_4$. Other resistant bacteria include coliforms (Sjogren and Port 1981), *Pseudomonas aeruginosa* (Marques et al. 1979), and *Klebsiella aerogenes* (Pickett and Dean 1976).

Silver-Resistant Bacteria The occurrence of Ag-resistant bacteria has been well documented (Belly and Kydd 1982). The soluble ionic form of Ag is the most toxic to bacteria and the toxicity is significantly reduced by adsorption on glass, precipitation with sulfides, chlorides, phosphates, and its interaction with organic matter. Bioaccumulation of Ag by a mixed culture of *Thiobacillus ferrooxidans* and *T. thiooxidans* occurs during the leaching of sulfide ore minerals.

Haefeli et al. (1984) isolated an Ag-resistant strain of *Pseudomonas stutzeri* from a silver mine in Utah, USA. The organism contained three plasmids. Other silver-resistant organisms include *Pseudomonas maltophilia* (Charley and Bull 1979), *Pseudomonas aeruginosa* (Marques et al. 1979), and *Staphylococcus aureus* (Charley and Bull 1979).

Tellurium-Resistant Bacteria Summers and Jacoby (1978) have investigated plasmid-determined resistance to Tellurium (Te) compounds. The Te resistance was unrelated to antibiotic resistance and distinct from As, Ag, and Hg resistance. In *Pseudomonas aeruginosa*, Te resistance was associated with incompatibility group P-2. The plasmids conferred resistance to 10^{-3}M tellurite. Marques et al. (1979) isolated a *P. aeruginosa* strain from soil that exhibited resistance to 64 μg/ml $K_2Te_2O_3$.

Copper-Resistant Bacteria Copper exerts toxic effects on many microorganisms. Only a few organisms have been reported to be copper-resistant. Copper-resistant Pseudomonads were isolated from a 100 ppm solution of $CuSO_4$. The organisms were tolerant to 10^{-3}M Cu^{2+} (Zevenhuizen et al. 1979).

Bacterial isolates from mine samples are resistant to 100 μg/ml $CuSO_4$. *Klebsiella aerogenes* has been found to be resistant to 10 ppm $CuCl_2$ (Bitton and Freibofer 1978) and capsulated strains survived better than non-capsulated strains.

Arsenic- and Antimony-Resistant Bacteria Hedges and Baumberg (1973) studied bacterial resistance to As compounds such as sodium arsenate, sodium arsenite, and arsenic trioxide, conferred by a plasmid (R773). Arsenic resistance in bacteria and its transmission by conjugation and phage has been studied by Smith (1978). *E. coli*, *Shigella*, *Proteus*, *K. pneumoniae*, and *Salmonella* were resistant to arsenite. Silver and Keach (1982) reported that energy-dependent arsenate efflux was the mechanism of plasmid-mediated resistance in *S. aureus* and *E. coli.* A 46-MDa conjugative plasmid (pDG101) harbored in *Corynebacterium flaccumfaciens* subsp. *oorti* encoded for resistance to arsenite, arsenate, and antimony (III) (Hendrick et al. 1984).

Boron- and Chromium-Resistant Bacteria Summers and Jacoby (1978) studied metals (As, Hg, Ag, and Te) resistance in Gram-negative bacteria. Certain plasmids in *P. aeruginosa*, belonging to the Inc P-2 group conferred resistance to boron and chromium. *Pseudomonas fluorescens* LB300 isolated from Cr-contaminated sediment in the upper Hudson river, USA (Bopp et al. 1983) was chromate-resistant.

Other Metal-Resistant Bacteria A number of bacteria resistant to metals such as Mo, Thallium (Tl), tungsten (W), uranium (U), and plutonium (Pu) have also been reported. *Pseudomonas aeruginosa* was found to be resistant to molybdenum, tungsten, thallium, and uranium (Marques et al. 1979). Wildung and Garland (1982) investigated the effects of Pu on soil microorganisms. Pu is found in low concentration (0.07 to 0.4 pg/g) in some surface soils as a result of fallout from nuclear testing (Wildung and Garland 1982).

6.3 Mechanisms of Metal Resistance

Microbial metal resistance is both basic and applied phenomenon as it is relevant to ecology, physiology, and genetics of the organisms in both polluted and unpolluted environments. Trevors et al. (1985) have reviewed mechanisms of metal resistances in bacteria. In many organisms, the genes controlling metal resistance are carried on plasmids but not all metal-resistant bacteria contain plasmids. The four basic mechanisms by which plasmids encode for metal resistance are (1) inactivation of the metal(s), (2) alteration of the site of inhibition, (3) impermeability of the metal(s), and (4) metal bypass mechanism. There are many other physical, chemical, and biological factors that influence metal resistance.

Nies (1999) has described mechanisms of resistance to heavy metals. Heavy metals are metals with a density above 5 g/cm^3. Of the over 90 naturally occurring elements, 21 are nonmetals, 16 are light metals, and 53 (including Arsenic) are heavy metals. Heavy metals have a relatively high atomic weight and density compared to water. Most heavy metals are transition elements.

6.3.1 Microbial Resistance to Heavy Metals

Chromium (Cr) Chromium mainly occurs as Cr (VI) in the divalent oxyanion chromate and as Cr (III), the trivalent cation. Reduction/oxidation reactions between the two states are thermodynamically possible under physiological conditions (Weast 1984), thus chromate and Cr^{3+} are both biologically important ions. Chromate is more toxic than Cr (III). In microorganisms, no beneficial influence of chromium was found. Chromate enters the cells of *Ralstonia* sp. strain CH34 (formerly *Alcaligenes eutrophus)* (Brim et al. 1999) by the sulfate-uptake system (Nies and Silver 1989). Chromate resistance is probably based on an interaction of chromate reduction and chromate reflux. *Pseudomonas fluorescens* strain LB800 was the first bacterium shown to reduce chromate (Bopp and Ehrlich 1988). In *Ralstonia* sp. CH34, both processes efflux and reduction are involved (Peitzsch et al. 1998).

Iron (Fe) Iron is the only macro-bioelement of the heavy metals. Because of its low solubility, Fe^{3+} is not toxic to aerobic bacteria. Microorganisms have evolved specific iron-binding complex compounds, the siderophores, which bind Fe (III) and shuttle it to the cell (Braun et al. 1998). Fe^{2+} is also transported into bacterial cells.

Cobalt (Co) Cobalt is found mainly in the Co^{2+} form. Co^{3+} is only stable in complex compounds. Co^{2+} has medium toxicity. Resistance to cobalt in Gram-negative bacteria is based on transenvelope efflux driven by a resistance, nodulation, cell division (RND) transporter. Cobalt resistance seems always to be the by-product of resistance to another heavy metal, either nickel or zinc.

Nickel (Ni) Free nickel occurs mostly in the Ni^{2+} cationic form. Nickel is detoxified by sequestration and/or transport. It is bound to phosphate in *S. aureus* (Gonzalez and Jensen 1998). In *Ralstonia* sp. CH34 and related bacteria, nickel resistance is based on nickel efflux driven by RND transporters.

Copper (Cu) Copper easily interacts with radicals, best with molecular oxygen. Its radical character makes copper very toxic and many organisms are more sensitive to copper (Gordon et al. 1994) than *E. coli.* Plasmid-encoded copper resistance in *E. coli* strongly interacts with chromosomally encoded functions and the actual resistance mechanism may depend on the growth phase (Brown et al. 1995). In copper-resistant *Pseudomonas* strains, copper is accumulated in the periplasm and outer membrane (Cooksey 1993, 1994).

Zinc (Zn) Zinc occurs as the divalent cation Zn^{2+}. Zinc is less toxic than copper. Zinc may be complexed by various cellular components and is transported by members of a variety of protein families. Unspecific and fast uptake of Zn^{2+} is mediated by Mg^{2+} transport systems, as shown in *Ralstonia* sp. CH34 (Nies and Silver 1989). Two systems are used for zinc detoxification in bacteria, P-type efflux ATPase, and RND-driven transporters. The P-type ATPases transport zinc only across the cytoplasmic membrane, whereas the RND systems are thought to transport across the complete cell wall of Gram-negative bacteria, including outer membrane. The driving force for the export of the heavy-metal cations is not ATP, but the proton-motive force (Nies 1995).

Arsenic (As) Arsenic is a heavy metalloid and acts sometimes as a metal, sometimes not. Mainly it occurs as As (V) in AsO_4, arsenate and as As (III), arsenite. Aerobic bacteria like *Alcaligenes faecalis* are able to oxidize arsenite again, thus a geomicrobial redox cycle of arsenic exists, similar to iron cycle. Arsenate detoxification involves an initial step to differentiate it from phosphate. This step is the reduction of arsenate to arsenite. For the resistant determinant in *E. coli*, arsenate reduction by the Ars C protein is coupled to glutathione (Oden et al. 1994) via glutaredoxin. Arsenite then leaves the bacterial cell since anion export from bacterial cells is always driven by the chemiosmotic gradient.

Cadmium (Cd) Cadmium is a toxic heavy metal. Resistance to cadmium in bacteria is based on cadmium efflux. Cyanobacteria, however, contain metallothioneins (Olafson et al. 1979). In Gram-negative bacteria, cadmium seems to be detoxified by RND-driven systems like Czc which is mainly a zinc exporter and Ncc which is mainly a nickel exporter.

Mercury (Hg) Mercury is the metal with strongest toxicity. Since bacteria are confronted with toxic Hg^{2+} concentrations, mercury resistance determinants, mer, are very widespread (Silver 1996, Silver and Phung 1996). Resistance to mercury is based on its unique peculiarities, its redox potential and the vapor pressure/melting/boiling point of metallic mercury, which is extraordinarily low for a metal. Thus, living cells are able to reduce Hg^{2+} to the metal, which does not remain inside the cell with the potential of becoming oxidized again, but leaves the cell by passive diffusion. To prevent toxic effects of Hg^{2+} on periplasmic proteins in Gram-negative bacteria, Hg^{2+} is transported in the cell via specific uptake systems. In Gram-negative bacteria, it is bound by the periplasmic Hg^{2+}-binding protein MerP as the first step of detoxification (Qian et al. 1998). MerP probably delivers the toxic cation to the mercury transporter MerT for transport into the cytoplasm (Hobman and Brown 1996).

Silver (Ag) Silver is a precious metal with medical use. It has been used as an antimicrobial agent in medicines (Slawson et al. 1992). Consequently, silver-resistant bacteria evolved. Resistance is catalyzed by a RND-type transporter with remarkable similarity to Czc system from *Ralstonia* sp. strain CH34. Thus, silver resistance may be based on RND-driven transenvelope efflux in Gram-negative bacteria, efflux by P-type ATPase in Gram-positive bacteria, and additional complexation by intracellular compounds.

Molybdenum (Mb) Molybdenum occurs mostly as Mo (VI) in molybdate. Molybdenum is an important trace element. Molybdate may also be transported by sulfate uptake systems. For most enzymes, molybdate is bound to a specific molybdate cofactor (Romao et al. 1995; Schindelin et al. 1996).

Manganese (Mn) Manganese exists in various oxidation states, from Mn (II) to Mn VII every state is possible with the Mn^{2+} cation being the predominant form. The toxicity of manganese is low. Mn is taken up by *Ralstonia* sp. by the Mg-uptake system (Nies and Silver 1989).

Vanadium (V) Vanadium mostly exists as V (V), the trivalent oxyanion vanadate. Vanadate is structurally similar to phosphate and is thus known as an inhibitor of ATPase, and it may be taken up by phosphate uptake systems (Mahanty et al. 1991; Rehder 1992). Bacteria like *Azotobacter chroococum* are able to form a vanadate-dependent nitrogenase for nitrogen fixation if molybdate is not present in the ecosystem (Chatterjee et al. 1997; Eady 1995). Work on vanadate resistance has been carried out in *Saccharomyces cerevisiae* (Nakamura et al. 1995).

Environmental and microbiological factors can influence heavy metal toxicity. Binding of metals to organic materials, precipitation, complexation, and ionic interactions are all important phenomena. Microbes possess a range of tolerance mechanisms most featuring some kind of detoxification (Gadd and Griffiths 1977). Some microorganisms have developed "strategies" for combating effects of toxic inorganics and several may prove useful for their removal from wastewater. Living organisms have influenced steady-state levels of elements in the atmosphere, in the oceans, and at the surface of the Earth throughout geological time. Both organic compounds and some inorganic complexes have been formed through biological activities that began about four billion years ago (Wood and Wang 1983).

Silver and Phung (1996) have reviewed studies in molecular genetics, environmental roles, and biochemical mechanisms of metal resistance. Bacterial plasmids encode resistance to toxic metal ions including Ag, AsO_2, $AsO^3{}_4$, Cd^2, Co^2, $CrO^2{}_4$, Cu^2, Hg^2, Ni^2, Pb^2, and Zn^2. Chromosomal determinants of toxic metal resistance are known but there is no clarity in the distinction between plasmid-mediated resistance and chromosomal genes resistance. In case of mercury and arsenic, plasmid and chromosomal determinants are the same. Copper transport ATPase and metallothionein cation-binding proteins are known from only chromosomal genes. Mostly, metal resistance systems function by energy-dependent efflux ions. Some of the efflux systems are ATPase and others are chemiosmotic cation/proton antiporters. The arsenic resistance efflux system transports arsenite, using alternatively either a two-component (Ars A and Ars B) ATPase or a single polypeptide (Ars B) functioning as a chemiosmotic transporter. The third gene in the arsenic resistance system, arsC encodes an enzyme that converts intracellular arsenate [As (VI)] to arsenite [As (III)], the substrate of the efflux system. The three-component CZC (Cd^2, Zn^2, Co^2) chemiosmotic efflux pump of soil microbes consists of inner membrane (czcA), outer membrane (czcC), and membrane spanning (czcB) proteins that together transport cations, from the cytoplasm across the periplasmic space to the outside of the cell. Metallothionein, which is a small protein that binds metal cations by means of numerous cysteine thiolates, has been characterized from cyanobacteria for the first time.

Rajwade et al. (1999) have studied biochemical basis of chromate reduction. A chromate reduction strain, *Pseudomonas mendocina* was found to transform hexavalent chromium to its trivalent form which precipitated as chromic hydroxide in the medium. Studies with respiratory inhibitors in conjunction with molar growth yield data indicated that chromate probably serves as an auxiliary oxidant that helps in generation of oxidized cofactors necessary in cellular metabolism. Chromate reduction by *P. mendocina* was found to be mediated by a periplasmic chromate reductase having temperature and pH optima of 70 °C and 8.5, respectively.

Molecular mechanisms involved in metal tolerance in halophilic archaea have been reviewed by Srivastava and Kowshik (2013). Haloarchaea live in hypersaline environments. Econiches like estuaries and solar crystallizer ponds may contain high concentrations of metals since they serve as ecological sinks for metal pollution and also as effective traps for river-borne metals. The availability of metals in these econiches depends upon the type of metal complexes formed and the solubility of the metal species at such high salinity. Haloarchaea have developed specialized mechanisms for the uptake of metals, required for various physiological processes and are not readily available at high salinity and evolved resistance mechanism for metal with high solubility. The authors have also discussed factors such as salinity and metal speciation that affect the bioavailability of metals to haloarchaea.

Domingues et al. (2020) have discussed role of extracellular polymeric substances (EPSs) in metal resistance of culturable heterotrophic bacteria. EPSs exhibit diverse properties such as surface modification, metal adsorption, and solubilization of hydrophobic substances, through surface tension reduction. Microbial communities are affected by ecosystems transformations due to mining activity. To overcome the problem of environmental stress, microorganisms adapt diverse metabolic and physiological strategies. One mechanism is the production of extracellular polymeric substances (EPSs) which potentially increase microbial resistance to high heavy metal concentrations through ion adsorption onto the molecule functional groups thus preventing metal penetration into the cytoplasm. The authors have studied EPS-producing bacteria from copper mine located in Brazilian Amazonia. *Actinomyces viscosus* E3.Pb5 and *Bacillus subtilis* group E3.As2 produced EPS that reduced the medium surface tension.

6.4 Environmental Pollution and Bioremediation

Heavy metal pollution is one of the most important environmental problems. Various industries, such as mining, metallurgy, electroplating, leather processing, photography, electric appliances manufacturing, metal surface treatment, produce and discharge waste containing various heavy metals in the environment. The metals are not biodegradable and tend to accumulate in living organisms, leading to various diseases and disorders. Thus, the metals can cause serious environmental pollution, threatening ecosystem and human health through their extreme toxicity. Physicochemical methods of removal of metals are expensive and result in incomplete removal of metals, have high reagent and energy requirements, and generate toxic sludge or other waste products. For development of low cost and easily available materials for adsorption of toxic heavy metals, microorganisms are considered as potent bioremediators, removing heavy metals via active or passive uptake mechanisms (Alam and Ahmad 2013).

Sinha et al. (2014) have emphasized on bioremediation potential of metallophiles. These microorganisms inherently possess properties that can influence the biomineralization process by changing the physicochemical behavior of the surrounding, metal speciation, mobility, and toxicity. Metallophiles possess an endogenous ability to exquisitely regulate their physiology to overcome the toxic effects of the external metal environment. Studies have shown that in combination with bioremediation and biomineralization, these metallophiles can be employed to convert environmentally problematic metals into valuable important and functional metallic materials. The authors have also described potential usage of these microbes in bioremediation of heavy metals, mechanism of metal bioremediation, and understanding metal-microbe interactions.

Afzal et al. (2017) have assessed heavy metal tolerance and biosorptive potential of *Klebsiella variicola* isolated from industrial effluents of Faisalabad, Pakistan. Thirteen isolates having tolerance to Ni and Co were studied for maximum tolerable concentration and multi-metal resistance. A native bacterial strain selected was identified as *Klebsiella variicola* and found to have heavy metal tolerance for Ni and Co.

6.5 Role of Metallophiles in Sustainability

Heavy metal contamination is one of the major global environmental concerns and the main sources of heavy metal contamination are either natural or anthropogenic. Depending upon soil pH and their speciation, these heavy metals can become mobile and a small part can leach to aquifer or become bioavailable to living microorganisms. Industrial waste waters contain cadmium, chromium, mercury, lead, nickel, cobalt, and copper and therefore their treatment is necessary (Afzal et al. 2017).

Bioremediation is a potential, cost-effective solution for the remediation of heavy metal-contaminated environment. The advantages of using microbes for bioremediation include natural occurrence, cheap production, easy availability to treat large volumes of wastewater, and removal and recovery of metals. Microorganisms that are able to survive well in high concentration of metals are of great interest as they can achieve different transformation and immobilization processes. They bring out bioaccumulation based on the incorporation of metals inside the living biomass or biosorption, in which metal ions are adsorbed at the cellular surface by different mechanisms. Some bacteria produce chelating agents that bound metals and reduce their toxicity. Many bacteria reduce or transform toxic metals into less toxic forms (Afzal et al. 2017).

Jain et al. (2017) have reviewed bacterial interactions with chromium. The rapidly increasing industrialization has mobilized chromium into the pristine environment. This has resulted in the concentration of chromium severalfold than usually found in nature. Due to mutagenic and carcinogenic effects in humans, hexavalent chromium has been considered as a priority, inorganic environmental pollutant by environmental agencies of different countries. Chromium is prevalent in several oxidation states in the environment but the two most ecologically stable forms are the Cr (III)-trivalent and Cr (VI)-hexavalent chromium. The toxicity of Cr (VI) is considered higher than Cr (III) due to its high solubility and mobility in the aqueous environment. Many indigenous bacterial species have been identified at chromium-contaminated site. These bacterial strains have evolved different strategies to evade chromium toxicity.

Pradhan et al. (2017) have reviewed work on bioreduction of hexavalent chromium in waste water treatment. Hexavalent chromium [Cr (VI)] in water is a proven carcinogen to living organisms. Due to different human activities, chromium reaches environment, at some sites to high concentration in the water system above the regulatory level. Chemical or biological techniques or a combination of the two have been used to remove Cr (VI) from polluted waters. Biological techniques include integrated bioremediation, such as the primary processes of direct bioreduction and biosorption and secondary processes of microbial fuel cell, biostimulation, surface-modified dry biomass and biochar adsorption, and engineered biofilm, and cell-free reductase.

Thus, metallophilic and metal-resistant/tolerant microorganisms can be employed to reduce environmental pollution caused by metals and contribute to sustainability of the environment and eventually human life.

6.6 Metallophilic, Metal-Resistant, and Metal-Tolerant Microorganisms Reported from India

Metal-rich environments exist in India mainly in mining areas in Madhya Pradesh and Chhattisgarh; arsenic-contaminated aquifers and groundwater in West Bengal; contaminated sites of chemical industries in Gujarat, Maharashtra, Kanpur, Uttar Pradesh; contaminated soils from Kashmir, Assam, and hot water spring of northeast India. Researchers have isolated metallophilic, metal-resistant, and metal-tolerant microorganisms from such metal-rich environments which are listed in Table 6.1.

6.6.1 Diversity

Some researchers from India have investigated diversity of metal-rich/contaminated sites and described the microbial species. Some have used culture-dependent approach while some have carried out culture-independent studies.

Dhakephalkar and Chopade (1994) studied multiple metal resistance of 40 strains of *Acinetobacter* isolated from different environmental sources. The strains were classified into four genospecies, namely, *A. baumannii* (33 isolates), *A. calcoaceticus* (three isolates), *A. junii* (three isolates), and *Acinetobacter* sp. 3 (one isolate). The strains were exposed to salts of 20 heavy metals and 18 antibiotics. All environmental isolates were resistant to multiple metal ions (at least 13 metal ions) and except for one strain to antibiotics (at least four antibiotics). Around 60% strains were sensitive to mercury. All strains were resistant to copper, lead, boron, and tungsten even at 10 mM concentration.

Wani et al. (2007) have reported hexavalent chromium Cr(VI)-reducing bacterial strain isolated from Lonar Lake, Maharashtra and identified as *Burkholderia cepacia* MCM B-821. The strain was resistant to 1000 ppm Cr (VI) and reduced 98% of the 75 ppm Cr (VI) in 36 h at pH 9.0 in presence of 2% salt and lactose as the donor. Electron paramagnetic resonance spectroscopy data suggested that the strain reduced Cr (VI) to Cr (III) via the formation of transient Cr (V).

Shivaji et al. (2005) have reported arsenic-resistant species, *Bacillus arsenicus* sp. nov. isolated from a concretion of arsenic ore obtained from a borehole, West Bengal, India. The organism could grow in presence of 20 mM of arsenate and 0.5 mM arsenite.

Suresh et al. (2004b) have isolated an arsenic- and radiation-resistant bacterium from an arsenic-contaminated aquifer water located in the Chakdah district in West Bengal, India. The organism was identified as *Deinococcus indicus* sp. nov.

Suresh et al. (2004a) have described an arsenic-resistant bacterium from a sand sample obtained from an arsenic-contaminated aquifer in Chakdah district in West Bengal, India (23° 3′ N 88° 35′ E). The bacterium was identified as *Bacillus indicus* sp. nov.

Nema et al. (2019) have reported for the first time the comprehensive culture-independent census of bacterial communities of deep subsurface of Krishna-Godavari (KG) basin, located in the eastern continental margin of India. The location

Table 6.1 Reports of metallophilic, metal-tolerant/resistant microorganisms from India

Sr. No.	Name of the organism/ group	Source/place	Diversity/ biotechnology	Reference
1	*Acinetobacter junii* ACN4	River water	Multiple metal resistance Plasmid-mediated mercury resistance	Dhakephalkar et al. 1993
2	*Acinetobacter baumannii*, *Acinetobacter* spp., *A. calcoaceticus*, *A. junii*	Different environmental sources	Multiple metal resistance	Dhakephalkar and Chopade 1994
3	*Pseudomonas mendocina* MCMB-180	MCM culture collection	Chromate-resistant Chromate reduction	Dhakephalkar et al. 1996
4	*Pseudomonas mendocina*	–	Hexavalent chromium bioremediation	Bhide et al. 1996
5	*Pseudomonas mendocina*	–	Chromate bioremediation	Salunkhe et al. 1998
6	*Citrobacter* strain MCM B-181	MCM culture collection	Biosorption for removal of lead, cadmium, and zinc	Puranik and Paknikar 1999
7	*Microbacterium lacticum*	Municipal sewage	Arsenic (III)-oxidizing Arsenic bioremediation	Mokashi and Paknikar 2002
8	*Thiobacillus thioparus*, *Cladosporium cladosporioides*	–	Silver-resistant thiosulfate-oxidizing Biosorption for recovery of silver	Pethkar and Paknikar 2003
9	*Pseudomonas mendocina*	MCM culture collection	Bioreduction of tellurite, treatment of waste photovoltaic modules	Rajwade and Paknikar 2003
10	*Bacillus indicus* sp. nov.	Sand sample from arsenic-contaminated aquifer in Chakdah district, West Bengal	Arsenic-resistant	Suresh et al. 2004a
11	*Deinococcus indicus* sp. nov.	Water from arsenic-contaminated aquifer, Chakdah district, West Bengal	Arsenic- and radiation-resistant	Suresh et al. 2004b

(continued)

Table 6.1 (continued)

Sr. No.	Name of the organism/group	Source/place	Diversity/biotechnology	Reference
12	*Bacillus arsenicus* sp. nov.	Concretion of arsenic ore from borehole, West Bengal	Arsenic-resistant	Shivaji et al. 2005
13	*Pseudomonas putida* KNP9	Panki Power Plant, Kanpur	Lead- and cadmium-resistant Siderophore production Plant growth promotion	Tripathi et al. 2005
14	*Ochrobacterium* sp. (DM1)	Contaminated sites of chemical industries, Gujarat	Hexavalent chromium reduction	Thacker and Madamwar 2005
15	*Providencia* sp.	Contaminated sites of chemical industries, Gujarat	Hexavalent chromium reduction Bioremediation of chromate	Thacker et al. 2006
16	*Pseudomonas aeruginosa*	–	Mercury bioremediation	Tupe et al. 2007
17	*Burkholderia cepacia* MCMB-821	Lonar Lake, Maharashtra	Hexavalent chromium-resistant	Wani et al. 2007
18	*Brucella* sp.	Chromate-contaminated sites	Chromate reduction, bioremediation	Thacker et al. 2007
19	*Bacillus cereus* (G1DM20), *Bacillus fusiformis* (G1DM22), *Bacillus sphaericus* (G1DM64)	Chromate-polluted landfill, Gujarat	Chromate reduction	Desai et al. 2008a
20	*Pseudomonas* sp. G1DM21	Chromate-polluted landfill, Gujarat	Chromate reduction	Desai et al. 2008b
21	Beta-proteobacteria and *Bacilli*	Selenium-contaminated sites in India	Selenite bioremediation	Ghosh et al. 2008
22	*Arthrobacter ramosus*	Mercuric salt-contaminated soil	Multiple metal-resistant characterization of mercuric reductase (Mer A) enzyme	Bafana et al. 2010
23	Enterobacter sp. EMB21, *Bacillus*, *Pseudomonas*	Oil-spilled soil	Bioaccumulation of mercury, plasmid-mediated mercury resistance	Sinha et al. 2013
24	*Pseudomonas* sp. W6	Hot water spring of northeast India	Lead-resistant, lead biosorption	Kalita and Joshi 2017

(continued)

Table 6.1 (continued)

Sr. No.	Name of the organism/ group	Source/place	Diversity/ biotechnology	Reference
25	*Rhizobium arsenicireducens* sp. nov.	Arsenic-rich groundwater, West Bengal	Diversity, new species	Mohapatra et al. 2017
26	*Pseudoxanthomonas arsenicir esistens* sp. nov.	Arsenate-contaminated groundwater, West Bengal	Diversity, new species	Mohapatra et al. 2018a
27	*Achromobacter* sp. Kas 3-5^T	Arsenate-contaminated groundwater, West Bengal	Diversity	Mohapatra et al. 2018b
28	*Staphylococcus equorum*, *Staphylococcus warneri*, *Bacillus safensis*, *Bacillus thuringiensis*	Saffron soils of Kashmir	Lead-tolerant, lead sequestration	Uqab et al. 2020
29	*Bacillus* sp. strain 111 J 3-1	Arsenic-contaminated groundwater of Brahmaputra River basin	Diversity, arsenic transformation	Ghosh et al. 2020

is a geological region well known for the abundance of economically important minerals. Elemental and mineralogical characterization of the sediments indicated the presence of carbon- and nitrogen-deprived conditions but with abundance of metalliferous minerals especially rich in valuable elements like zirconium, vanadium, cesium, and rare Earth elements. Diversity analysis based on Illumina MiSeq high-throughput sequencing platform revealed presence of Firmicutes (44.24%), Proteobacteria (34.17%), Bacteroidetes (15.18%).

Uqab et al. (2020) have isolated lead-tolerant bacteria from saffron soils of Kashmir and examined for their response to different concentrations of lead ranging from 100 ppm to 1000 ppm. Four isolates *Staphylococcus equorum*, *Staphylococcus warneri*, *Bacillus safensis* and *Bacillus thuringiensis* were found to have the efficacy of sequestration of lead at varying MIC. *Bacillus thuringiensis* exhibited highest (900 ppm) tolerance and 65.3% sequestration while *Staphylococcus warneri* showed the lowest (500 ppm) lead tolerance and 52.8% sequestration of lead. The potential of these organisms to tolerate and sequester lead could be exploited for reducing toxicity of lead-contaminated environment.

Desai et al. (2009) have investigated influence of long-term chromium pollution on soil bacterial community structures of highly chromium-polluted industrial landfill sites (G1 and G2) and nearby control site (G3), from Gujarat. Cultivation-dependent and cultivation-independent analyses of the bacterial community structures were assessed. 16S rRNA gene sequencing was carried out. Twelve different bacterial phyla were represented amongst 35, 34, and 72 different bacterial genera retrieved from sites G1, G2, and G3 respectively. The bacterial community of site G1 consisted of *Firmicutes* (52.75%), *Gamma-proteobacteria* (18%), *Actinobacteria* (14.5%), Bacteroidetes (9.5%) and *Deinococcus-Thermus* (5.25%) and that of site G2 consisted of *Firmicutes* (31.25%), *Alpha-proteobacteria* (7%), *Beta-proteobacteria* (8%), *Gamma-proteobacteria* (19%), *Delta-proteobacteria* (9.5%), *Epsilon-proteobacteria* (3%), Actinobacteria (13%), *Bacteroidetes* (7.75%), and *Deinococcus-Thermus* (1.5%).

Nandre et al. (2017) have investigated effect of carbon starvation on rate of arsenite oxidation. The authors isolated arsenic-oxidizing cultures from arsenic-devoid environment. A total of 40 heterotrophic aerobic, arsenic-transforming bacterial strains representing 18 different genera were identified, of which 19 strains were isolated from tannery effluent and 21 from tannery soil. It was observed that a strong correlation between the carbon starvation and arsenic oxidation potential of the isolates existed across different genera in Proteobacteria obtained.

Mohapatra et al. (2017) have characterized an arsenate-reducing novel bacterium *Rhizobium arsenicireducens* sp. nov., isolated from arsenic-rich groundwater of West Bengal. Cells of the strain were Gram-negative, rod-shaped, motile, and facultative anaerobic. Optimum pH for growth 6.0–7.0 and temperature 30 °C. The organism showed 98.4% similarity with genus *Rhizobium* based on 16S rRNA gene sequencing.

Mohapatra et al. (2018a) have described arsenate-reducing novel gamma proteobacterium, isolated from highly arsenate-contaminated groundwater, West Bengal. The strain showed 16S rRNA gene sequence similarity with the members of the genus *Pseudoxanthomonas*. Based on DNA-DNA hybridization and phylogenetic analysis, the strain was found to be a novel species. New species name *arseniciresistens* was proposed. The strain is identified as *Pseudoxanthomonas arseniciresistens* sp. nov.

Mohapatra et al. (2018b) have carried out molecular and ecophysiological characterization of arsenic (As) transforming *Achromobacter* sp. KAs 3-5T isolated from As-contaminated groundwater of West Bengal.

Ghosh et al. (2020) have carried out molecular and taxonomic characterization of arsenic (As) transforming *Bacillus* sp. strain III J3-1 isolated from As-contaminated groundwater of Brahmaputra river basin India.

Nath et al. (2019) have described metal-tolerant bacteria from soil samples collected from different locations of Cachar district, Assam, India. The isolates were identified as *Bacillus megaterium*, *Bacillus cereus*, *Pseudomonas aeruginosa*, and *Chromobacterium pseudoviolaceum*. The isolates were tested for tolerance to metals Cd, Pb, Fe, and Cu and resistance to antibiotics. *B. cereus* strain GCC-SO2 showed high degree of tolerance to lead and copper with MIC of 2050 ± 50 and 2533.33 ug/ml, respectively. *Pseudomonas aeruginosa* strain GCC-S03 showed maximum tolerance towards cadmium (1516.67 μg/ml), iron (1633.33 μg/ml), and copper 1716.66 μg/ml; *Bacillus megaterium* GCC-S01 to lead (1650 μg/ml) and copper (2583.33 μg/ml) and *Chromobacterium pseudoviolaceum* to cadmium 1166.67 μg/ml, lead 1350 μg/ml, copper 1683.33 μg/ml. Thus, metal contamination represents a long-standing widespread and recalcitrant selective pressure, and microbial communities in soil respond by exhibiting metal tolerance.

Sushma and Amudan (2012) isolated bacteria from Mithi river, Mumbai, and screened for tolerance to metal chromium (Cr), copper (Cu), and lead (Pb). The bacteria showed tolerance to concentration ranging from 3000 to 5000 ppm. They were identified as *Enterobacter aerogenes* M5 (growing in presence of copper and lead); *Bacillus pumilus* X8 (only chromium), *Bacillus alvei* X9 (only copper), and *Bacillus cereus* X11 (only lead). Mithi river is known to be polluted due to industrial effluents and hence the bacteria isolated from this river water were found to be metal-tolerant.

Pal et al. (2005) examined metal content and microbial population in serpentine soils collected from Saddle Hills, Chidya Tapu and Rutland of Andaman Islands, India. The serpentine soils contained high levels of nickel (1740–8033.4 mg/kg dry soil); cobalt (93.2–533.4 mg/kg dry soil), and chromium (302.9–4437 mg/kg dry soil), iron (62–152 g/kg dry soil), and 37–60 g of magnesium per kg dry soil. The microbial density was $6.2–11.3 \times 10^6$ colony forming unit/g of soil, lower than that of non-serpentine soils. Total 342 isolates (200 from serpentine and 142 from non-serpentine soils) were tested for Ni, Co, and Cr resistance. Serpentine soil microbes were, in general, highly resistant than the non-serpentine soil microbes. From serpentine soils, eight bacterial isolates tolerated >12.0 mM Ni and 11 bacterial isolates tolerated >16.0 mM Cr. The serpentine soil isolates also showed co-resistance to Cu, Zn, and Mn but were sensitive to Hg and Cd. The isolates also exhibited antibiotic resistance.

Pal et al. (2007) studied rhizospheric microorganisms from two plants (*Rinorea bengalensis*) [wall] O. K. and *Dichapetalum gelonioides* ssp. *andamanicum* endemic to serpentine outcrops of Andaman Islands, India. The soils were rich in Ni along with Co, Cr, Fe, and Mg but poor in microbial density. Ni-tolerant bacterial isolates were found to belong to *Pseudomonas*, *Bacillus*, and *Cupriavidus* and accumulate nickel 209.5–224 μM Ni/g protein.

Thacker and Madamwar (2005) have reported hexavalent chromium-reducing bacterial culture (DM 1) from contaminated sites of chemical industries, Gujarat. The culture was examined for its ability to reduce hexavalent chromium to trivalent chromium, a detoxification process in cell suspension and cell extract. The isolate was identified as *Ochrobactrum* sp. and found to tolerate chromium up to a maximum concentration of 300 ppm, at optimum temperature 35 °C and pH 7.0. Assay with the permeabilized cells (treated with toluene and Triton x-100) and cell-free extract demonstrated that the hexavalent chromium reduction is mainly associated with the soluble fraction of the cells.

Desai et al. (2008a) have evaluated in vitro Cr (VI) reduction potential of *Bacillus* sp. Three strains G1DM 20, G1DM 22, and GM1DM 64 were isolated from Cr (VI)-polluted landfill, Gujarat, and identified as closely related to *Bacillus cereus*, *Bacillus fusiformis*, and *Bacillus sphaericus*, respectively. The suspended cultures of the isolates exhibited more than 85% reduction at 1000 μM Cr (VI) within 30 h. Assays with the permeabilized cells and cell-free extracts from each isolate demonstrated that the hexavalent chromate reductase activity was mainly associated with the soluble fraction of cells and expressed constitutively.

6.6.2 Biotechnology

Applications of metallophiles in biotechnology, for example, recovery of metals from metal rich environments/industrial wastes have been described. Different mechanisms such as biosorption, bioaccumulation, reduction/oxidation of metals for their removal from contaminated sites are investigated.

Dhakephalkar et al. (1993) have discussed mechanism of plasmid-encoded mercury volatilization in *Acinetobacter junii* ACN4 isolated from river water. The organism exhibited resistance to multiple metals and antibiotics. The strain contained a single 78 kb plasmid designated as pup 1102 which encoded broad spectrum resistance to mercurial compounds such as mercuric chloride, mercuric acetate, fluoresceine acetate, and merbromin. Qualitative volatilization studies revealed that volatilization of mercuric chloride was the mechanism of pup1-encoded mercury resistance in ACN1. The plasmid pup 1102 also mediated volatilization of organo-mercurial in ACN4 but less efficiently than that of $HgCl_2$.

Dhakephalkar et al. (1996) have reported plasmid-mediated chromate reduction and resistance in *Pseudomonas mendocina* MCM B-180.

Pethkar and Paknikar (2003) have investigated thiosulfate biodegradation-silver biosorption process for the treatment of photo film processing wastewater which contains both silver and thiosulfate. A chemoautotrophic bacterium *Thiobacillus thioparus* tolerant to 350 mg silver/L oxidized thiosulfate to sulfate and sulfur. *Thiobacillus*-treated wastewater was then subjected to biosorption by fungal culture *Cladosporium cladosporioides* for recovery of silver.

Sinha et al. (2013) have described mechanism of bioaccumulation of metals like mercury. Five bacterial strains belonging to genera *Enterobacter*, *Bacillus*, and *Pseudomonas* were isolated from oil-spilled soil. One of the strains *Enterobacter* sp. EMB21 exhibited mercury bioaccumulation inside the cells. The bioaccumulation of mercury was confirmed by transmission electron microscopy and energy dispersive X-ray. The mercury-resistant loci in the *Enterobacter* sp. EMB21 cells were found to be plasmid-mediated. The organism seems promising for mercury remediation.

Puranik and Paknikar (1999) have described biosorption process for removal of lead, cadmium, and zinc by *Citrobacter* strain MCM B-181. Metal sorption by *Citrobacter* strain was found to be influenced by the pH of the solution, initial metal concentration, biomass concentration, and type of growth medium. Kinetic studies showed that metal uptake by *Citrobacter* strain MCM B-181 was a fast process achieving $>90\%$ adsorption efficient in <20 min.

Kalita and Joshi (2017) have reported *Pseudomonas* sp. W6 isolated from hot water spring of northeast India and having lead biosorption ability. The bacterium exhibited resistance to 1.0 mM lead in both solid and liquid minimal media, and 65% and 61.2% removal of lead from the synthetic Bangladesh groundwater medium in batch and column study, respectively. The bacterium produced exopolysaccharide which influenced biosorption of lead. The organism could be used for bioremediation of lead-contaminated wastewater.

Tupe et al. (2007) studied optimization of bioremediation process for removal of mercury by *Pseudomonas aeruginosa*. The organism was found to reduce 10 mg/L Hg (II) to Hg with 70% efficiency in 24 h. For optimization of process parameters, Taguchi design of experiments was used. The analysis of results predicted Hg (II) reduction efficiency to be 95% in 24 h under optimized process parameters. The validation experiment yielded 99.29% efficiency in Hg (II) removal.

Bhide et al. (1996) have described microbiological process for removal of Cr (VI) from cooling tower effluent using *Pseudomonas mendocina.* Scale-up process in 20 L continuous stirred tank reactor resulted in removal of 25–100 mg chromate/L in 4.5–8 h with 99.9% efficiency when sugarcane molasses was used as nutrient. The process could work in wide range of pH (6.5–9.5) and temperature (25–40 °C).

Rajwade and Paknikar (2003) have described microbiological process using *Pseudomonas mendocina* MCM B-180 for bioreduction of tellurite to elemental tellurium. Bench scale process in 2 L continuously stirred tank reactor with sucrose and diammonium hydrogen phosphate as nutrients removed 10 mg/L tellurium in 12 h with 99.8% efficiency. The process worked in pH range of 5.5–8.5 and temperature range of (25–45 °C) and initial tellurite concentration (10–100 mg/L). An integrated biosorption-bioreduction process for the treatment of waste photovoltaic modules using microbial tellurite reduction was developed.

Bafana et al. (2010) have described a multiple metal-resistant *Arthrobacter ramosus* strain G2 isolated from mercuric salt-contaminated soil. The organism was found to tolerate metals such as cadmium, cobalt, zinc, chromium, and mercury at maximum concentration of 37, 525, 348, 1530 and 369 μM, respectively. The isolate could reduce and detoxify redox-active metals like chromium and mercury indicating its potential in bioremediation of heavy metal-contaminated sites. The organism produced chromate reductase and mercury reductase. The authors have characterized enzyme mercury reductase (MerA) and reported for the first time from *Arthrobacter* sp.

Alam and Ahmad (2013) have described biosorption performance of dried and nongrowing biomasses of *Exiguobacterium* sp. ZM-2 isolated from soil contaminated with tannery effluents, Kanpur, Northern India for removal of Cd^{2+}, Ni^{2+}, Cu^{2+}, and Zn^{2+} from aqueous solution at 25, 50, 100, 150, and 200 mg/L concentration. The biosorption capacity was significantly altered by pH of the solution. Total sorption of metal ions occurred within 30 min. Recovery of metals (Cd^{2+}, Zn^{2+}, Cu^{2+}, and Ni^{2+}) was found to be better with dried biomass as compared to nongrowing biomass. Metal removal through bioaccumulation was determined by growing the strain in nutrient medium-amended different concentrations of metal ions.

Trihadiningrum et al. (2014) has conducted study to measure the viability of two strains of *Bacillus cereus* (ATCC 1178 and ATCC 9632) in chromium, copper, and zinc containing cultures, and bioremoval efficiencies of the cultures for the heavy metals. Both the strains were viable in culture containing Cr, Cu, and Cd at different concentrations. Chromium (102.4 mg/L) removal efficiency of 51%, 100% for copper (2 mg/L) and 77% in case of cadmium (15 to 25 mg/L) was achieved.

Pradhan et al. (2019) have described biosorption of hexavalent chromium from solution using the biomass of microalgae *Scenedesmus* sp. as adsorbent for the removal of hexavalent chromium. Different relevant parameters like initial pH, contact time, initial concentration of Cr (VI), biosorbant dosage, particle size, and temperature were examined to evaluate their effect. They were found to be effective with the maximum of 92.89% chromium loading on to the biomass. The FTIR analysis revealed the presence of easily hydrolyzed functional groups like aldehydes, amides, carboxylic acids, phosphates, and halides, which charge positively below the point of zero charge at pH 2.65. The biosorption proceeded through the anionic adsorption mechanism and followed the pseudo-first-order kinetic model. Its feasibility was confirmed as the experimental data closely fitted to both Langmuir and Freundlich isotherms. The biosorption was spontaneous.

Desai et al. (2008b) have reported hexavalent chromate reductase activity in cytosolic fractions of *Pseudomonas* sp. G1DM 21isolated from Cr (VI)-contaminated industrial landfill, Gujarat. The suspended culture of the bacterium reduced 99.7 % of 500 μM Cr (VI) and 93.06% of 1000 μM Cr(VI) in 48 h. The permeabilized cells of the culture reduced 90 % of 100 μM Cr (VI) in 120 min. The enzyme chromate reductase was characterized and found to be maximum at 30 °C and pH 7.0.

6.6.3 Sustainability

Metallophilic and metal-resistant microorganisms are employed for bioremediation of metal-contaminated sites and thus contribute to sustainability of environment.

Pal and Paknikar (2012) have described microbially mediated bioremediation of arsenic from contaminated water. Microbial remediation of arsenic from aquatic environments presents an interesting option because of its high efficiency, low cost, and eco-friendly nature.

Salunkhe et al. (1998) have developed a bioremediation method for chromate-contaminated sites. The culture of *Pseudomonas mendocina* when added in soil microcosms could immobilize 100 μg (2 mM) chromate/g soil in 8 h by converting it into trivalent form. The chromate-contaminated soil, after microbiological treatment, supported growth of wheat seedlings without exerting any toxic effects.

Ghosh et al. (2008) isolated and characterized selenite and selenite-tolerant microorganisms from selenium-contaminated sites (soil and water samples) in India. The isolates showed very high MIC values ranging from 300 to 750 mM for selenite and selenate. The organisms were found to belong to the classes β-Proteobacteria and Firmicutes (*Bacilli*). The isolates could be used for bioremediation of selenium-contaminated sites.

Mokashi and Paknikar (2002) developed a microbially assisted process for removal of arsenic from contaminated groundwater. Arsenic (III)-oxidizing (up to 50 mmol/L) culture of *Microbacterium lacticum* was isolated from municipal sewage (Pune, Maharashtra) by an enrichment technique. The culture was immobilized on brick pieces and packed in a glass column. The system could achieve complete oxidation of As (III) from groundwater at neutral pH, ambient temperature, and methanol as substrate. The oxidized As species were removed from groundwater using zerovalent iron, activated charcoal, and ferric chloride.

Tripathi et al. (2005) have reported siderophore-producing lead- and cadmium-resistant *Pseudomonas putida* KNP9, isolated from Panki Power Plant, Kanpur, India. Heavy metals inhibit growth of plants and cause plant death. Hence, the isolate was explored and found to simulate significant ($P > 5\%$) root and shoot growth of Mung bean (green gram) to the extent of 16.48% and 28.80%, respectively in presence of $CdCl_2$ (110 μM), while 20% and 19.5%, respectively, in presence of lead acetate (660 μM). The concentration of accumulated lead and cadmium in root and shoot was also reduced in presence of this isolate, ranging from 37.5% to 93.19%. The isolate thus could be used as plant growth promoting in presence of cadmium and lead.

Thacker et al. (2007) studied reduction of chromate by cell-free extract of *Brucella* sp. isolated from Cr (VI)-contaminated sites. The organism reduced chromate to 100%, 94.1%, 93.2%, 66.9%, and 41.6% at concentrations of 50, 100, 150, 200, and 300 mg/L, respectively, at pH 7.0 and 37 °C temperature. The strain also showed multiple heavy (Ni, Zn, Hg, Pb, Co) tolerance and resistance to various antibiotics. Assay with crude cell-free extracts demonstrated that the hexavalent chromium reduction was mainly associated with the soluble fraction of the cell. The strain with resistance to high concentration of Cr (VI) and high chromium (VI)-reducing ability could be suitable candidate for bioremediation of arsenic-contaminated sites.

Thacker et al. (2006) have reported hexavalent chromium reduction by *Providencia* sp. isolated from contaminated sites of chemical industries. The isolate could grow and reduce chromate 100% at a concentration ranging from 100–300 mg/L and 99.3% at 400 mg/L, at pH 7.0 and temperature 37 °C and exhibited tolerance to heavy metals, Ni, Zn, Hg, Pb, and Co. Assay with resting cells, permeabilized cells (treated with toluene and triton X 100) and crude cell-free extracts demonstrated that the hexavalent chromium reduction is mainly associated with the soluble fraction of the cell. The organism could be used for bioremediation of Cr (VI)-containing wastes and specific environmental cleanup operations.

6.6.4 Contributions to the Knowledge of Metallophilic Microbial Species, New to Science, from India

Researchers from India have isolated and identified metallophilic, metal-resistant, and metal-tolerant microorganisms from various metal-rich environments in India. Some of these microbial species are novel species, and thus these findings contribute to the knowledge of metallophilic microorganisms all over the world. These species, new to science, from India are listed in Table 6.2.

Table 6.2 Metallophilic microbial species, new to science, from India

Sr. No.	Name of the organism	Source	Metal tolerance/ resistance	Reference
1	*Bacillus indicus* sp. nov.	Sand sample from arsenic-contaminated aquifer in Chakdah district, West Bengal	Arsenic-resistant	Suresh et al. 2004a
2	*Deinococcus indicus* sp. nov.	Water from arsenic-contaminated aquifer, Chakdah district, West Bengal	Arsenic- and radiation-resistant	Suresh et al. 2004b
3	*Bacillus arsenicus* sp. nov.	Concretion of arsenic ore from borehole, West Bengal	Arsenic-resistant	Shivaji et al. 2005
4	*Rhizobium arsenicireducens* sp. nov.	Arsenic-rich groundwater, West Bengal	Arsenite-reducing	Mohapatra et al. 2017
5	*Pseudoxanthomonas arsenicirresistens* sp. nov.	Arsenate-contaminated groundwater, West Bengal	Arsenite-reducing	Mohapatra et al. 2018a

With the ever-increasing interest and curiosity, the researchers from India will further explore different pristine environments for metallophilic microorganisms and add to the world data.

References

Afzal AM, Rasool MH, Waseem M, Aslam B (2017) Assessment of heavy metal tolerance and biosorptive potential of *Klebsiella variicola* isolated from industrial effluents. AMB Express 7: 184. https://doi.org/10.1186/s13568-017-0482-2

Alam MZ, Ahmad S (2013) Multi-metal biosorption and bioaccumulation by *Exiguobacterium* sp. ZM-2. Ann Microbiol 63:1137–1146. https://doi.org/10.1007/s13213-012-0571-z

Austin B, Allen DA, Mills AL, Colwell RR (1977) Numerical taxonomy of heavy metal-tolerant bacteria isolated from an estuary. Can J Microbiol 23:1433–1447

Bafana A, Krishnamurthi K, Patil M, Chakrabarti T (2010) Heavy metal resistance in *Arthrobacter ramosus* strain G2 isolated from mercuric salt-contaminated soil. J Hazard Mater 177(1-3): 481–486. https://doi.org/10.1016/j.jhazmat.2009.12.058

Belly RT, Kydd GC (1982) Silver resistance in microorganisms. Develop Indust Microbiol 23:567–577

Bhide JV, Dhakephalkar PK, Paknikar KM (1996) Microbiological process for the removal of Cr (VI) from chromate-bearing cooling tower effluent. Biotechnol Lett 18(6):667–672

Bitton G, Freihofer V (1978) Influence of extracellu1at polysaccharides on the toxicity of copper and cadmium toward *Klebsiella aerogenes*. Microb Ecol 4:119–125

Bopp LH, Ehrlich HL (1988) Chromate resistance and reduction in *Pseudomonas fluorescens* strain LB300. Arch Microbiol 150:426–431

Bopp LH, Chakrabarty AM, Ehrlich HL (1983) Chromate resistance plasmid in *Pseudomonas fluorescens*. J Bacteriol 155:1105–1109

Braun V, Hantke K, Koster W (1998) Bacterial iron transport: mechanisms, genetics, and regulation. Metal Ions Biol Syst 35:67–145

Brim H, Heyndrickx M, De Vos P, Wilmotte A, Springael D, Schlegel HG, Mergeay M (1999) Amplified rDNA restriction analysis and further genotypic characterisation of metal-resistant soil bacteria and related facultative hydrogenotrophs. Syst Appl Microbiol 22(2):258–268

Brown NL, Barrett SR, Camakaris J, Lee BT, Rouch DA (1995) Molecular genetics and transport analysis of the copper-resistance determinant (pco) from *Escherichia coli* plasmid pRJ1004. Mol Microbiol 17:1153–1166

Charley RC, Bull AT (1979) Bioaccumulation of silver by a multi-species community of bacteria. Arch Microbiol 123:239–244

Chatterjee R, Ludden PW, Shah VK (1997) Characterization of VNFG, the delta subunit of the vnf-encoded apodinitrogenase from *Azotobacter vinelandii*. Implications for its role in the formation of functional dinitrogenase 2. J Biol Chem 272:3758–3765

Chopra I (1971) Decreased uptake of cadmium by a resistant strain of *Staphylococcus aureus*. J Gen Microbiol 63:265–267

Cooksey DA (1993) Copper uptake and resistance in bacteria. Mol Microbiol 7:1–5

Cooksey DA (1994) Molecular mechanisms of copper resistance and accumulation in bacteria. FEMS Microbiol Rev 14:381–386

Desai C, Jain K, Madamwar D (2008a) Evaluation of in vitro Cr (VI) reduction potential in cytosolic extracts of three indigenous *Bacillus* sp. isolated from Cr (VI) polluted industrial landfill. Bioresour Technol 99(14):6059–6069

Desai C, Jain K, Madamwar D (2008b) Hexavalent chromate reductase activity in cytosolic fractions of *Pseudomonas* sp. G1DM21 isolated from Cr (VI) contaminated industrial landfill. Process Biochem 43(7):713–721

Desai C, Parikh RY, Vaishnav T, Shouche YS, Madamwar D (2009) Tracking the influence of long-term chromium pollution on soil bacterial community structures by comparative analyses of 16S rRNA gene phylotypes. Res Microbiol 160(1):1–9

Dhakephalkar PK, Chopade BA (1994) High levels of multiple metal resistance and its correlation to antibiotic resistance in environmental isolates of *Acinetobacter*. Biometals 7(1):67–74

Dhakephalkar PK, Kumbhar AS, Padhye SB, Chopade BA (1993) Mechanism of plasmid encoded mercury volatilization in *Acinetobacter junii* ACN4. J Inorg Biochem 51(1-2):388

Dhakephalkar PK, Bhide JV, Paknikar KM (1996) Plasmid mediated chromate reduction and resistance in *Pseudomonas mendocina* MCM B-180. Biotechnol Lett 18:1199–1202

Domingues VS, de Souza MA, Júlio ADL, Queiroz ALL, dos Santos VL (2020) Diversity of metal-resistant and tensoactive-producing culturable heterotrophic bacteria isolated from a copper mine in Brazilian amazonia. Sci Rep 10:6171. https://doi.org/10.1038/s41598-020-62780-8

Duxbury T, Bieknell B (1983) Metal-tolerant bacterial populations from natural and metal-polluted soils. Soil Biol Biochem 15:243–250

Eady RR (1995) Vanadium nitrogenases of Azotobacter. Metal Ions Biol Syst 31:363–405

Gadd GM, Griffiths AJ (1977) Microorganisms and heavy metal toxicity. Microb Ecol 4:303–317. https://doi.org/10.1007/BF02013274

Ghosh A, Mohod AM, Paknikar KM, Jain RK (2008) Isolation and characterization of selenite-and selenate-tolerant microorganisms from selenium-contaminated sites. J World J Microbiol Biotechnol 24(8):1607–1611

Ghosh S, Mohapatra B, Satyanarayana T, Sar P (2020) Molecular and taxonomic characterization of arsenic (As) transforming Bacillus sp. strain IIIJ3–1 isolated from As-contaminated groundwater of Brahmaputra river basin, India. BMC Microbiol 20(1):1–20

Gonzalez H, Jensen TE (1998) Nickel sequestering by polyphosphate bodies in *Staphylococcus aureus*. Microbios 93:179–185

Gordon AS, Howell LD, Harwood V (1994) Response of diverse heterotrophic bacteria to elevated copper concentrations. Can J Microbiol 40:408–411

Haefeli C, Franklin C, Hardy K (1984) Plasmid-determined silver resistance in *Pseudomonas stutzeri* isolated from a silver mine. J Bacteriol 158:389–392

Hedges RW, Baumberg S (1973) Resistance to arsenic compounds conferred by a plasmid transmissible between strains of *Escherichia coli*. J Bacteriol 115:459–460

Hendrick CA, Haskins WP, Vidaver AK (1984) Conjugative plasmid in *Corynebacterium flaccumfaciens* subsp, *oorti* that confers resistance to arsenite, arsenate and antimony (Ill). Appl Environ Microbiol 48:56–60

Hobman JL, Brown NL (1996) Overexpression of MerT, the mercuric ion transport protein of transposon Tn501, and genetic selection of mercury hypersensitivity mutations. Mol Gen Genet 250:129–134

Jain K, Desai C, Madamwar D (2017) Bacterial interactions with chromium and strategies for the remediation of hexavalent chromium pollution. In: Handbook of metal-microbe interactions and bioremediation. CRC Press, Boca Raton, pp 719–729

John MK (1971) Lead contamination of some agricultural soils in western Canada. Environ Sci Technol 5:1199–1203

Kalita D, Joshi SR (2017) Study on bioremediation of Lead by exopolysaccharide producing metallophilic bacterium isolated from extreme habitat. Biotechnol Rep 16:48–57

Kondo I, Ishikawa T, Nakahara H (1974) Mercury and cadmium resistances mediated by the penicillinase plasmid in *Staphylococcus aureus*. J Bacteriol 117:1–7

Mahanty SK, Khaware R, Ansari S, Gupta P, Prasad R (1991) Vanadate-resistant mutants of Candida albicans show alterations in phosphate uptake. FEMS Microbiol Lett 68:163–166

Marques AM, Congregado F, Simon-Pujol DM (1979) Antibiotic and heavy metal resistance of *Pseudomonas aeruginosa* isolated from soils. J Appl Bacteriol 47:347–350

Mergeay M (2007) Metallophiles and acidophiles in metal-rich environments. Extremophiles III:1–24

Mohapatra B, Sarkar A, Joshi S, Chatterjee A, Kazy SK, Maiti MK, Satyanarayana T, Sar P (2017) An arsenate-reducing and alkane-metabolizing novel bacterium, *Rhizobium arsenicireducens* sp. nov., isolated from arsenic-rich groundwater. Arch Microbiol 199(2):191–201

Mohapatra B, Sar P, Kazy SK, Maiti MK, Satyanarayana T (2018a) Taxonomy and physiology of *Pseudoxanthomonas arseniciresistens* sp. nov., an arsenate and nitrate-reducing novel gammaproteobacterium from arsenic contaminated groundwater, India. PloS One 13(3): e0193718

Mohapatra B, Satyanarayana T, Sar P (2018b) Molecular and eco-physiological characterization of arsenic (As)-transforming *Achromobacter* sp. KAs 3–5T from As-contaminated groundwater of West Bengal, India. J Environ Sci Health, Part A, 53(10):915–924

Mokashi SA, Paknikar KM (2002) Arsenic (III) oxidizing *Microbacterium lacticum* and its use in the treatment of arsenic contaminated groundwater. Lett Appl Microbiol 34:258–262

Nakahara H, lshikawa T, Saral Y, Kondo I, Kozukue H, Mitsuhashi S (1977) Mercury resistance and R plasmids in *Escherichia coli* isolated from clinical lesions in Japan. Antimicrob Agents Chemother 11:999–1003

Nakamura T, Namba H, Ohmoto T, Liu Y, Hirata D, Miyakawa T (1995) Cloning and characterization of the Saccharomyces cerevisiae SVS1 gene which encodes a serine- and threonine-rich protein required for vanadate resistance. Gene 165:25–29

Nandre VS, Bachate SP, Salunkhe RC, Bagade AV, Shouche YS, Kodam KM (2017) Enhanced detoxification of arsenic under carbon starvation: a new insight into microbial arsenic physiology. Curr Microbiol 74(5):614–622

Nath S, Paul P, Roy R, Bhattacharjee S, Deb B (2019) Isolation and identification of metal-tolerant and antibiotic-resistant bacteria from soil samples of Cachar district of Assam, India. Appl Sci 1: 727. https://doi.org/10.1007/s42452-019-0762-3

Nelson JD, Blair W, Brinckman FE, Colwell RR, lverson WP (1973) Biodegradation of phenylmercuric acetate by mercury-resistant bacteria. Appl Microbiol 26:321–326

Nema P, Paul D, Karmalkar NR, Shouche YS (2019) Bacterial diversity in the metal-rich terrestrial deep subsurface sediments of Krishna Godavari Basin, India. Geomicrobiol J 36(10):917–932

Nies DH (1995) The cobalt, zinc, and cadmium efflux system CzcABC from *Alcaligenes eutrophus* functions as a cation-proton-antiporter in *Escherichia coli*. J Bacteriol 177:2

Nies DH (1999) Microbial heavy-metal resistance. Appl Microbiol Biotechnol 51:730–750

Nies DH, Silver S (1989) Metal ion uptake by a plasmid-free metal-sensitive *Alcaligenes eutrophus* strain. J Bacteriol 171:4073–4075

Novick RP, Roth C (1968) Plasmid-linked resistance to inorganic salts in *Staphylococcus aureus*. J Bacteriol 95:1335–1342

Oden KL, Gladysheva TB, Rosen BP (1994) Arsenate reduction mediated by the plasmid-encoded ArsC protein is coupled to glutathione. Mol Microbiol 12:301–306

Olafson RW, Abel K, Sim RS (1979) Prokaryotic metallothionein: preliminary characterization of a blue-green algae heavy-metal binding protein. Biochem Biophys Res Commun 89:36–43

Olson GJ, Porter FD, Rubinstein J, Silver S (1982) Mercuric reductase enzyme from a mercury-volatilizing strain of *Thiobacillus ferrooxidans*. J Bacteriol 151:1230–1236

Pal A, Paknikar KM (2012) Bioremediation of arsenic from contaminated water. In: Microorganisms in environmental management. Springer, Dordrecht, pp 477–523

Pal A, Dutta S, Mukherjee PK, Paul AK (2005) Occurrence of heavy metal-resistance in microflora from serpentine soil of Andaman. J Basic Microbiol. https://doi.org/10.1002/jobm.200410499

Pal A, Wauters G, Paul AK (2007) Nickel tolerance and accumulation by bacteria from rhizosphere of nickel hyperaccumulators in serpentine soil ecosystem of Andaman, India. Plant Soil 293(1): 37–48. https://doi.org/10.1007/s11104-007-9195-7

Peitzsch N, Eberz G, Nies DH (1998) *Alcaligenes eutrophus* as a bacterial chromate sensor. Appl Environ Microbiol 64:453–458

Pethkar AV, Paknikar KM (2003) Thiosulfate biodegradation–silver biosorption process for the treatment of photofilm processing wastewater. Process Biochem 38(6):855–860

Pickett AW, Dean ACR (1976) Cadmium and zinc sensitivity and tolerance in *Klebsiella (Aerobacter) aerogenes*. Microbios 15:79–91

Pradhan D, Sukla LB, Sawyer M, Rahman PKSM (2017) Recent bioreduction of hexavalent chromium in wastewater treatment: A review. J Ind Eng Chem 55:1–20. https://doi.org/10.1016/j.jiec.2017.06.040

Pradhan D, Sukla LB, Mishra BB, Devi N (2019) Biosorption for removal of hexavalent chromium using microalgae *Scenedesmus* sp. J Clean Prod 209:617–629. https://doi.org/10.1016/j.jclepro10.288

Puranik PR, Paknikar KM (1999) Biosorption of lead, cadmium, and zinc by *Citrobacter* strain MCM B-181: characterization studies. Biotechnol Prog 15(2):228–237

Qian H, Sahlman L, Eriksson PO, Hambraeus C, Edlund U, Sethson I (1998) NMR solution structure of the oxidized form of MerP, a mercuric ion binding protein involved in bacterial mercuric ion resistance. Biochemistry 37:9316–9322

Rajwade JM, Paknikar KM (2003) Bioreduction of tellurite to elemental tellurium by *Pseudomonas mendocina* MCM B-180 and its practical application. Hydrometallurgy 71(1–2):243–248

Rajwade JM, Salunkhe PB, Paknikar KM (1999) Biochemical basis of chromate reduction by *Pseudomonas mendocina*. Process Metallurgy 9:105–114

Rehder D (1992) Structure and function of vanadium compounds in living organisms. Biometals 5:3–12

Romao MJ, Archer M, Moura I, Moura JJG, LeGall J, Engh R, Schneider MRH (1995) Crystal structure of the xanthinoxidase-related aldehyde oxidoreductase from *D. gigas*. Science 270:1170–1176

Salunkhe PB, Dhakephalkar PK, Paknikar KM (1998) Bioremediation of hexavalent chromium in soil microcosms. Biotechnol Lett 20(8):749–751

Schindelin H, Kisker C, Hilton J, Rajagopalan KV, Reese DC (1996) Crystal structure of DMSO reductase: redox-linked changes in molybdopterin coordination. Science 272:1615–1621

Shivaji S, Suresh K, Chaturvedi P, Dube S, Sengupta S (2005) *Bacillus arsenicus* sp. nov., an arsenic-resistant bacterium isolated from a siderite concretion in West Bengal, India. Int J Syst Evol Microbiol 55(3):1123–1127

Silver S (1996) Bacterial resistances to toxic metal ions - a review. Gene 179:9–19

Silver S, Keach D (1982) Energy-dependent arsenate efflux: the mechanism of plasmid-mediated resistance. Proc Natl Acad Sci U S A 79:6114–6118

Silver S, Phung LT (1996) Bacterial heavy metal resistance: new surprises. Annu Rev Microbiol 50: 753–789. https://doi.org/10.1146/annurev.micro.50.1.753

Sinha A, Kumar S, Khare SK (2013) Biochemical basis of mercury remediation and bioaccumulation by *Enterobacter* sp. EMB21. Appl Biochem Biotechnol 169(1):256–267. https://doi.org/10.1007/s12010-012-9970-7

Sinha A, Sinha R, Khare SK (2014) Heavy metal bioremediation and nanoparticle synthesis by metallophiles. In: Parmar N, Singh A (eds) Geomicrobiology and biogeochemistry. Soil biology, vol 39. Springer, Berlin, pp 101–118. https://doi.org/10.1007/978-3-642-41837-2_6

Sjogren RE, Port J (1981) Heavy metal-antibiotic resistant bacteria in a lake recreational area. Water Air Soil Pollut 15:29–44

Slawson RM, Van Dyke MI, Lee H, Trevors JT (1992) Germanium and silver resistance, accumulation, and toxicity in microorganisms. Plasmid 27:72–79

Smith HW (1978) Arsenic resistance in Enterobacteria: its transmission by conjugation and by phage. J Gen Microbiol 109:49–56

Srivastava P, Kowshik M (2013) Mechanisms of metal resistance and homeostasis in haloarchaea. Archaea 2013:732864. https://doi.org/10.1155/2013/732864

Summers AO, Jacoby GA (1978) Plasmid-determined resistance to boron and chromium compounds in *Pseudornonas aeruginosa*. J Bacteriol 13:637–640

Suresh K, Prabagaran SR, Sengupta S, Shivaji S (2004a) *Bacillus indicus* sp. nov., an arsenic-resistant bacterium isolated from an aquifer in West Bengal, India. Int J Syst Evol Microbiol 54(4):1369–1375

Suresh K, Reddy GSN, Sengupta S, Shivaji S (2004b) *Deinococcus indicus* sp. nov., an arsenic-resistant bacterium from an aquifer in West Bengal, India. Int J Syst Evol Microbiol 54(Pt 2):457–461. https://doi.org/10.1099/ijs.0.02758-0

Sushma M, Amudan R (2012) Screening for heavy metal bioaccumulating bacteria from an urban wetland. Proceeding of International Conference SWRDM-2012, pp 17–21

Thacker U, Madamwar D (2005) Reduction of toxic chromium and partial localization of chromium reductase activity in bacterial isolate DM1. World J Microbiol Biotechnol 21(6):891–899

Thacker U, Parikh R, Shouche Y, Madamwar D (2006) Hexavalent chromium reduction by Providencia sp. Process Biochem 41(6):1332–1337

Thacker U, Parikh R, Shouche Y, Madamwar D (2007) Reduction of chromate by cell-free extract of *Brucella* sp. isolated from Cr(VI) contaminated sites. Bioresour Technol 98(8):1541–1547. https://doi.org/10.1016/j.biortech.2006.06.011

Tornabene TG, Edwards HW (1972) Microbial uptake of lead. Science 196:1334–1335

Trevors JT, Oddie KM, Belliveau BH (1985) Metal resistance in bacteria. FEMS Microbiol Rev 32: 39–54

Trihadiningrum Y, Arinda T, Sholikah U, Shovitri M, Wilujeng SA, Pandebesie ES (2014) Bioremoval of chromium, copper and cadmium by *Bacillus cereus* in simulated electroplating wastewater, IPTEK. J Proc Ser 1:31–35

Tripathi M, Munot HP, Shouche Y, Meyer JM, Goel R (2005) Isolation and functional characterization of siderophore-producing lead-and cadmium-resistant *Pseudomonas putida* KNP9. Curr Microbiol 50(5):233–237

Tuovinen OH, Niemela SI, Gyllenberg HG (1971) Tolerance of *Thiobacillus ferroxidans* to some metals. Antonie Van Leeuwenhoek 37:489–396

Tupe SG, Rajwade JM, Paknikar KM (2007) Taguchi approach significantly increases bioremediation process efficiency: a case study with Hg (II) removal by *Pseudomonas aeruginosa*. Lett Appl Microbiol 45(1):36–41

Tynecka Z, Gos Z, Zajac J (1981a) Energy-dependent efflux of cadmium coded by a plasmid resistance determinant in *Staphylococcus aureus*. J Bacteriol 147:313–319

Tynecka Z, Gos Z, Zajac J (1981b) Reduced cadmium transport determined by a resistance plasmid in *Staphylococcus aureus*. J Bacteriol 147:305–312

Uqab B, Nazir R, Ganai BA, Rahi P, Rehman S, Farooq S, Dar R, Parray JA, Al-Arjani A-BF, Al-Arjani TB, Allah EFA (2020) MALDI-TOF-MS and 16S rRNA characterization of lead tolerant metallophile bacteria isolated from saffron soils of Kashmir for their sequestration potential. Saudi J Biol Sci 27(8):2047–2053. https://doi.org/10.1016/j.sjbs.2020.04.021

Wani R, Kodam KM, Gawai KR, Dhakephalkar PK (2007) Chromate reduction by *Burkholderia cepacia* MCMB-821, isolated from the pristine habitat of alkaline crater lake. Appl Microbiol Biotechnol 75(3):627–632

Weast RC (1984) CRC handbook of chemistry and physics, 64th edn. CRC, Boca Raton

Wildung RE, Garland TR (1982) Effects of plutonium on soil microorganisms. Appl Environ Microbiol 43:418–423

Wood JM, Wang HK (1983) Microbial resistance to heavy metals. Environ Sci Technol 17(12): 582–590. https://doi.org/10.1021/es00118a002

Zevenhuizen LPTM, Dolfing J, Eshuis EJ, Scholten-Koerselman I J (1979) Inhibitory effects of copper on bacteria related to the free ion concentration. Microb Ecol 5:139–146

7 Psychrophilic, Psychrotrophic, and Psychrotolerant Microorganisms

Abstract

Based on temperature-dependent definition, psychrophiles are described as cold-loving microorganisms and grow optimally at about 15 °C, maximum temperature for growth at about 20 °C and minimum 0 °C or lower. Psychrotrophic microorganisms have a maximum temperature for growth about 20 °C and widespread in natural environments. Psychrotolerant microorganisms can survive at low temperature and can grow at elevated temperatures. Psychrophilic microorganisms occur in cold environments like polar surface, ice-cap regions, glaciers, cold deserts, permafrost, and deep oceans. Cold-loving bacteria include the genera *Colwellia*, *Achromobacterium*, *Alcaligenes*, *Alteromonas*, *Bacillus*, and many more. Physiological basis of cold adaptation includes protein synthesizing mechanisms, alteration in cellular structures, and inactivation of enzymes. Psychrophilic microorganisms produce antifreeze proteins, cryoprotectants, cold-shock proteins, and cold-active enzymes. They have applications in food industry, waste processing, agriculture, medicines, and environmental bioremediation.

In India, glaciers from Indian Himalayan Ranges, cold lakes like Pangong, and high-altitude soils serve as cold environments for isolation of psychrophilic microorganisms. Cold-adapted hydrolytic enzymes have been reported from *Staphylococcus lipolyticus* (lipase), *Actinomycetes*, and *Alcaligenes* (amylase, xylanase, cellulase, protease). Over 20 novel species of psychrophilic microorganisms, new to Science, have been reported from India.

Keywords

Psychrophilic · Psychrotolerant · Cold lakes · Glaciers · Cold-adapted enzymes · Antifreeze proteins · Cryoprotectants · *Colwellia* · *Alcaligenes* · Indian Himalayan Ranges · Pangong lake

P. P. Kanekar, S. P. Kanekar, *Diversity and Biotechnology of Extremophilic Microorganisms from India*, Microorganisms for Sustainability 41,
https://doi.org/10.1007/978-981-19-1573-4_7

There were debates and confusion about the definition of 'psychrophile' (Hamdan 2018) in the past. In 1887, Foster isolated microorganisms from fish preserved by cold temperature (Georlette et al. 2004). He then reported in 1892 other bacterial isolates from various environments that were able to grow at 0 °C (Morita 1975). Schmidt-Nielson (1902) first mentioned the term "psychrophile" for bacteria capable of growing at 0 °C (Feller and Gerday 1997). Muller (1903) criticized this term since these organisms were able to grow rapidly at elevated temperatures (Morita and Moyer 2001). Kruse (1910) suggested these bacteria as "psychrotolerant." Morita (1975) proposed a definition to the term "psychrophile" for cold-adapted or cold-loving microorganisms having minimum temperature 0 °C, optimum 15 °C, and maximum 20 °C. The term psychrotrophic is used for the organisms that do not meet the definition of psychrophilic. Gounot (1986) described psychrophilic and psychrotrophic microorganisms as those that have the ability to grow at 0 °C. Psychrotrophic microorganisms have a maximum temperature for growth about 20 °C and widespread in natural environments and in foods. Psychrophilic microorganisms are restricted to permanently cold habitats.

Cavicchioli (2016) has discussed at length the concept of a psychrophile. If microorganisms grow in naturally cold environments, a good assumption is that they are cold-adapted and "like" the cold. The term "psychrophile" reflects this—from Greek, psukhros meaning "cold" and philein "to love." However, temperature is not the only factor that controls their existence; like all environmental systems, the abundance of individual taxa and the ecology of the system is determined by the lifestyle of each taxon, their responses to all biotic and abiotic parameters that define the system and biogeographic constraints, that is, the effects of physical barriers and distance on dispersal and colonization. Problems with the concept of a psychrophile are due to the dependency of many in the field to use temperature-dependent growth rate as a measure for assessing how well adapted a microorganism is to the cold. The term "psychrophile" had been used since the early 1900s, and like mesophiles and thermophiles, psychrophiles were also defined based on their optimum temperature. There are limitations to this approach. The author (Cavicchioli) has suggested that the singular term "psychrophile" is both appropriate and sufficient for describing microbes indigenous to cold environments in contrast to the terms "psychrotolerant," "psychrotroph," and "facultatively psychrophile."

Again, Moyer et al. (2017) have described psychrophiles and psychrotrophs. Psychrophiles are extremophilic bacteria or archaea which are cold-loving and have an optimal temperature for growth at about 15 °C, a maximal temperature for growth at about 20 °C, and a minimal temperature for growth at about 0 °C or lower. Psychrotrophs are cold-tolerant bacteria or archaea and have ability to grow at low temperature, with optimal and maximal growth temperatures above 15 and 20 °C, respectively. Psychrophiles and psychrotrophs are adapted to cold environments like polar surface, ice-cap regions, permafrost, and deep oceans and often capable of enduring extended period of cryptobiosis.

Thus, for practical purposes, the temperature-dependent definition of psychrophiles will be considered.

7.1 Ecological Distribution of Psychrophiles

The Earth's biosphere is predominantly cold and permanently exposed to temperatures below 5 °C; such low mean temperature mainly arises from the fact that 71% of the Earth's surface is covered by oceans that have a constant temperature of 2–4 °C below 1000 m depth irrespective of the latitude. The polar regions account for another 15%, to which the glacier and alpine regions must be added as well as permafrost representing more than 20% of the terrestrial soils. Although inhospitable, all these low-temperature biotopes have been successfully colonized by cold-adapted organisms. Psychrophiles thrive in permanently cold environments and even at subzero temperatures in supercooled liquid water (Feller 2013).

7.2 Habitats of Psychrophiles

The oceans are the natural habitat for psychrophiles (Morita 1975). The largest coverage of these cold environments is colonized by diverse groups of microorganisms such as bacteria, archaea, yeasts, filamentous fungi, and algae. Cold habitats dominate the vast majority of Earth, covering 75% of the Earth's surface and span from the arctic to the Antarctic and from high mountain regions to the deep oceans. The major fraction of this low-temperature environment is represented by the deep sea (90% of the ocean volume) followed by snow (35% of the land surface), permafrost (24% of land surface), sea ice (13% of the Earth's surface), and glaciers (10% of the land surface). Other cold environments are cold-water lakes, cold soils, cold deserts, and caves (Hamdan 2018). Chattopadhyay et al. (2014) have mentioned the Earth as a cold biosphere with a major part (>85%) permanently exposed to temperature below 5 °C. The cold biosphere provides habitats such as snow, permafrost, sea ice, glaciers, oceanic waters and sediments, cold water lakes, soil, and caves.

Farooq et al. (2016) have reviewed habitats of psychrophiles. Cold soils play important biogeochemical roles in primary biomass production and nutrient cycling. Cryosphere is not only the integral part of the global climate system but also one of the major habitable ecosystems of Earth's biosphere. Glacier ice and snow are interrelated ecosystems because glacier ice is formed from snow as a result of gradual compression and burial for millions of years. Glacier is a unique ecosystem preserving microbial life and past climate changes chronologically for millions of years. Another specific habitat colonizing microorganisms and existing on the surface of the glacier ice are cryoconite holes ("cryo" meaning ice and "conite" meaning dust). Most cryoconites holes are open systems to the surrounding environment and contain abundant population of active living organisms.

7.3 Diversity of Psychrophiles

Psychrophilic microorganisms have been studied by culture-dependent and culture-independent methods. The isolation procedure for psychrophilic bacterial cells calls for the usual microbial methods except that the source material, pipettes, media etc. are kept cold at all times. This emphasizes the abnormal thermal lability of the psychrophilic bacteria. The enrichment technique, with the desired medium and incubation between 0 and 5 °C before proceeding for isolation procedures, is recommended. Psychrophiles are not rare in permanently cold environments and can be easily isolated if the correct isolation procedures are applied (Morita 1975).

Chattopadhyay et al. (2014) have reviewed microbial diversity of psychrophiles from Antarctica, Arctic, and Himalayan glaciers. Diverse microorganisms such as bacteria, archaea, yeast, fungi, and algae are known to survive, divide, and colonize these cold habitats.

While ocean waters, permafrost (frozen soils), glaciers and ice sheets, polar sea ice, and snow have been studied for psychrophilic microbial diversity, the high-altitude cold habitats of the Himalayas are little explored for bacterial diversity. Soil samples collected from different Himalayan habitats showed that majority of cultured bacterial strains belonged to *Proteobacteria*, *Actinobacteria*, and *Bacteroidetes*. Gradual decline in the number of genera was observed with increasing altitude. Some commonly found bacterial phyla in glacial soils include *Actinobacteria*, *Proteobacteria*, *Cyanobacteria*, *Firmicutes*, *Acidobacteria*, and *Bacteriodetes* (Farooq et al. 2016). Soils of alpine regions undergo dramatic temporal changes in their microclimatic properties resulting in bacterial selection gradient. The abundance of total and culturable bacteria deposited into the East Rongbuk Glacier, Mt. Everest were investigated (Farooq et al. 2016).

In ocean depths, psychrophiles face extremely high pressure in addition to cold temperature and thus are called as piezo-psychrophiles or baro-psychrophiles. The microbial communities that are found in sea ice and exposed to high salt concentration are called as halo-psychrophiles (Farooq et al. 2016).

Upadhyay (2016) has reviewed diversity of psychrophilic pathogens as potential risk for food-borne illness. Contrary to common belief that hot environment favors growth of pathogenic microorganisms, refrigerators also support growth of pathogens. Minimum temperature for growth of some potential pathogenic microorganisms is reported as 5 °C for *Acetobacter,* 3.3–10 °C for *Clostridium botulinum,* 0 °C for *Clostridium putrefaciens*, 5–10 °C for *E. coli* 1 °C for *Listeria*, 5 to 10 °C for *Staphylococcus*, 4 °C for *Vibrio*, *Pseudomonas* and 0–4 °C for *Yersinia enterocolitica.* The psychrotrophic microorganisms reported to survive in refrigerated foods include *Campylobacter jejuni* (in raw chicken, unpasteurized milk), *Listeria monocytogenes* (raw milk, cheese, ice cream, raw vegetables, raw poultry, and meats), *Yersinia enterocolitica* (raw milk, chocolate, raw meats), *Aeromonas hydrophila* (seafoods) and *Pseudomonas* spp. (meat, fish, and dairy products).

Hamdan (2018) has reviewed diversity of psychrophiles. Bacteria represent very important members of the sea ice habitat. Heterotrophic gas-vacuolate bacteria, not reported in other marine habitats, have been discovered in and near sea ice. Among the cold-adapted bacteria, the genus *Colwellia* is an unusual example. Members of this genus play an important role in carbon and nutrient cycling. A wide range of phylogenetic diversity within the genera *Achromobacterium*, *Alcaligenes*, *Alteromonas*, *Aquaspirillum*, *Arthrobacter*, *Bacillus*, *Bacteroides*, *Brevibacterium*, *Clostridium*, *Colwellia*, *Cytophaga*, *Flavobacterium*, *Gelidibacter*, *Methanococcoides*, *Methanogenium*, *Methanosarcina*, *Microbacterium*, *Micrococcus*, *Moritella*, *Octadecabacter*, *Phormidium*, *Phytobacterium*, *Polaribacter*, *Polaromonas*, *Pseudomonas*, *Psychroserpens*, *Shewanella*, *Psychrobacter*, and *Vibrio* have been found to be psychrophilic across the domain Bacteria (Staley and Gosink 1999; Deming 2002; Miteva 2008; Zhang et al. 2008).

Yadav et al. (2019) have reviewed diversity of psychrotrophic microorganisms. The extreme cold environments harbor novel psychrotrophic microbes. The psychrotrophic microorganisms have been reported from worldwide cold habitats and belong to all three domains archaea, bacteria, and eukarya including different phyla such as *Actinobacteria*, *Firmicutes*, *Verrucomicrobia*, *Bacteroidetes*, *Chloroflexi*, *Cyanobacteria*, *Euryarchaeota*, *Proteobacteria*, *Thaumarchaeota*, and *Nitrospirae*. The most dominant genera belonging to *Arthrobacter*, *Bacillus*, *Exigobacterium*, *Paenibacillus*, *Providencia*, *Pseudomonas*, and *Serratia* have been reported from the cold habitats.

Junge et al. (2019) have reviewed diversity of psychrophilic bacteria in sea and glacial ice environments. Psychrophilic bacteria and microbial communities play important biogeochemical roles in the vast majority of Earth environments. They are phylogenetically diverse and exhibit a host of molecular tracts that confer functionality at low temperatures for growth. Recent advances in genomic, metagenomic, proteomic, and metaproteomic technologies have provided a detailed view of the phylogenetic diversity and a better way to describe the ecological function of this interesting group of microorganisms, a better understanding of low-temperature microbial ecosystem function and potential changes in an increasingly warmer Earth.

Some researchers from India have described psychrophilic microorganisms and novel species from Antarctica and Arctic.

Shivaji et al. (2013) have reported draft genome sequence of *Arthrobacter gangotriensis* strain Lz1y^{T} isolated from a Penguin Rookery soil sample collected in Antarctica, near the Indian Station Dakshin Gangotri.

Shivaji et al. (2012) have described a novel Gram-stain negative, nonmotile strain M12-11BT, isolated from arctic sediment sample collected at a depth of 200 m from Kongsfjorden, Svalbard. The organism was found to be psychrotolerant and belonging to the genus *Cyclobacterium.* Based on phenotypic and chemotaxonomic characters, 16S rRNA gene sequencing, phylogenetic analysis, and DNA-DNA hybridization studies, the strain was assigned a new species as *Cyclobacterium quasimii* sp. nov.

Sharma et al. (2018) isolated a Gram-negative, yellowish-orange pigmented, rod-shaped, motile bacterium, strain ARC 111 T, from sediment of Arctic permafrost at Midtre Lovenbreen glacier, Svalbard. 16S rRNA gene sequencing showed highest similarity (97.8%) to *Subsaxibacter broadyi* P7T. Phylogenetic analysis revealed distinct positioning of strain ARC 111 T within the genus *Subsaxibacter.* Based on DNA-DNA hybridization studies, the strain was found to be a novel species and a new species as *Subsaxibacter sediminis* sp. nov. was proposed.

Jadhav et al. (2013) isolated an extracellular cold active lipase-producing psychrotrophic bacterium (BRI8) from the Antarctic seawater sample. The 16S rRNA gene sequencing revealed that the isolate belonged to genus *Halomonas.* Production of the enzyme lipase was optimized. The enzyme was stable in various organic solvents and showed 100% activity in presence of isoamyl alcohol. Optimum pH and temperature for enzyme activity were 7.0 and 10 °C, respectively.

Shivaji et al. (2005a, b) isolated 12 strains of psychrophilic bacteria from cyanobacterial mat samples collected from various water bodies in the McMurdo Dry Valley region of Antarctica. Based on phenotypic and chemotaxonomic studies, the strain was assigned to genus *Psychrobacter.* Two strains were found to be novel species and new species as *Psychrobacter vallis* sp. nov. and *Psychrobacter aquaticus* sp. nov. were proposed.

Reddy et al. (2003) isolated 11 psychrophilic bacteria from a solid layer of fast ice in the middle of Pointe-Geologie Archipelago, Adelie Land, Antarctica. All the 11 strains were found to belong to genus *Halomonas* based on phenotypic, chemotaxonomic characteristics and phylogenetic analysis. DNA-DNA homology was found to be >85%. New species as *Halomonas glaciei sp. nov.* was identified.

Singh et al. (2014) studied culturable psychrophilic bacterial diversity from cryoconite holes at Midre Lovenbreen (ML), Austre Broggerbreen (AB), and Vestre Broggerbreen (VB) glaciers. The culturable bacterial count ranged from 2.7×10^3 to 8.8×10^4 cfu/g. Thirty-five bacterial isolates were obtained and found to belong to genera *Pseudomonas*, *Polaromonas*, *Micrococcus*, *Subtercola*, *Agreia*, *Leifsonia*, *Cryobacterium*, and *Flavobacterium.*

Singh et al. (2015) have reported draft genome sequence of obligate psychrophilic, *Cryobacterium* sp. MLB-32 isolated from cryoconite holes of high Arctic glaciers.

Kapse et al. (2017) have reported draft genome sequence of *Marinobacter* sp. strain AC-23 isolated from Kongsfjorden in the Arctic. The draft genome sequence will aid in understanding the psychrophilic and sea ice-specific lifestyle.

Antony et al. (2012) examined bacterial abundance, diversity, and glaciochemical composition from three depths of an ice core from coastal Dronning Maud Land, East Antarctica. Fourteen bacterial isolates belonging to genera *Methylobacterium*, *Brevundimonas*, *Paenibacillus*, *Bacillus*, and *Micrococcus* were retrieved. With high in situ sea salt Na^+ and dust concentration suggested that the bacteria might have been transported and deposited into ice along with dust particles and marine aerosols. Frequent isolation of similar bacterial genera from different cold environments suggests that they possess features that enable survival and metabolism for extended periods of time at subzero temperatures.

Antony et al. (2016) investigated the diversity and functional capabilities of microorganisms from three regions of East Antarctica. A diverse assemblage of bacteria (Proteobacteria, Actinobacteria, Firmicutes, Bacteroidetes,

Deinococcus-Thermus, Planctomycetes, Verrucomicrobia), archaea (Euryarchaeota) were detected through culture-dependent and culture-independent methods.

Singh et al. (2017) have reported draft genome sequence of *Nesternkonia* sp. strain PF2819, a psychrophilic bacterium isolated from 44,800-year-old permafrost. The draft genome sequence of this strain revealed the presence of genes involved in the production of cold active genomes, carotenoid biosynthesis, fatty acid biosynthesis, and resistance to heavy metals.

Reddy et al. (2000) isolated psychrophilic bacterium CMS19YT from a cyanobacterial mat sample from a pond in McMurdo Dry Valley, Antarctica. The bacterium was aerobic, Gram-positive, non-spore-forming, nonmotile, exhibited rod-coccus cycle and produced a water-insoluble yellow pigment soluble in methanol. The isolate grew at temperature 5 to 30 °C, pH 6 to 9 and tolerated up to 11.5% NaCl. Based on 16S rDNA analysis, the isolate was found be closely related to group *Arthrobacter* species and highest (97.91%) similarity with *Arthrobacter agilis.* Based on DNA-DNA hybridization studies, a new species was proposed for the isolate CMS19YT as *Arthrobacter flavus* sp. nov.

Shivaji et al. (1988) isolated five cultures from soil samples collected in Schirmacher oasis, Antarctica, and identified as members of the family Micrococcaceae, three belonging to genus *Micrococcus* and two to *Planococcus.* Three isolates (37R, 45R, and 49R) were red-pigmented and identified as *Micrococcus roseus.* Two strains (30y and Lz30R) were yellow and orange in color and identified as *Planococcus* sp.

Shivaji et al. (1992) reported two pure cultures of bacteria isolated from soil sample collected in Schirmacher oasis, Antarctica, as belonging to genus *Sphingobacterium.* Based on DNA-DNA hybridization studies, the *Sphingobacterium antarticus* sp. nov. was proposed for the two Antarctic strains. The type strain is 4BY (MTCC67S).

Reddy et al. (2004) isolated 31 strains belonging to *Pseudomonas* from cyanobacterial mat samples collected from various water bodies in Antarctica. All 31 isolates were psychrotrophic and divided in three groups CMS35T, CMS38T, and CMS64T. Based on phenotypic, chemotaxonomic characteristics and DNA-DNA hybridization studies, three new species were proposed as *Pseudomonas antarctica* sp. nov., *Pseudomonas meridian asp*. nov., and *Pseudomonas proteolytic asp*. nov.

Reddy et al. (2003) isolated a strain CMS760rT, an orange pigment from a cyanobacterial mat sample from a pond located in McMurdo Dry Valley, Antarctica. Based on chemotaxonomic and phylogenetic properties, the strain CMS760rT was identified as a member of the genus Kocuria, and a new species as *Kocuria polaris* sp. nov. was proposed.

7.4 Cold Adaptation: Physiology and Mechanisms

The effects of temperature are of prime importance with regard to psychrophiles and their activities both under controlled conditions in laboratory and in natural environments. Physiological basis of psychrophiles includes protein-synthesizing mechanisms, ribosomes, soluble fractions, whole cell protein synthesis, alteration in cellular structures, and inactivation of enzymes (Inniss 1975).

Gounot (1986) has mentioned that the ability to grow at low temperature may be correlated to an increasing proportion of unsaturated fatty acids in the lipid phase of the cell membrane, which makes it more fluid, and a protein conformation functional at low temperature. The relatively low maximum temperature of growth of these microorganisms is often considered to be due to the thermolability of one or more essential cellular components, particularly enzymes.

Hamdan (2018) has reviewed metabolic adaptation of psychrophiles. The psychrophilic microorganisms use a vast variety of unique adaptive features of their cellular components for survival in harsh and extremely cold environments. Chintalapati et al. (2004) have categorized three phases of cold-shock response. Phase I is an acclimatization phase which follows immediately to cold exposure, leads to reduced growth rate as a result of reduced membrane fluidity and several cold-shock proteins are produced. During Phase II, a recovery phase, cells are considered "cold-adapted" and resume growth and restart bulk protein synthesis. Phase III is a stationary phase wherein non-cold-shock proteins are synthesized allowing cells to proceed to slow-rate growth at low temperature (Weber and Marahiel 2003). Bacteria possess the ability to cope with such challenges to complex strategies. One important strategy is directed towards their extremely efficient DNA repair mechanism sustained under frozen conditions (Christner 2002).

Ray et al. (1994) have depicted role of phosphorylation of lipopolysaccharides in the psychrophilic bacteria, in temperature adaptation. Phosphorylation of lipopolysaccharides (LPS) from a psychrotrophic bacterium *Pseudomonas syringae* from the Antarctica was studied. The bacterium was found to possess an LPS kinase which could phosphorylate more LPS photosynthetically at higher temperatures. The phosphorylation was low at a low temperature and also found to occur in vivo. After phosphorylation of LPS in vitro, it was found that major part of the radioactivity (>85%) was associated with the core oligosaccharide region of the LPS. The phosphate group of this region are probably involved in the binding of the metal ions, which could be removed by EDA. The cells grown at lower temperature probably contained fewer divalent cations because of the smaller amount of phosphate and thereby were more sensitive to EDTA.

Chintalapati et al. (2004) have reviewed various strategies by which bacteria modulate the fluidity of the membrane for cold adaptation. Psychrophilic and psychrotolerant bacteria evolve different strategies to adapt to low temperature. One important strategy relates to the ability of the cell to modulate the fluidity of the membrane. Bacteria modulate membrane fluidity by altering their fatty acid composition, the lipid head group, the protein content of the membrane, the type of carotenoids synthesized, the fatty acid chain length, and the proportion of *cis* to *trans* fatty acids. Bacteria also have a two-component signal transduction pathway consisting of a membrane-bound sensor and a soluble cytoplasmic response regulator involved in the perception and transduction of low-temperature signals.

Shivaji and Prakash (2010) have discussed how bacteria respond to low temperature. Rigidification of the membrane appears to be the primary signal perceived by a bacterium when exposed to low temperature. The perception and transduction of the signal then occurs through a two-component signal transduction pathway consisting of a membrane-associated sensor and a cytoplasmic response regulator, and as a consequence, a set of cold-regulated genes are activated. Changes in DNA topology due to change in temperature may also trigger cold-responsive mechanisms. Inducible proteins thus accumulated repair the damage caused by cold stress. The fluidity of the rigidified membrane is restored by altering the levels of saturated and unsaturated fatty acids, by altering the fatty acid chain length, by changing the proportion of *cis* to *trans* fatty acids, and by changing the proportion of *anteiso* to *iso* fatty acids.

7.4.1 Regulation of Membrane Fluidity

The most frequent adaptive strategy relates to the ability of the cell to regulate or modulate the fluidity of the membrane in freezing environments. Shivaji and Prakash (2010) reported that membranes become more rigid at cold temperatures, which activated a membrane-associated sensor resulting in upregulation of genes involved in membrane fluidity modulation for exchange of metabolites from and to the cell. Changes in lipid composition facilitate this process, by modifications in the lipid's fatty acyl chains that maintain optimum membrane fluidity (Russell 2008).

7.4.2 Carotenoid Pigments

Carotenoid pigments in bacteria isolated from Antarctic Sea ice have been reported to play an important role in the maintenance of membrane structure, fluidity, and protection from UV radiation (Armstrong 1997). Accumulation of C-50 carotenoid, Bacterioruberin, found in the psychrotrophic strain *Arthrobacter agilis* was postulated to play a crucial role in the regulation of membrane fluidity at low temperature (Fong et al. 2001). Several polar and nonpolar carotenoid pigments synthesized by *Micrococcus roseus* and *Sphingobacterium antarcticus* Antarctic strains were found to bind vesicles made of both synthetic and natural lipids and to rigidify them (Chattopadhyay and Jagannadham 2001).

7.4.3 Antifreeze Proteins

Antifreeze proteins (AFPs) are ice-binding proteins that have the ability to decrease the freezing point of water and show extracellular ice recrystallization inhibition activity during latter stages of warming cycle (Gilbert et al. 2004; Christner 2010).

Duman and Olsen (1993) first demonstrated the presence of AFPs in cold-adapted bacteria. The bacterium *Moraxella* sp. was the first reported Antarctic strain to produce AFP (Yamashita et al. 2002). Some psychrophilic AFPs have been purified from cell extracts of *Micrococcus cryophilus*, *Pseudomonas putida*, and *Rhodococcus erythropolis* (Muryoi et al. 2004).

7.4.4 Cryoprotectants

Cryoprotectants are exopolymeric substances produced in high amounts and believed to prevent cold-induced aggregation of proteins and maintain optimum membrane fluidity under unfavorable low temperature (Krembs et al. 2002; Mancuso et al. 2005). Ko et al. (1994) demonstrated the growth-enhancing effect of glycine betaine on *Listeria monocytogenes* at low temperature. Trehalose was found to prevent protein denaturation and aggregation in psychrophilic bacteria (Phadtare 2004).

7.4.5 Cold-Shock Proteins

Cold-shock proteins (CSPs) have been extensively studied and are considered the most prominent response of cells to cold shock in order to counteract the detrimental effect of temperature downshift. CSPs play a critical role in cold adaptation (Hebraud and Potier 1999). They had been functionally linked to the regulation of cellular protein synthesis. They also act as chaperones by preventing the formation of mRNA secondary structures (mRNA "folding") and maintenance of chromosome structure (Chaikam and Karlson 2010; Phadtare 2011). These stress proteins together with cold-acclimation proteins have been detected and overexpressed during the entire growth of several Antarctic bacteria (Phadtare 2004).

7.4.6 Cold-Active Enzymes

Cold-active or cold-adapted enzymes display high catalytic efficiency at low temperature compared to their mesophilic counterparts (Samie et al. 2012). Struvay and Feller (2012) reported multiple adaptive features developed by constrained psychrophiles to design enzymes perfectly compatible to the given environment. Psychrophiles produce enzymes that are highly flexible in structure having up to tenfold higher specific activity than their mesophilic homologues, thus providing better access to the active site of substrates at lower temperature (Gerday et al. 2000). These cold-active enzymes offset the inhibitory effect of low temperatures on reaction rates and maintain adequate metabolic fluxes to the growing organism. The apparent maximal activity of cold-active enzymes is shifted towards low temperature to cope with the increased viscosity of the aqueous environment induced under cold conditions.

Feller (2013) has reviewed psychrophilic enzymes. Psychrophiles thriving permanently at near-zero temperatures synthesize cold-active enzymes to sustain their cell cycle. Genome sequences, proteomic and transcriptomic studies suggest various adaptive features to maintain adequate translation and proper folding under cold conditions. Most psychrophilic enzymes optimize a high activity, therefore reducing the free energy barrier of the transition state. A weak temperature dependence of activity ensures moderate reduction of the catalytic activity in the cold. In these naturally evolved enzymes, the optimization to low-temperature activity is reached via destabilization of the whole molecule. This involves a reduction in the number and strength of all types of weak interactions or the disappearance of stability factors, resulting in improved dynamics of active site residues in the cold.

Farooq et al. (2016) have reviewed cold-adapted enzymes. Psychrophiles possess ability to synthesize enzymes that catalyze reactions at temperatures close the freezing point of water. Low temperatures tend to increase the compactness of proteins and also lead to exponential decrease in chemical reaction rates. The cold-adapted enzymes have been observed to possess high specific activity or catalytic efficiency at low and moderate temperatures.

7.4.7 Lipid Adaptation

Hassan et al. (2020) have described temperature-driven lipid adaptation in glacial psychrophilic bacteria. Bacteria inhabiting nonpolar glaciers are exposed to large variations in temperature, which affects the fluidity of bacterial cell membrane. To maintain normal functions of the cell membranes, psychrophilic bacteria adapt by changing the composition of the cell membrane fatty acids. The authors isolated 42 bacterial strains from the Ghulmet, Ghulkin, and Hopar glaciers of the Hunza valley in the Karakoram Mountain range, Pakistan, and studied their cell membrane fatty acid distribution. Gram-negative and Gram-positive isolates were grown at different temperatures (5, 15, 25 and 35 °C) to determine effect of temperature on cell membrane (CM) fatty acid distribution. The major group groups of cell membrane fatty acids (FA) were found as straight-chain mono-unsaturated fatty acids (n-MUFAs) and branched fatty acids (br-FAs). The level of br-FAs decreased relative to n-FAs with increasing temperatures. The production of polyunsaturated fatty acids (PUFAs) was observed only at low temperatures. The authors claim this study to be for the first time reporting the role of br-FAs in the maintenance of cell membrane fluidity of bacteria inhabiting nonpolar habitats.

7.5 Biotechnological Applications of Psychrophiles

Psychrophiles are being extensively studied for their applications in food industry, waste processing, mining, environmental bioremediation, agriculture, medicine, and molecular diagnostics (Hamdan 2018).

Kasana and Gulati (2011) have reviewed cellulases from psychrophiles with a focus on their activity, molecular characteristics, gene cloning, crystal structure, and strategies for the improvement of psychrophilic cellulases. The cellulases from psychrophiles find applications in laundry additives, environmental bioremediation, food industry, and molecular biology.

Farooq et al. (2016) has mentioned that the biotechnological application of cold-adapted enzymes comes from their high K_{cat} at low to moderate temperatures, high thermostability at elevated temperatures, and their ability to function inorganic solvents. Cold-active enzymes find applications in starch and food processing industries due to their high catalytic activity at temperatures that minimize spoilage and alterations in taste and nutritional values. In pharmaceutical industry, cold-active α-amylases can be used as a digestive aid in food industry for the reduction of haze formation in juices. In brewing industry, cold-active α-amylases could be used to speed the mashing phase at low temperatures. In food industry, the lactose from milk can be removed by use of β-galactosidase from psychrophiles. In baking industry, psychrophilic glycosidases can be used to avoid the retention of residual activity after cooking that alters the structure of the final product during storage. Cold-active pectinases can be used in reducing the viscosity and to clarify fruit juices at low temperatures. Cold-adapted proteases can be used for tenderization and taste improvement of refrigerated meat products, removal of undesirable tissues from seafood, such as descaling and removal of fish skin. Cold-active proteases are used in industrial peeling of leather, which requires heating of water to 37 °C for the process to be performed by mesophilic enzymes. This can be carried out using cold-active enzymes at ordinary temperatures of water.

7.5.1 Applications in Food Industry

There is an increasing industrial trend to treat food stuffs under low-temperature conditions in order to avoid detrimental effects on taste, texture, and nutritional value and also to save energy. Cold-active β-galactosidase, which hydrolyses lactose-free milk-derived foods for lactose-intolerant people, who represent approximately 30% of the world population (Nam and Ahn 2011). Hoyoux et al. (2001) patented a cold-active β-galactosidase from an Antarctic psychrophile, *Pseudoalteromonas haloplanktis* for its capacity to hydrolyze lactose during milk storage at low temperatures. Cold-active xylanase from the same Antarctic organism effectively maintains the mechanical dough properties and final bread quality with a positive effect on loaf volume.

7.5.2 Medical and Pharmaceutical Applications

Several promising psychrophilic strains were reported by Tomova et al. (2015) as a valuable source of new active antimicrobial compounds at low temperatures. Poly-unsaturated fatty acids, produced to improve membrane permeability and nutrient transport in psychrophilic bacteria, may constitute an economic alimentary source for aquaculture industries with favorable activities on cholesterol and triglyceride transport for optimal nervous system and cardiovascular health (Jadhav et al. 2010).

7.5.3 Detergent and Fabric Industry

The detergent industry requires enzymes capable of working at low temperature in the context of energy-saving. Cold-adapted enzymes actively used in detergent formulations—lipases, proteases, and α-amylases are systematically introduced to improve the efficiency of detergents and reduce the amount of chemicals used in order to protect the texture and colors of fabrics and reduce wear and tear during washing (Feller and Gerday 2003; Gerday et al. 2000; Maiangwa et al. 2015). The application of cold-adapted cellulases in fabric production and denim finishing could increase the smoothness and softness of tissues, decolorization of textile effluents, and textile bleaching and allows the development of environmentally friendly technologies in fiber processing (Araujo et al. 2008).

Parvizpour et al. (2021) have reviewed psychrophilic enzymes, their adaptation, pharmaceutical and industrial applications. Psychrophiles are cold-loving microorganisms synthesizing enzymes that are permanently active at almost near-zero temperatures. Psychrozymes are supposed to be structurally more flexible than their homologous proteins. This structural flexibility enables these proteins to undergo confirmational changes during catalysis and improve catalytic efficiency at low temperatures. The outstanding characteristics of the psychrophilic enzymes have attracted the attention of the scientific community to utilize them in a wide variety of industrial and pharmaceutical applications.

7.6 Role of Psychrophiles in Bioremediation and Sustainability

Psychrophiles hold tremendous potential as "environmental cleaners" by degrading pollutants of petroleum hydrocarbons in extreme cold conditions (Margesin and Schinner 2001; Zhou et al. 2008).

Whyte et al. (1999) reported that the microbial catalytic pathways responsible for the degradation of petroleum hydrocarbons, including n-alkanes and polycyclic aromatic hydrocarbons, are widespread in cold regions. Several indigenous cold-adapted microbial populations that use petroleum hydrocarbons include *Rhodococcus* (Ruberto et al. 2005), *Pseudomonas* (Aislabie et al. 2006), and *Pseudoalteromonas* (Lin et al. 2009).

7.7 Psychrophilic Microorganisms Reported from India

India is blessed with a high mountain area, the Indian Himalayan Ranges (IHR). Many glaciers exist in these ranges, to name a few, Hamta, Pindari, Roopkund, Kafni, and Thajwas. There are also a number of cold lakes in these regions, for example, two subglacial lakes, Gurudonmar and Pangong. High-altitude soils also serve as cold environments. All these cold environments are populated by psychrophilic, psychrotrophic, and psychrotolerant microorganisms. Researchers from India have reported a number of psychrophiles from various cold habitats which are listed in Table 7.1.

The researchers from India have investigated a rich diversity of psychrophiles using both culture-dependent and culture-independent methods. Some findings are described.

7.7.1 Diversity

Reddy et al. (2014) have reported draft genome sequence of *Cyanobacterium roopkundensis* strain RuGI7 isolated from a soil sample collected in the periphery of Roopkund Lake, Himalaya, India.

Ajaz et al. (2018) have isolated and characterized psychrophilic bacteria from soil samples from agricultural waste disposal sites of Ladakh (latitude 34°10′12″ N and longitude 77°34′48″ E). Ten waste (soil mixed with waste) samples were processed for isolation of bacteria growing at 10 °C. Total nine isolates were obtained and characterized biochemically, studied for optimization of growth conditions and screened for extracellular enzymes. The isolates were identified as belonging to genera *Actinomycetes*, *Alcaligenes* sp., lactic acid bacteria, and *Bacillus*. The isolates produced extracellular enzymes amylase, xylanase, cellulase, and protease. Agricultural wastes are residues from the growing and processing of raw agricultural products and may contain beneficial material. The extracellular hydrolases produced by the psychrophilic bacteria would be involved in solid waste management.

Suyal et al. (2014) have described the response of a Himalayan psychrophilic diazotroph to low-temperature diazotrophy. Seven cold-adaptive N_2-fixing bacteria were isolated and identified as *Bacillus* sp., *Arthrobacter* sp., *Rhodococcus* sp., and *Pseudomonas* sp. In order to examine physiological response to cold temperature diazotrophy, differential proteomic analysis of *Pseudomonas migulae* S10724 strain was carried out using two-dimensional electrophoresis and MALDI-TOF-MS. Functional assessment of 66 differentially expressed proteins revealed several mechanisms thought to be involved in low-temperature adaptation and nitrogen fixation, general stress adaptation, protein and nucleic acid synthesis, energy metabolism, cell growth/maintenance, etc. Major fraction of the upregulated proteins was stress proteins.

Soni et al. (2015) studied differential proteomic expression of Himalayan psychrotrophic diazotroph *Pseudomonas palleroniana* N26 under low-temperature diazotrophic conditions. Proteomic analysis of *Pseudomonas palleroniana* N26 was carried out using two-dimensional electrophoresis technique. Fifty-three proteins spots were found to be differentially expressed revealing several mechanisms thought to be involved in low-temperature adaptation and nitrogen fixation.

Pandey et al. (2019) have performed comparative assessment and bioprospection of psychrotolerant bacteria isolated from high-altitude soil samples from Indian

Table 7.1 Psychrophilic, psychrotrophic, and psychrotolerant microorganisms reported in India

Sr. No.	Name of organism/ group	Source	Diversity/ biotechnology	Reference
1	*Pedobacter himalayensis* sp. nov. strain HHS22T	Glacier water from snout of the Hamta glacier in Himalayan Mountain ranges	Diversity, new species	Shivaji et al. (2005a, b)
2	*Dyadobacter hamtensis* sp. nov. strain HHS 11T	Water from snout of the Hamta glacier in Himalayan Mountain ranges	Diversity, new species	Chaturvedi et al. (2005)
3	*Planococcus stackebrandtii* sp. nov. strain K22-03T	Cold desert of Himalaya	Diversity, new species	Mayilraj et al. (2005)
4	*Bacillus aerius* sp. nov. strain 24KT *Bacillus aerophilus* sp. nov. strain 28 KT *Bacillus stratosphericus* sp. nov. strain 41KF2aT *Bacillus altitudinis* sp. nov. strain 41KF2bt	Cryogenic tubes used to collect air samples at altitudes 24, 28, and 41 km	Diversity, new species	Shivaji et al. (2006)
5	*Exiguobacterium indicum* sp. nov. strain HHS 31T	Melt water of Hamta glacier of Himalayan Mountain ranges	Diversity, new species	Chaturvedi and Shivaji (2006)
6	*Leifsonia pindariensis* sp. nov. strain PON10T	Pindari glacier of Indian Himalayas	Diversity, new species	Reddy et al. (2008a, b)
7	*Bacillus cecembensis* sp. nov.	Pindari glacier of Indian Himalayas	Diversity, new species	Reddy et al. (2008a, b)
8	*Janibacter hoylei* sp. nov. type strain PVAS-1T *Bacillus isronensis* sp. nov. Type strain B3W22T *Bacillus aryabhatti* sp. nov. Type strain	Cryotubes used to collect air samples at altitudes between 27 and 41 km	Diversity, new species	Shivaji et al. (2009)
9	*Bacillus* *Pseudomonas* *Arthrobacter* *Proteobacteria* 45% *Firmicutes* 32% *Actinobacteria* 16% *Bacteroidetes* 6%	Soils from Western Himalayas	Diversity	Gangwar et al. (2009)

(continued)

Table 7.1 (continued)

Sr. No.	Name of organism/ group	Source	Diversity/ biotechnology	Reference
10	Three 16S rRNA gene clone libraries, 798 clones belonging to 25 classes. *Actinobacteria* predominant class	Soil samples from the periphery of Roopkund glacier in the Himalayan ranges	Diversity	Pradhan et al. (2010)
11	*Paenibacillus glacialis* sp. nov. strain KFC 91T	Kafni glacier of the Himalayas	Diversity	Kishore et al. (2010)
12	*Cryobacterium roopkundense* sp. nov., strain RuGI7T	Soil from periphery of the glacial lake Roopkund in the Himalayan Mountain ranges	Diversity, new species	Reddy et al. (2010)
13	Three 16S rRNA gene clone libraries (P1L, P4L, and P8L) 703 clones *Actinobacteria* *Firmicutes* *Proteobacteria* *Acidobacteria* *Bacteroidetes* *Fibrobacteres* *Nitrospirae* *Verrucomicrobia*	Soil in the vicinity of Pindari glacier, Himalayan Mountain ranges	Diversity	Shivaji et al. (2011a, b)
14	Two 16S rRNA gene clone libraries (KF and KS) 648 clone libraries *Actinobacteria* *Bacteroidetes* *Chloroflexi* *Firmicutes* *Verrucomicrobia* *Acidobacteria* *Nitrospirae* *Lentisphaerae*	Proglacial soil from Kafni glacier, Himalayan Mountain ranges	Diversity	Srinivas et al. (2011)
15	*Tsukamurella* sp. *Alishewanella* sp. *Sphingomonas* sp. *Ochrobacterium* sp. *Brevundimonas* sp. *Paucibacter* sp. *Duganella zoogloeoides*	Two different subglacial lakes, Gurudonmar (GDM) Pangong (PNG) from Himalayas	Diversity	Sahay et al. (2012)
16	*Staphylococcus lipolyticus* sp. nov. strain PN 2T	Sea sediment from Bay of Bengal	Diversity, new species Biotechnology cold-adapted lipase	Arora (2013)

(continued)

Table 7.1 (continued)

Sr. No.	Name of organism/ group	Source	Diversity/ biotechnology	Reference
17	*Paenisporosarcina indica* sp. nov. strain PN 2T	Soil near Pindari glacier of Indian Himalayas	Diversity, new species	Reddy et al. (2013)
18	*Arthrobacter gangotriensis* sp. nov. strain LZ1yT	Penguin Rookery soil near Indian Station Dakshin Gangotri in Antarctica	Diversity, new species	Shivaji et al. (2013)
19	*Cryobacterium roopkundensis* sp. nov. strain RuGI7	Soil from periphery of Roopkund Lake, Himalaya	Draft genome sequence	Reddy et al. (2014)
20	*Bacillus* sp. *Arthrobacter* sp. *Rhodococcus* sp. *Pseudomonas migulae* S10724	Himalayan psychrophilic, diazotrophic, seven cold-adaptive N_2-fixing bacteria	Diversity, differential proteomic analysis of *Pseudomonas migulae* S10724 66 proteins	Suyal et al. (2014)
21	*Pseudomonas palleroniana* N26	Himalayan psychrotrophic diazotroph	Proteomic analysis of 53 proteins	Soni et al. (2015)
22	*Rufibacter immobilis* sp. nov.	Psychrophilic bacterium from water of high-altitude (4350 m) saline lake, Pangong lake in Western Himalayas	Diversity, new species	Polkade et al. (2015)
23	*Dyadobacter psychrophilus* B2	Western Himalayan agroecosystems	Diversity, cold stress adaptation studies	Suyal et al. (2017)
24	*Actinomycetes* *Alcaligenes* sp. Lactic acid bacteria *Bacillus*	Soil samples (soil mixed with waste) from agricultural waste disposal site of Ladakh	Diversity, extracellular enzymes amylase, xylanase, cellulase, protease involved in waste disposal	Ajaz et al. (2018)
25	*Bacillus* *Pseudomonas*	High-altitude soil samples from Indian Himalayas Region (IHR)	Diversity	Pandey et al. (2019)
26	*Pseudomonas helmanticensis*	High-altitude cold climate Gangotri soil Western Indian Himalayas (WIH)	Proteome analysis Cold stress resistance Chaperons and cold-shock proteins	Kumar et al. (2020)
27	Psychrophilic and psychrotrophic bacteria	Himalayan Thajwas glacial soil	Diversity, cold-active enzymes	Farooq et al. (2022)

Himalayan Region (IHR), based on 16S rRNA gene sequencing and MALDI-TOF mass spectrometry. Results of both the methods revealed maximum number of *Bacillus* followed by *Pseudomonas* species.

Fig. 7.1 Gurudonmar lake, Sikkim (Google image)

Pote et al. (2014) have isolated psychrotrophic microorganisms from marine water samples (including four samples from Antarctic). Fourteen bacterial isolates were explored for their biotechnological potential.

Suyal et al. (2017) have analyzed proteomes of two psychrotrophs *Dyadobacter psychrophilus* B2 and *Pseudomonas jessenii* MP1 using two-dimensional electrophoresis and MALDI-TOF-MS, to determine the physiological response of altitudinally different but indigenous microorganisms in response to cold stress under N depleting conditions. Nitrogen (N) deficiency and low-temperature conditions are the prominent facet of Western Himalayan agroecosystems. A slight change in the environment alters the protein expression of the microorganisms. Hence, the authors carried out these studies. Functional assessment of 150 differentially expressed proteins from both the psychrotrophs revealed that several mechanisms might be involved in cold stress adaptation, protein synthesis/ modifications, energy metabolism, and cell growth/maintenance.

Sahay et al. (2012) have reported cold-active hydrolases producing bacteria from two different subglacial Himalayan lakes. The lake Gurudonmar (GDM) (Fig. 7.1) is situated at 5250 m above mean sea level at 28° N 88° E which is at the footfall of Kanchenjunga peak in eastern part of the Himalayas, India. The annual mean temperature of this lake was varying between −10 and 4 °C while pH varies from 7.2 to 8.0. The lake Pangong (PNG) is one of the largest brackish water (salinity 25 ppt) snow-capped lake and is located at 4250 m above mean sea level at 33° N 78° E. The lake of 134 km long and 5 km wide is completely frozen during winter. Out of total 44 isolates obtained from lake PNG, nine were psychrophilic, 16 psychrotrophic, and 19 psychrotolerant, and in case of GDM lake, out of

52 isolates obtained 12 were psychrophilic, 15 psychrotrophic, and 25 psychrotolerant. The snow-capped surface waters of PNG lake have unique culturable bacterial species such as *Tsukamurella* sp., *Alishewanella* sp., *Sphingomonas* sp., *Ochrobacterium* sp., and *Brevundimonas* sp. The studies showed highly diverse populations, belonging to *Proteobacteria* (α, β, and γ), *Firmicutes*, and *Actinobacteria* in these lakes. Two isolates, NBGD4 and NBGD39 from GDM lake showed 99% and 98% similarity with *Paucibacter* sp. and *Duganella zoogloeides*, respectively. Sixteen isolates exhibited cold-active protease, amylase, xylanase, and cellulase activity.

Gangwar et al. (2009) have investigated bacterial diversity of soils from Western Himalayas, India. High-altitude cold habitats of the Himalayas are little explored for bacterial diversity. Diverse bacterial species and phylotypes were obtained by culture-dependent and culture-independent approaches. Psychrophilic and psychrotolerant bacteria dominated the Himalayan habitats, accounting for 60% of the cultivated strains. The isolates produced hydrolytic enzymes, lipase being secreted by most strains (62%). Partial 16S rRNA gene sequences were obtained for 99 bacterial strains and 74 clones obtained from soil samples from the Western Himalayas. Among the cultured bacterial strains, 45% belonged to the *Proteobacteria*, 32% *Firmicutes*, 16% *Actinobacteria*, and 6% *Bacteroidetes*. Most of the strains belonged to the genus *Bacillus* (30%) followed by *Pseudomonas (24%)* and *Arthrobacter* (12%). In culture-independent studies, phylotypes belonging to the *Proteobacteria* were dominant (73%). The bacterial diversity exhibited an altitude gradient with a gradual decline in the number of genera with increase in altitude.

Farooq et al. (2022) have described psychrophilic and psychotropic bacterial diversity and distribution in Himalayan Thajwas glacial soil. The total bacterial count of seven soil samples collected near glacier snout at different altitudes was investigated through a culture-dependent method. The highest colony-forming units, cfu/g ($6.8 \pm 2 \times 10^7$) was recorded at sight I and the lowest cfu/g ($4.6 \pm 2 \times 10^3$) at sight VII. The 31 isolated strains were identified and grouped as psychrophiles (optimum growth temperature $\geq$15 °C) and psychrotrophs. (Optimum growth $\geq$20 °C). The strains were screened for cold-active enzymes. Variation in microbial diversity may be related to soil physicochemical properties since high positive correction was observed with water content total nitrogen and organic carbon. The authors claim this as the first research on microbial communities from the Thajwas Glacier (Fig. 7.2) adding to knowledge of microbial diversity found in Himalayan cold ecosystem.

Kumar et al. (2020) have reported psychrophilic *Pseudomonas helmanticensis* proteome under simulated cold stress. The organism was isolated from high-altitude cold climatic Gangotri soil (altitude 3415 m 30.98° N, 78.93° E) (Fig. 7.3) while growing on nitrogen-deficient solid medium (Burk's medium) at 2 °C for 48 h (Kumar 2018; Kumar et al. 2019). Optimum growth temperature of this bacterium was reported 10 °C in the Nutrient Broth medium. For proteome extraction, the culture was grown separately at 2 °C (T, treatment) and 20 °C (C, control). The proteome was extracted from mid-log phase culture and analyzed by LC-MS.

Fig. 7.2 Thajwas glacier (Google image)

Fig. 7.3 Gangotri glacier (Google image)

The results revealed that most of the differentially expressed proteins providing cold stress resistance were molecular chaperons and cold-shock proteins.

High-altitude terrestrial ecosystems of Western Indian Himalaya (WIH) are one of such cold-stressed ecosystems which undergo frequent freezing-thawing cycles (Awasthi et al. 2019), Microorganisms in these Himalayan regions are mostly psychrophilic or psychrotrophic in nature and play a critical role in biogeochemical cycle operating under low temperature (Joshi et al. 2017).

Western Indian Himalaya (WIH) is well known for its rich diversity of cold-adapted microorganisms. Cold-adapted *Arthrobacter humicola*, *Brevibacillus invocatus*, *Dyadobacter*, *Pseudomonas palleroniana*, *Pseudomonas jensenii*, *Pseudomonas mandelii*, *Pseudomonas migulae*, *Pseudomonas helmanticensis*, and *Rhodococcus qingshengii* have been reported from WIH (Kumar et al. 2018; Suyal et al. 2017, 2019).

Gangotri, being the second largest glacier of Himalaya, experiences severe fluctuating cold stress with maximum and minimum annual temperature of 11.1 ± 0.7 °C and -2.3 ± 0.4 °C respectively (Kumar et al. 2019).

Arora (2013) has reported a new cold-adapted lipase producing marine species. The bacterium, SS-33^T was isolated from sea sediments collected from the Bay of Bengal, India, and subjected to polyphasic taxonomic study. Phylogenetic analysis based on the 16S rRNA gene sequences showed that the strain SS-33^T belonged to the genus *Staphylococcus.* Based on phenotypic, chemotaxonomic, and DNA-DNA hybridization studies, a novel species as *Staphylococcus lipolyticus* sp. nov. was proposed. The type strain is SS-33^T (=MTCC 10101^T = JCM 16560). The strain hydrolyzed various substrates including tributyrin, olive oil, Tween 20, Tween 40, Tween 60, and Tween 80 at low temperatures as well as mesophilic temperatures. Lipase from the strain was purified and characterized. Lipase exhibited 90% activity of its optimum temperature (30 °C) at 10 °C and was more stable at 10 °C as compared to 30 °C. As such, the enzyme has wide range of applications in industrial processes.

Reddy et al. (2008a, b) isolated a bacterial strain PNS T from the Pindari glacier (Fig. 7.4) of the Indian Himalayas. The organism is a Gram-positive, aerobic, motile, rod-shaped bacterium. Based on the phenotypic and chemotaxonomic characteristics, the bacterial strain was assigned the genus *Bacillus.* Phylogenetic analysis and DNA-DNA relatedness indicated that the strain was novel and hence a new species as *Bacillus cecembensis* sp. nov. was proposed for the strain PN5T.

Kishore et al. (2010) isolated a novel strain of *Paenibacillus*, KFC91T from Kafni glacier of the Himalayas. Based on the phylogenetic analysis, chemotaxonomic characteristics and other phenotypic traits, strain KFC91T was assigned to the genus *Paenibacillus.* DNA-DNA hybridization studies indicated that the strain was a novel species, and hence a new species as *Paenibacillus glacialis* sp. nov. was proposed.

Reddy et al. (2013) isolated a Gram-positive, aerobic spore forming rod-shaped bacterium, PN2T, from a soil sample collected near Pindari glacier of Indian Himalayas. Based on chemotaxonomic studies, phylogenetic analysis, and 16S rRNA gene sequence, the strain PN2T was assigned to the genus *Paenisporosarcina.* Based on DNA-DNA relatedness, the strain was found to be a novel species and a new species as *Paenisporosarcina indica* sp. nov. was proposed.

Fig. 7.4 Pindari glacier, Indian Himalayas (Google image)

Fig. 7.5 Hamta glacier located in the Himalayan Mountain ranges of India (Google image)

Chaturvedi et al. (2005) isolated a strain HHS11T from a water sample collected from the snout of Hamta glacier (Fig. 7.5) located in the Himalayan Mountain ranges of India. Based on phenotypic, chemotaxonomic, and phylogenetic analyses, the strain was assigned the genus *Dyadobacter*. However, the strain showed differences with the existing species of *Dyadobacter*. Hence, a new species as *Dyadobacter hamtensis* sp. nov. was proposed for the strain HHS11T.

Mayilraj et al. (2005) isolated a bacterium strain K22-03T from a cold desert of the Himalayas, India. Based on phenotypic characters and 16S rRNA gene sequencing, the strain was confirmed to belong to the genus *Planococcus*; however, the strain was found to differ from the known species of *Planococcus*. Based on DNA-DNA hybridization studies, a novel species as *Planococcus stackebrandtii* sp. nov. was proposed for the strain K22-03T.

Reddy et al. (2010) isolated a strain RuG17T from a soil sample collected at the periphery of the glacial lake Roopkund in the Himalayan Mountain range, India. The organism was Gram-positive aerobic, rod-shaped, and grew optimally between 15 to 18 °C. Based on phenotypic and chemotaxonomic studies, the strain was assigned to the genus *Cryobacterium.* Based on DNA-DNA hybridization, a new species as *Cryobacterium roopkundense* sp. nov. was proposed.

Chaturvedi and Shivaji et al. (2006) have reported a psychrophilic bacterium from the melt water of Hamta glacier of the Himalayan Mountain ranges of India. The strain isolated, HHS31T was Gram-positive, motile, non-spore-forming rod. Based on phenotypic and chemotaxonomic characteristics, the strain was found to be related to the genus *Exiguobacterium.* However, DNA-DNA relatedness indicated the strain to be novel species. A new species *Exiguobacterium indicum* sp. nov. was proposed.

Shivaji et al. (2011a, b) studied bacterial diversity of soil in the vicinity of Pindari glacier, Himalayan Mountain ranges, India. Three 16S rRNA gene clone libraries (P1L, P4L and P8L) were constructed using three soil samples (P1S, P4S and P8S) collected near Pindari glacier, Himalayas which yielded 703 clones. *Actinobacteria*, *Firmicutes*, and *Proteobacteria* were common to the three libraries. Libraries P1L and P8L also shared phyla *Acidobacteria*, *Bacteroidetes*, *Gemmatimonadetes*, and *Planctomycetes*. Phyla *Chlamydiae*, *Chlorobi*, *Chlorfexi*, *Dictyoglomi*, *Fibrobacteres*, *Nitrospirae*, *Verrucomicrobia* were present only in P1L.

Shivaji et al. (2005a, b) isolated a strain HH522T from a glacier water sample from the snout of the Hamta glacier located in the Himalayan Mountain ranges of India. Phenotypic, chemotaxonomic, and phylogenetic analysis established the affiliation of the isolate to the genus *Pedobacter.* DNA-DNA relatedness indicated the strain to be a novel species for which the name *Pedobacter himalayensis* sp. nov. was proposed. The type strain is HHS 22T (=JCM12171T = MTCC 6384T).

Pradhan et al. (2010) studied bacterial diversity of two soil samples collected from the periphery of the Roopkund glacial lake and one soil sample from the surface of the Roopkund glacier (Fig. 7.6) in the Himalayan ranges by constructing three 16S rRNA gene clone libraries. The three libraries yielded 798 clones belonging to 25 classes. *Actinobacteria* was the most predominant class (>10% of the clones) in the three libraries. In the library from glacial soil, class Betaproteobacteria was the most predominant (24.2%).

Srinivas et al. (2011) studied diversity in proglacial soil from Kafni glacier (Fig. 7.7), Himalayan Mountain ranges, India. Two 16S rRNA gene clone libraries (KF and KS) were constructed using two soil samples (K7s and K8s) collected near Kafni glacier. The two libraries yielded 648 clones. Phyla *Actinobacteria*, *Bacteroidetes*, *Chlorflexi*, *Firmicutes*, *Proteobacteria*, *Spirochaetae*, *Tenericutes*, and *Verrucomicrobia* were common to two libraries. Phyla *Acidobacteria*, *Chlamydiae*, and *Nitrospirae* were present only in KF library whereas *Lentisphaerae* in the KS only. In the two libraries, clones belonging to phyla *Bacteroidetes* and *Proteobacteria* were the most predominant.

Shivaji et al. (2006) isolated four novel bacterial strains from cryogenic tubes used to collect air samples at altitudes 24, 28, and 41 km. The four strains 24KT, 28KT, 41KF2aT, and 41KF2bT were identified as members of the genus *Bacillus.* Based on phenotypic chemotaxonomic characteristics, phylogenetic analysis based

Fig. 7.6 Rookund glacier (Google image)

Fig. 7.7 Kafni glacier (Google image)

on 16S rRNA gene sequencing and DNA-DNA hybridization studies, the four strains are proposed as new species *Bacillus aerius* sp. nov., *Bacillus aerophilus* sp. nov., *Bacillus stratosphericus* sp. nov. and *Bacillus altitudinis* sp. nov.

Shivaji et al. (2009) isolated three novel bacterial strains, PVAS-1T, B3W22T, and B8W22T from cryotubes used to collect air samples at altitudes between 27 and 41 km. Based on phenotypic characteristics, chemotaxonomic features, DNA-DNA hybridization with the nearest phylogenetic neighbors, and phylogenetic analysis based on 16S rRNA gene sequences, the three strains were identified as representing

Fig. 7.8 Pangong lake (Google image)

novel species as *Janibacter hoylei* sp. nov. (type strain PVAS-1T = MTCC8307T = DSM 21601T = CCUG56714T), *Bacillus isronensis* sp. nov. (type strain B3W22T = MTCC7902 = JCM13838T) and *Bacillus aryabhattai* sp. nov. (type strain B8W22T = MTCC775ST = JCM 13839T).

Reddy et al. (2008a, b) isolated a bacterial strain PON10T from the Pindari glacier of the Indian Himalayas. The strain is a yellow-pigmented, Gram-positive, aerobic, motile rod-shaped bacterium. Based on the phenotypic and chemotaxonomic characteristics, the strain PON10T was assigned to the genus *Leifsonia.* The strain differed from existing species of *Leifsonia* and was proposed as a novel species as *Leifsonia pindariensis* sp. nov.

Polkade et al. (2015) have reported psychrophilic bacteria from water of a high-altitude (4350 m) saline lake, Pangong lake (Fig. 7.8) located in Western Himalayas of India extending to Tibet in China. The organisms were aerobic, Gram-stain-negative, capable of growing at low temperature (15 °C), in presence of 0–2% NaCl and at pH 6.5–9.0. Cellular fatty acid analysis was performed. Predominant polar lipids were phosphatidylethanolamine, phosphatidylglycerol, and diphosphatidylglycerol. The organisms were found to belong to genus *Rufibacter* but based on morphological, chemotaxonomical, and molecular characteristics, a new species was proposed as *Rufibacter immobilis* sp. nov.

Mehta et al. (2021) have reported genome sequencing and analysis of a psychrotrophic methanogen *Methanosarcina* sp. nov. MSH10X1 isolated from sub-seafloor methane hydrate deposits of Krishna Godavari Basin on India's east coast. The strain could grow from 5 to 40 °C and follows methylotrophic, acetoclastic, and hydrogenotrophic modes of methanogenesis using different substrates. The KEGG analysis highlighted the presence of genes responsible for all three modes of methanogenesis.

7.7.2 Biotechnology

Researchers from India have reviewed work on biotechnological and industrial applications of psychrophiles.

Ramana et al. (2000) have described biotechnological applications of psychrophiles. Due to their unique biochemical features, the psychrophiles can be used in manufacturing food enzymes and value-added pharmaceutical products. Recent developments show that psychrophiles are good source of novel catalysts of industrial interest. Some of the enzymes have been isolated, and the genes are successfully cloned and expressed in target hosts. Biopolymer degrading enzymes like amylases, pullulanases, xylanases, and proteases play an important role in food, detergent, and pulp industries. Cell membranes of psychrophiles contain surfactants bearing unique stability at low temperatures and that can be used in pharmaceutical formulations.

Kasana (2010) has reviewed proteases from psychrotrophs. Although proteases can be obtained from animals and plants, microorganisms are preferred source for industrial applications. Among various groups of microorganisms, psychrotrophs become ideal candidates for enzyme production since proteases from psychrotrophs are active at low temperature and stable under alkaline conditions. Thus, they have application as laundry detergent additive and also in environmental bioremediation, food industry, and in molecular diagnostics. The author has focused on aspects as production, optimization, molecular characteristics, applications, substrate specificity, and crystal structure of psychrotrophic proteases.

Furhan (2020) has reviewed production and biotechnological potential of cold-adapted proteases from psychrophiles and psychrotrophs. Proteases are important hydrolytic enzymes hydrolyzing large proteins into smaller peptides. The cold-adapted proteases have high catalytic capacity at low temperatures. Thus, interest in cold-adapted microorganisms producing cold-adapted proteases is rising day by day. Psychrophiles and psychrotrophs inhabiting permanently or partially cold environments appear as rich sources of cold-adapted proteases. The author (Furhan 2020) has also reviewed recent advances in cold-adapted protease gene cloning and structural studies and prospective applications of cold-adapted proteases.

Annapure and Nair (2022) have reviewed psychrozymes as a novel and promising resource for industrial applications. The Earth's biosphere is dominated by low-temperature regions, which include deep oceans, alpine reaches, and polar regions. The microorganisms that inhabit them, the psychrophiles are abundant in terms of biomass, diversity, and distribution. These indigenous psychrophilic microorganisms provide an enormous resource of enzymes (psychrozymes) that function effectively in the cold. Such enzymes constitute excellent biocatalysts owing to their high catalytic activity at temperatures below 25 °C, eliminating the need for heating processes that hamper the quality, sustainability, and cost-effectiveness of industrial production. This also minimizes the likelihood of chemical side reactions that generally occur at higher temperatures. Cold-adapted xylanases characterized from the deep-sea psychrophilic bacterium *Flammeovirga pacifica* can be used as an additive to wheat flour to improve dough rheology and quality of baked products. Likewise, cold-adapted proteases can be used in detergents that allow the washing of textiles at room temperature.

Shivaji et al. (2011a, b) used cell-free culture supernatants of five psychrophilic bacteria *Pseudomonas antarctica*, *Pseudomonas proteolytic*, *Pseudomonas meridiana*, *Arthrobacter kerquelensis*, and *Arthrobacter gangotriensis* to synthesize silver nanoparticles (AgNPs). The sizes of the AgNPs ranged from 6 to 13 nm and were stable for 8 months in dark. The synthesis and stability appeared to depend on the temperature, pH, and bacterial species. It was observed that *A. kerquelensis* supernatant could not produce AgNPs at the temperature where the *P. antarctica* could synthesize AgNPs.

7.7.3 Role of Psychrophiles in Sustainability

The psychrophilic microorganisms play an important role in bioremediation of waste disposal sites at high altitude, for example, in Ladakh. Some psychrophiles produce plant-growth promoting substance and thus enhance agriculture productivity. Bioremediation of contaminated sites contributes to environmental sustainability.

Farooq et al. (2016) have described role of psychrophiles in bioremediation and sustainability. Psychrophilic microorganisms have been used as a bioremediation option to treat petroleum-contaminated sites in cold regions, where it is generally thought oil spills are more damaging and that ecosystem recovery is slower than in warmer climates. Bacteria able to degrade high amounts of organic compounds within a short time at low temperatures represent a promising source of technology for accelerated wastewater treatment. Cold-adapted *Arthrobacter psychrolactophilus* strain was found to be promising to use as microbial starter for bioremediation process. The state of Jammu and Kashmir, India, has a great potential for cold-active enzymes extraction from psychrophilic microorganisms since the state experiences cold climate and has a number of glaciers.

Yadav et al. (2019) have reviewed role of psychrotrophic microbes in sustainable agriculture. The extreme cold environments harbor novel psychrotrophic microbes which have been reported as plant growth promoters and biocontrol agents for sustainable agriculture and in industry as cold-adapted hydrolytic enzymes. Inoculation with psychrotrophic/psychrotolerant strains significantly enhanced root/shoot biomass and nutrients uptake as compared to non-bacterized control. Psychrotrophic microorganisms play an important role in alleviation of cold stress in plants growing at high hill and low-temperature conditions.

7.7.4 Contribution to the Knowledge of Psychrophilic Microbial Species, New to Science, from India

With the advent of molecular tools, it has been possible to identify and describe new species of psychrophilic microorganisms from cold environments in India. Over

20 new species of psychrophilic microorganisms have been reported from India and are listed in Table 7.2. They are added to global microbial data collection.

Further investigations of different cold environments in India would enrich the diversity of psychrophilic microorganisms, with new species.

Table 7.2 Psychrophilic microbial species, new to science, from India

Sr. No.	Name of the organism	Source	Reference
1	*Pedobacter himalayensis* sp. nov. strain HHS22T	Hamta glacier in Himalayan Mountain ranges	Shivaji et al. (2005a, b)
2	*Dyadobacter hamtensis* sp. nov. strain HHS11T	Hamta glacier in Himalayan Mountain ranges	Chaturvedi et al. (2005)
3	*Planococcus stackebrandtii* sp. nov. strain K22-03T	Cold desert of Himalayas	Mayilraj et al. (2005)
4	*Bacillus aerius* sp. nov. strain 24 KT *Bacillus aerophilus* sp. nov. strain 28KT *Bacillus stratosphericus* strain 41KF2aT *Bacillus altitudinis* sp. nov. strain 41KF2bT	Cryogenic tubes used to collect air samples at altitudes 24, 28, and 41 km	Shivaji et al. (2006)
5	*Exiguobacterium indicum* sp. nov. strain HHS31T	Hamta glacier of Himalayan Mountain ranges	Chaturvedi and Shivaji et al. (2006)
6	*Leifsonia pindariensis* sp. nov. strain PON10T	Pindari glacier of Indian Himalayas	Reddy et al. (2008a, b)
7	*Bacillus cecembensis* sp. nov.	Pindari glacier of Indian Himalayas	Reddy et al. (2008a, b)
8	*Janibacter hoylei* sp. nov. Type strain PVAS-1T *Bacillus isronensis* sp. nov., Type strain $B_3W_{22}T$ *Bacillus aryabhattai* sp. nov.	Cryotubes used to collect air samples at altitudes between 27 and 41 km	Shivaji et al. (2009)
9	*Paenibacillus glacialis* sp. nov. strain KFC91T	Kafni glacier of Himalayas	Kishore et al. (2010)
10	*Cryobacterium roopkundense* sp. nov. Type strain SS-33T	Roopkund in the Himalayan Mountain ranges	Reddy et al. (2010)
11	*Staphylococcus lipolyticus* sp. nov. Type strain SS-33T	Sea sediment from Bay of Bengal	Arora (2013)
12	*Paenisporosarcina indica* sp. nov. strain PN_2T	Soil near Pindari glacier in Indian Himalayas	Reddy et al. (2013)
13	*Arthrobacter gangotriensis* sp. nov. strain LZ1YT	Penguin Rookery soil near Indian Station Dakshin Gangotri in Antarctica	Shivaji et al. (2013)
14	*Cryobacterium roopkundensis* sp. nov. strain RuGI7	Soil from periphery of Roopkund Lake, Himalaya	Reddy et al. (2014)
15	*Rufibacter immobilis* sp. nov.	High-altitude (4350 m) saline lake, Pangong lake in Western Himalayas	Polkade et al. (2015)

References

Aislabie J, Saul DJ, Foght JM (2006) Bioremediation of hydrocarbon-contaminated polar soils. Extremophiles 10:171–179. https://doi.org/10.1007/s00792-005-0498-4

Ajaz M, Zargar MY, Baba ZA, Dar ZM, Malik MA, Rasool K, Maqbool S, Khan IJ, Asif M (2018) Characterization of psychrophilic bacteria involved in solid waste decomposition under temperate conditions. Int J Curr Microbiol App Sci 7(9):3788–3794. https://doi.org/10.20546/ijcmas.2018.709.468

Annapure US, Nair P (2022) Microbial extremozymes, novel sources and industrial applications. In: Chapter 14 - Psychrozymes: a novel and promising resource for industrial applications, pp 185–195. https://doi.org/10.1016/B978-0-12-822945-3.00018-X

Antony R, Krishnan KP, Laluraj CM, Thamban M, Dhakephalkar PK, Engineer AS, Shivaji S (2012) Diversity and physiology of culturable bacteria associated with a coastal Antarctic ice core. Microbiol Res 167(6):372–380

Antony R, Sanyal A, Kapse N, Dhakephalkar PK, Thamban M, Nair S (2016) Microbial communities associated with Antarctic snow pack and their biogeochemical implications. Microbiol Res 192:192–202

Araujo R, Casal M, Cavaco-Paulo A (2008) Application of enzymes for textile fibres processing. Biocatal Biotransformation 26:332–349. https://doi.org/10.1080/10242420802390457

Armstrong GA (1997) Genetics of eubacterial carotenoid biosynthesis: a colourful tale. Annu Rev Microbiol 51:629–659. https://doi.org/10.1146/annurev.micro.51.1.629

Arora PK (2013) *Staphylococcus lipolyticus* sp. nov., a new cold-adapted lipase producing marine species. Ann Microbiol 63:913–922. https://doi.org/10.1007/s13213-012-0544-2

Awasthi S, Sharma A, Saxena P, Yadav J, Pandiyan K, Kumar M, Singh A, Chakdar H, Bhowmik A, Kashyap PL, Srivastava AK (2019) Molecular detection and in silico characterization of cold shock protein coding gene (cspA) from cold adaptive Pseudomonas koreensis. J Plant Biochem Biotechnol 28(4):405–413

Cavicchioli R (2016) On the concept of a psychrophile. ISME J 10:793–795. https://doi.org/10.1038/ismej.2015.160

Chaikam V, Karlson DT (2010) Comparison of structure, function and regulation of plant cold shock domain proteins to bacterial and animal cold shock domain proteins. BMB Rep 43:1–8. https://doi.org/10.5483/BMBRep.2010.43.1.001

Chattopadhyay MK, Jagannadham MV (2001) Maintenance of membrane fluidity in Antarctic bacteria. Polar Biol 24:386–388. https://doi.org/10.1007/s003000100232

Chattopadhyay MK, Reddy GS, Shivaji S (2014) Psychrophilic bacteria: biodiversity, molecular basis of cold adaptation and biotechnological implications. Curr Biotechnol 3(1):100–116. https://doi.org/10.2174/22115501113026660039

Chaturvedi P, Shivaji S (2006) *Exiguobacterium indicum* sp. nov., a psychrophilic bacterium from the Hamta glacier of the Himalayan Mountain ranges of India. Int J Syst Evol Microbiol 56(12): 2765–2770

Chaturvedi P, Reddy GSN, Shivaji S (2005) *Dyadobacter hamtensis* sp. nov., from Hamta glacier, located in the Himalayas, India. Int J Syst Evol Microbiol 55(5):2113–2117

Chintalapati S, Kiran MD, Shivaji S (2004) Role of membrane lipid fatty acids in cold adaptation. Cell Mol Biol (Noisy-le-Grand, France) 50(5):631–642

Christner BC (2002) Incorporation of DNA and protein precursors into macromolecules by bacteria at −15°C. Appl Environ Microbiol 68:6435–6438. https://doi.org/10.1128/AEM.68.12.6435-6438.2002

Christner BC (2010) Bioprospecting for microbial products that affect ice crystal formation and growth. Appl Microbiol Biotechnol 85:481–189. https://doi.org/10.1007/s00253-009-2291-2

Deming JW (2002) Psychrophiles and polar regions. Curr Opin Microbiol 5:301–309. https://doi.org/10.1016/S1369-5274(02)00329-6

Duman JG, Olsen TM (1993) Thermal hysteresis protein activity in bacteria, fungi and phylogenetically diverse plants. Cryobiology 30:322–328. https://doi.org/10.1006/cryo.1993.1031

Farooq S, Nazir R, Uqab B (2016) Psychrophiles: their habitat and applications. J Himalayan Ecol Sustain Dev 11:27–36

Farooq S, Nazir R, Ganai BA, Mushtaq H, Dar GJ (2022) Psychrophilic and psychrotrophic bacterial diversity of Himalayan Thajwas glacial soil, India. Biologia 77:203–213. https://doi.org/10.1007/s11756-021-00915-6

Feller G (2013) Psychrophilic enzymes: from folding to function and biotechnology. Scientifica 2013:512840. https://doi.org/10.1155/2013/512840

Feller G, Gerday C (1997) Psychrophilic enzymes: molecular basis of cold adaptation. Cell Mol Life Sci 53(10):830–841. https://doi.org/10.1007/s000180050103

Feller G, Gerday C (2003) Psychrophilic enzymes: hot topics in cold adaptation. Nat Rev Microbiol 1:200–208. https://doi.org/10.1038/nrmicro773

Fong NJC, Burgess ML, Barrow KD, Glenn DR (2001) Carotenoid accumulation in the psychrotrophic bacterium *Arthrobacter agilis* in response to thermal and salt stress. Appl Microbiol Biotechnol 56:750–756. https://doi.org/10.1007/s002530100739

Furhan J (2020) Adaptation, production, and biotechnological potential of cold-adapted proteases from psychrophiles and psychrotrophs: recent overview. J Genet Eng Biotechnol 18:36. https://doi.org/10.1186/s43141-020-00053-7

Gangwar P, Alam SI, Bansod S, Singh L (2009) Bacterial diversity of soil samples from the western Himalayas, India. Can J Microbiol 55(5). https://doi.org/10.1139/W09-011

Georlette D, Blaise V, Collins T, D'Amico S, Gratia E, Hoyoux A, Marx J-C, Sonan G, Feller G, Gerday C (2004) Some like it cold: Biocatalysis at low temperatures. FEMS Microbiol Rev 28:25–42. https://doi.org/10.1016/j.femsre.2003.07.003

Gerday C, Aittaleb M, Bentahier M, Chessa JP, Claverie P, Collins T, D'Amico S, Dumont J, Garsoux G, Georlette D, Hoyoux A, Lonhienne T, Meuwis M-A, Feller G (2000) Cold-adapted enzymes: from fundamentals to biotechnology. Trends Biotechnol 18:103–107. https://doi.org/10.1016/S0167-7799(99)01413-4

Gilbert JA, Hill PJ, Dodd CE, Laybourn-Parry J (2004) Demonstration of antifreeze protein activity in Antarctic Lake bacteria. Microbiology 150:171–180. https://doi.org/10.1099/mic0.26610-0

Gounot AM (1986) Psychrophilic and psychrotrophic microorganisms. Experientia 42(11–12): 1192–1197. https://doi.org/10.1007/BF01946390

Hamdan A (2018) Psychrophiles: ecological significance and potential industrial application. S Afr J Sci 114(5–6). https://doi.org/10.17159/sajs.2018/20170254

Hassan N, Anesio AM, Rafiq M, Holtvoeth J, Bull I, Haleem A, Shah AA, Hasan F (2020) Temperature driven membrane lipid adaptation in glacial psychrophilic bacteria. Front Microbiol 11:824. https://doi.org/10.3389/fmicb.2020.00824

Hebraud M, Potier P (1999) Cold shock response and low temperature adaptation in psychrotrophic bacteria. J Mol Microbiol Biotechnol 1:211–219

Hoyoux A, Jennes I, Dubois P, Genicot S, Dubail F, Francois JM, Baise E, Feller G, Gerday C (2001) Cold-adapted β-galactosidase from the Antarctic psychrophile *Pseudoalteromonas haloplanktis*. Appl Environ Microbiol 67:1529–1535. https://doi.org/10.1128/AEM.67.4.1529-1535.2001

https://serc.carleton.edu/microbelife/extreme/cold/index.html

Inniss WE (1975) Interaction of temperature and psychrophilic microorganisms. Annu Rev Microbiol 29:445–466. https://doi.org/10.1146/annurev.mi.29.100175.002305

Jadhav V, Jamle M, Pawar P, Devare M, Bhadekar R (2010) Fatty acid profiles of PUFA producing Antarctic bacteria: Correlation with RaPd analysis. Ann Microbiol 60:693–699. https://doi.org/10.1007/s13213-010-0114-4

Jadhav VV, Pote SS, Yadav A, Shouche YS, Bhadekar RK (2013) Extracellular cold active lipase from the psychrotrophic *Halomonas* sp. BRI 8 isolated from the Antarctic Sea water, Songklanakarin. J Sci Technol 35(6):623–630

Joshi D, Kumar S, Suyal DC, Goel R (2017) The Microbiome of the Himalayan ecosystem. In: Kalia V, Shouche Y, Purohit H, Rahi P (eds) Mining of microbial wealth and meta genomics. Springer, Singapore, pp 110–116

Junge K, Cameron K, Nunn B (2019) Chapter 12 - Diversity of psychrophilic bacteria in sea and glacier ice environments—insights through genomics, metagenomics, and proteomics approaches. In: Microbial diversity in the genomic era. Academic Press, pp 197–216. https://doi.org/10.1016/B978-0-12-814849-5.00012-5

Kapse N, Singh P, Roy U, Singh SM, Dhakephalkar PK (2017) Insights into the psychrophilic and sea ice-specific lifestyle of *Marinobacter* sp. strain AC-23: a genomic approach. Genome Announc 5(15):e00134–e00117

Kasana RC (2010) Proteases from psychrotrophs: an overview. Crit Rev Microbiol 36(2):134–145. https://doi.org/10.3109/10408410903485525

Kasana RC, Gulati A (2011) Cellulases from psychrophilic microorganisms: a review. J Basic Microbiol 51(6):572–579. https://doi.org/10.1002/jobm.201000385

Kishore KH, Begum Z, Pathan AAK, Shivaji S (2010) *Paenibacillus glacialis* sp. nov., isolated from the Kafni glacier of the Himalayas, India. Int J Syst Evol Microbiol 60(8):1909–1913

Ko R, Smith LT, Smith GM (1994) Glycine betaine confers enhanced osmotolerance and cryotolerance on *Listeria monocytogenes*. J Bacteriol 176:426–431. https://doi.org/10.1128/jb.176.2.426-431.1994

Krembs C, Eicken H, Junge K, Deming JW (2002) High concentrations of exopolymeric substances in Arctic winter sea ice: Implications for the polar ocean carbon cycle and cryoprotection of diatoms. Deep Sea Res 49:2163–2181. https://doi.org/10.1016/S0967-0637(02)00122-X

Kruse W (1910) Allegemeine Mikrobiologie. FCW Vogel

Kumar S (2018) Documentation of the bacterial and diazotrophic diversity from Garhwal Himalaya through culturable and unculturable approaches. G.B.P.U.A.&T, Pantnagar

Kumar S, Suyal DC, Bhoriyal M, Goel R (2018) Plant growth promoting potential of psychrotolerant *Dyadobacter* sp. for pulses and finger millet and impact of inoculation on soil chemical properties and diazotrophic abundance. J Plant Nutr 41(8):1035–1046. https://doi.org/10.1080/01904167.2018.1433211

Kumar S, Suyal DC, Yadav A, Shouche Y, Goel R (2019) Microbial diversity and soil physiochemical characteristic of higher altitude. PLoS One 14(3):e0213844

Kumar S, Suyal DC, Yadav A, Shouche Y, Goel R (2020) Psychrophilic *Pseudomonas helmanticensis* proteome under simulated cold stress. J Cell Stress Chaper 25(6):1025–1032. https://doi.org/10.1007/s12192-020-01139-4

Lin X, Yang B, Shen J, Du N (2009) Biodegradation of crude oil by an Arctic psychrotrophic bacterium *Pseudoalteromomas* sp. P29. Curr Microbiol 59:341–345. https://doi.org/10.1007/s00284-009-9440-9

Maiangwa J, Ali M, Salleh A, Abd RR, Shariff F, Leow T (2015) Adaptational properties and applications of cold-active lipases from psychrophilic bacteria. Extremophiles 19:235–247. https://doi.org/10.1007/s00792-014-0710-5

Mancuso NCA, Guezennec J, Bowman JP (2005) Bacterial exopolysaccharides from extreme marine environments with special consideration of the Southern Ocean, sea ice, and deep-sea hydrothermal vents: a review. Mar Biotechnol 7:253–271. https://doi.org/10.1007/s10126-004-5118-2

Margesin R, Schinner F (2001) Biodegradation and bioremediation of hydrocarbons in extreme environments. Appl Microbiol Biotechnol 56:650–663. https://doi.org/10.1007/s002530100701

Mayilraj S, Prasad GS, Suresh K, Saini HS, Shivaji S, Chakrabarti T (2005) *Planococcus stackebrandtii* sp. nov., isolated from a cold desert of the Himalayas, India. Int J Syst Evol Microbiol 55(1):91–94

Mehta P, Deshmukh K, Dagar SS, Dhakephalkar PK, Lanjekar VB (2021) Genome sequencing and analysis of a psychrotrophic methanogen *Methanosarcina* sp. nov. MSH10X1 cultured from methane hydrate deposits of Krishna Godavari Basin of India. Mar Genomics 100864

Miteva V (2008) Bacteria in snow and glacier ice. In: Margesin R, Schinner F, Marx J-C, Gerday C (eds) Psychrophiles: from biodiversity to biotechnology. Springer, Berlin, pp 31–50. https://doi.org/10.1007/978-3-540-74335-4_3

Morita RY (1975) Psychrophilic bacteria. Bacteriol Rev 39(2):144–167

Morita RY, Moyer CL (2001) Psychrophiles. In: Encyclopedia of biodiversity, 2nd edn, pp 298–303. https://doi.org/10.1016/B978-0-12-384719-5.00176-3

Moyer CL, Eric R, Richard C, Morita Y (2017) Psychrophiles and psychrotrophs. In: Reference module in life sciences. https://doi.org/10.1016/B978-0-12-809633-8.02282-2

Muller M (1903) Uber das Wachstum und die Lebenstatigkeit von Bakterien sowie den Ablauf fermentativer Prozesse bei niederen Temperaturen unter spezieller Berutcksichtigung des Fleisches als Nahrungsmittel. Arch Hyg 47:127–193

Muryoi N, Sato M, Kaneko S, Kawahara H, Obata H, Yaish MW, Griffith M, Glick BR (2004) Cloning and expression of *afpA,* a gene encoding an antifreeze protein from the arctic plant growth-promoting rhizobacterium *Pseudomonas putida* GR12-2. J Bacteriol 186:5661–5671. https://doi.org/10.1128/JB.186.17.5661-5671.2004

Nam E, Ahn J (2011) Antarctic marine bacterium *Pseudoalteromonas* sp. kncuc808 as a source of cold adapted lactose hydrolyzing enzyme. Braz J Microbiol 42:927–936

Pandey A, Jain R, Sharma A, Dhakar K, Kaira GS, Rahi P, Dhyani A, Pandey N, Adhikari P, Shouche YS (2019) 16S rRNA gene sequencing and MALDI-TOF mass spectrometry based comparative assessment and bioprospection of psychrotolerant bacteria isolated from high altitudes under mountain ecosystem. SN Appl Sci 1(3):1–12

Parvizpour S, Hussin N, Shamsir MS, Razmara J (2021) Psychrophilic enzymes: structural adaptation, pharmaceutical and industrial applications. Appl Microbiol Biotechnol 105(3):899–907. https://doi.org/10.1007/s00253-020-11074-0

Phadtare S (2004) Recent developments in bacterial cold-shock response. Curr Issues Mol Biol 6:125–136

Phadtare S (2011) Unwinding activity of cold shock proteins and RNA metabolism. RNA Biol 8:394–397. https://doi.org/10.4161/rna.8.3.14823

Polkade AV, Ramana VV, Joshi A, Pardesi L, Shouche YS (2015) *Rufibacter immobilis* sp. nov. isolated from a high-altitude saline lake. Int J Syst Evol Microbiol 65(Pt5):1592–1597. https://doi.org/10.1099/ijs.0.000144

Pote S, Chaudhary Y, Upadhayay S, Tale V, Walujkar S, Shouche Y, Bhadekar R (2014) Identification and biotechnological potential of psychrotrophic marine isolates. Eur Asian J Bio Sci 8:51–60

Pradhan S, Srinivas TNR, Pindi PK, Kishore KH, Begum Z, Singh PK, Singh AK, Pratibha MS, Yasala AK, Reddy GSN, Shivaji S (2010) Bacterial biodiversity from Roopkund glacier, Himalayan Mountain ranges, India. Extremophiles 14(4):377–395

Ramana KV, Singh L, Dhaked RK (2000) Biotechnological application of psychrophiles and their habitat to low-temperature. J Sci Ind Res 59(2):87–101

Ray MK, Kumar GS, Shivaji S (1994) Phosphorylation of lipopolysaccharides in the Antarctic psychrotroph Pseudomonas syringae: a possible role in temperature adaptation. J Bacteriol 176(14):4243–4249

Reddy GS, Aggarwal RK, Matsumoto GI, Shivaji S (2000) *Arthrobacter flavus* sp. nov., a psychrophilic bacterium isolated from a pond in McMurdo Dry Valley, Antarctica. Int J Syst Evol Microbiol 50(4):1553–1561

Reddy G, Raghavan P, Sarita N, Prakash J, Nagesh N, Delille D, Shivaji S (2003) *Halomonas glaciei* sp. nov. isolated from fast ice of Adelie Land, Antarctica. Extremophiles 7(1):55–61

Reddy GSN, Matsumoto GI, Schumann P, Stackebrandt E, Shivaji S (2004) Psychrophilic pseudomonads from Antarctica: *Pseudomonas antarctica* sp. nov., *Pseudomonas meridiana* sp. nov. and *Pseudomonas proteolytica* sp. nov. Int J Syst Evol Microbiol 54(3):713–719

Reddy GSN, Prabagaran SR, Shivaji S (2008a) *Leifsonia pindariensis* sp. nov., isolated from the Pindari glacier of the Indian Himalayas, and emended description of the genus *Leifsonia*. Int J Syst Evol Microbiol 58(9):2229–2234

Reddy GSN, Uttam A, Shivaji S (2008b) *Bacillus cecembensis* sp. nov., isolated from the Pindari glacier of the Indian Himalayas. Int J Syst Evol Microbiol 58(10):2330–2335

Reddy GSN, Pradhan S, Manorama R, Shivaji S (2010) *Cryobacterium roopkundense* sp. nov., a psychrophilic bacterium isolated from glacial soil. Int J Syst Evol Microbiol 60(4):866–870

Reddy GSN, Manasa BP, Singh SK, Shivaji S (2013) *Paenisporosarcina indica* sp. nov., a psychrophilic bacterium from a glacier, and reclassification of *Sporosarcina antarctica* Yu et al., 2008 as *Paenisporosarcina antarctica* comb. nov. and emended description of the genus *Paenisporosarcina*. Int J Syst Evol Microbiol 63(Pt_8):2927–2933

Reddy GS, Sreenivas A, Shivaji S (2014) Draft genome sequence of *Cryobacterium roopkundensis* strain RuGl7, isolated from a soil sample in the vicinity of Roopkund Lake, Himalayas, India. Genome Announc 2(6):e01206–e01214

Ruberto L, Vazquez S, Lobalbo A, Mac CW (2005) Psychrotolerant hydrocarbon-degrading *Rhodococcus* strains isolated from polluted Antarctic soils. Antarct Sci 17:47–56. https://doi.org/10.1017/S0954102005002415

Russell NJ (2008) Psychrophily and resistance to low temperature. In: Extremophiles – vol II, encyclopedia of life support systems (EOLSS), pp 1–34

Sahay H, Babu BK, Singh S, Kaushik R, Saxena AK, Arora DK (2012) Cold-active hydrolases producing bacteria from two different sub-glacial Himalayan lakes. J Basic Microbiol 52:1–13

Samie N, Noghabi K, Gharegozloo Z, Zahiri H, Ahmadian G, Sharafi H, Behrozi R, Hojatollah V (2012) Psychrophilic α-amylase from *Aeromonas veronii* NS07 isolated from farm soils. Process Biochem 47:1381–1387. https://doi.org/10.1016/j.procbio.2012.05.007

Sharma A, Jani K, Feng G-D, Karodi P, Vemuluri VR, Zhu H-H, Shivaji S, Thite V, Kajale S, Rahi P, Shouche Y (2018) *Subsaxibacter sediminis* sp. nov., isolated from Arctic glacial sediment and emended description of the genus *Subsaxibacter*. Int J Syst Evol Microbiol 68(5):1678–1682

Shivaji S, Prakash J (2010) How do bacteria sense and respond to low temperature? Arch Microbiol 192:85–95. https://doi.org/10.1007/s00203-009-0539-y

Shivaji S, Rao NS, Saisree L, Sheth V, Reddy GSN, Bhargava PM (1988) Isolation and identification of *Micrococcus roseus* and *Planococcus* sp. from schirmacher oasis, Antarctica. J Biosci 13(4):409–414

Shivaji S, Ray MK, Rao NS, Saisree L, Jagannadham MV, Kumar GS, Reddy GSN, Bhargava PM (1992) *Sphingobacterium antarcticus* sp. nov., a psychrotrophic bacterium from the soils of Schirmacher Oasis, Antarctica. Int J Syst Evol Microbiol 42(1):102–106

Shivaji S, Chaturvedi P, Reddy GSN, Suresh K (2005a) *Pedobacter himalayensis* sp. nov., from the Hamta glacier located in the Himalayan Mountain ranges of India. Int J Syst Evol Microbiol 55(3):1083–1088

Shivaji S, Reddy GSN, Suresh K, Gupta P, Chintalapati S, Schumann P, Stackebrandt E, Matsumoto GI (2005b) *Psychrobacter vallis* sp. nov. and *Psychrobacter aquaticus* sp. nov., from Antarctica. Int J Syst Evol Microbiol 55(2):757–762

Shivaji S, Chaturvedi P, Suresh K, Reddy GSN, Dutt CBS, Wainwright M, Narlikar JV, Bhargava PM (2006) *Bacillus aerius* sp. nov., *Bacillus aerophilus* sp. nov., *bacillus stratosphericus* sp. nov. and *Bacillus altitudinis* sp. nov., isolated from cryogenic tubes used for collecting air samples from high altitudes. Int J Syst Evol Microbiol 56(7):1465–1473

Shivaji S, Chaturvedi P, Begum Z, Pindi PK, Manorama R, Padmanaban DA, Shouche YS, Pawar S, Vaishampayan P, Dutt CBS, Datta GN, Manchanda RK, Rao UR, Bhargava PM, Narlikar JV (2009) *Janibacter hoylei* sp. nov., *Bacillus isronensis* sp. nov. and *Bacillus aryabhattai* sp. nov., isolated from cryotubes used for collecting air from the upper atmosphere. Int J Syst Evol Microbiol 59(12):2977–2986

Shivaji S, Madhu S, Singh S (2011a) Extracellular synthesis of antibacterial silver nanoparticles using psychrophilic bacteria. Process Biochem 46(9):1800–1807

Shivaji S, Pratibha MS, Sailaja B, Kishore KH, Singh AK, Begum Z, Anarasi U, Prabagaran SR, Reddy GSN, Srinivas TNR (2011b) Bacterial diversity of soil in the vicinity of Pindari glacier,

Himalayan Mountain ranges, India, using culturable bacteria and soil 16S rRNA gene clones. Extremophiles 15(1):1–22

Shivaji S, Reddy PVV, Rao SSSN, Begum Z, Manasa P, Srinivas TNR (2012) *Cyclobacterium qasimii* sp. nov., a psychrotolerant bacterium isolated from Arctic marine sediment. Int J Syst Evol Microbiol 62(Pt_9):2133–2139

Shivaji S, Ara S, Bandi S, Singh A, Pinnaka AK (2013) Draft genome sequence of *Arthrobacter gangotriensis* strain Lz1yT, isolated from a penguin rookery soil sample collected in Antarctica, near the Indian Station Dakshin Gangotri. Genome Announc 1(3):e00347–e00313

Singh P, Singh SM, Dhakephalkar P (2014) Diversity, cold active enzymes and adaptation strategies of bacteria inhabiting glacier cryoconite holes of High Arctic. Extremophiles 18(2): 229–242

Singh P, Kapse N, Arora P, Singh SM, Dhakephalkar PK (2015) Draft genome of *Cryobacterium* sp. MLB-32, an obligate psychrophile from glacier cryoconite holes of high Arctic. Mar Genomics 21:25–26

Singh P, Kapse N, Roy U, Singh SM, Dhakephalkar PK (2017) Draft genome sequence of permafrost bacterium *Nesterenkonia* sp. strain PF2B19, revealing a cold adaptation strategy and diverse biotechnological potential. Genome Announc 5(15):e00133–e00117

Soni R, Suyal DC, Agrawal K, Yadav A, Shouche Y, Goel R (2015) Differential proteomic expression of Himalayan psychrotrophic diazotroph *Pseudomonas palleroniana* N26 under low temperature diazotrophic conditions. CryoLetters 36(2):74–82

Srinivas TNR, Singh SM, Pradhan S, Pratibha MS, Kishore KH, Singh AK, Begum Z, Prabagaran SR, Reddy GSN, Shivaji S (2011) Comparison of bacterial diversity in proglacial soil from Kafni Glacier, Himalayan Mountain ranges, India, with the bacterial diversity of other glaciers in the world. Extremophiles 15(6):673–690

Staley JT, Gosink JJ (1999) Poles apart: biodiversity and biogeography of sea ice bacteria. Annu Rev Microbiol 53:189–215. https://doi.org/10.1146/annurev.micro.53.1.189

Struvay C, Feller G (2012) Optimization to low temperature activity in psychrophilic enzymes. Int J Mol Sci 13:11643–11665. https://doi.org/10.3390/ijms130911643

Suyal DC, Yadav A, Shouche Y, Goel R (2014) Differential proteomics in response to low temperature diazotrophy of Himalayan psychrophilic nitrogen fixing Pseudomonas migulae S10724 strain. Curr Microbiol 68(4):543–550

Suyal DC, Kumar S, Yadav A, Shouche Y, Goel R (2017) Cold stress and nitrogen deficiency affected protein expression of psychrotrophic *Dyadobacter psychrophilus* B2 and *Pseudomonas jessenii* MP1. Front Microbiol 8:430

Suyal DC, Joshi D, Kumar S, Soni R, Goel R (2019) Differential protein profiling of soil diazotroph *Rhodococcus qingshengii* S10107 towards low-temperature and nitrogen deficiency. Sci Rep 9 (1):1–9

Tomova I, Stoilova-Disheva M, Lazarkevich I, Vasileva-Tonkova E (2015) Antimicrobial activity and resistance to heavy metals and antibiotics of heterotrophic bacteria isolated from sediment and soil samples collected from two Antarctic islands. Front Life Sci 8:348–357. https://doi.org/10.1080/21553769.2015.1044130

Upadhyay G (2016) Psychrophilic pathogens: potential risk for food borne illness. Res J Recent Sci 5(7):50–52

Weber MH, Marahiel MA (2003) Bacterial cold shock responses. Sci Progr 86:9–75. https://doi.org/10.3184/003685003783238707

Whyte LG, Bourbonnière L, Bellerose C, Greer CW (1999) Bioremediation assessment of hydrocarbon-contaminated soils from the high Arctic. Biorem J 3:69–79. https://doi.org/10.1080/10889869991219217

Yadav AN, Yadav N, Sachan SG, Saxena AK (2019) Biodiversity of psychrotrophic microbes and their biotechnological applications. J Appl Biol Biotechnol 7(04):99–108. https://doi.org/10.7324/JABB.2019.70415

Yamashita Y, Nakamura N, Omiya K, Nishikawa J, Kawahara H, Obata H (2002) Identification of an antifreeze lipoprotein from *Moraxella* sp. of Antarctic origin. Biosci Biotechnol Biochem 66:239–247. https://doi.org/10.1271/bbb.66.239

Zhang X, Ma X, Wang N, Yao T (2008) New subgroup of *Bacteroidetes* and diverse microorganisms in Tibetan plateau glacial ice provide a biological record of environmental conditions. FEMS Microbiol Ecol 67:21–29. https://doi.org/10.1111/j.1574-6941.2008.00604.x

Zhou Z, Qing J, Jinlai M, Fangming L (2008) Antarctic psychrophile bacteria screening for oil degradation and their degrading characteristics. Mar Sci Bull 10:50–57

Radiophilic, Radioresistant, and Radiotolerant Microorganisms

8

Abstract

Radiophilic microorganisms exhibit an extraordinary ability to withstand the lethal and mutagenic effects of ionizing radiations. *Deinococcus radiodurans* is the bacterial species which is highly resistant to radiation levels up to 15,000 GY of acute ionizing radiation and 60 GY/h of chronic radiation (GY is Gray and 1 GY = 100 rad, rad is radiation absorbed dose). The organism is found in radiation contaminated sites, hot springs, deserts, and oceans. The organism also exhibits resistance to desiccation, hypertonic stress, and oxidation. The highly efficient radiation-protection mechanisms include self-repair of DNA damage, efficient cellular damage clearance mechanisms like hydrolysis of damaged proteins, overexpression of repair proteins, and effective clearance of reactive oxygen species.

In India, different species of *Deinococcus* have been described from marine fish, arsenic-contaminated aquifer. In addition to *Deinococcus*, nitrogen-fixing cyanobacterium *Anabaena* sp., *Micrococcus*, *Bacillus*, and *Actinobacteria* have been reported as radiation-resistant bacteria. Four novel species of radiation-resistant bacteria, new to Science, have been reported from India.

Keywords

Radiation resistant · GY · *Deinococcus radiodurans* · Self-repair of DNA damage · Hydrolysis of damaged proteins · Radiation-contaminated sites · Arsenic-contaminated aquifer · Desiccation resistant

Resistance to radiation was first discovered by scientists who were interested in food sterilization. After multiple experiments using high doses of radiation, certain bacterial communities persisted, one which was named *Deinococcus radiodurans*. Bacteria of the genus *Deinococcus* exhibit an extraordinary ability to withstand the

P. P. Kanekar, S. P. Kanekar, *Diversity and Biotechnology of Extremophilic Microorganisms from India*, Microorganisms for Sustainability 41,
https://doi.org/10.1007/978-981-19-1573-4_8

lethal and mutagenic effects of DNA-damaging agents of particularly the effects of ionizing radiation. These bacteria are the most DNA damage-tolerant organisms ever identified. These unusual bacteria are very highly resistant to what scientists had considered to be absolutely lethal radiation levels. *Deinococcus radiodurans* is remarkable for its ability to withstand radiation levels 100–1000 times higher than that which would completely debilitate any human or Earth (Battista 1997). It has been discovered that *Deinococcus* is as resistant to complete dehydration as it is to radiation and shows similar response to extremely dry conditions as in highly radioactive environments. In facts, researchers believe the resistance of *Deinococcus* to radiation is only incidental to the discovery, and development of radiation emitting technology and that dehydration resistance may have evolutionary preceded radiation resistance. (Battista 1997).

After the Chernobyl Nuclear Power Plant (ChNPP) accident and the negative ecological aftereffects on life in this area, researchers were interested in radiation-resistant bacteria as the most dynamic model of the given ecosystem and mechanism which provide resistance to bacteria against ionizing radiation. The extreme radiation-resistant bacteria are represented by the *Deinococcus* species which form a group phylogenetically close to the line *Thermus-Meiothermus*. Other radioresistant bacteria include members of the genera *Rubrobacter*, *Methylobacterium*, *Kokuria*, *Bacillus*, and some archaebacteria. In a number of cases, it is difficult to distinguish their natural habitats. They are isolated from the samples which were previously exposed to X-ray or gamma irradiation or from the ecosystems having naturally raised radioactivity (Romanovskaia et al. 2000). The authors have reviewed the mechanism of action of various species of radiation on cells and macromolecules, physiological sign of the cell damage caused by radiation. The extreme resistance of *D. radiodurans* to the DNA-damaging factor is defined by 1) repair mechanisms which fundamentally differ from those in other prokaryotes; (2) ability to increase efficiency of a standard set of the DNA-repairing proteins. The ecological consequences of the ChNPP accident for soil bacteria and recession of microbial diversity under the effect of anthropogenic radiation was shown.

8.1 Mechanisms of Radiation Resistance

Jin et al. (2019) have reviewed radiation resistance mechanisms of *Deinococcus*. *Deinococcus* is an extremophilic microorganism found in wide range of habitats including radiation-contaminated sites, hot springs, Antarctic soils, deserts, and oceans. It exhibits some of the highest level of resistance to ionizing radiation known in nature tolerating up to 15,000 GY of acute ionizing radiation and 60 Gy/h of chronic radiation (Daly 2000). In addition, the resistance of *Deinococcus* to drought and hypertonic stress in relatively high. The highly efficient radiation-protection mechanisms of *Deinococcus* depend upon combination of passive and active defense mechanisms, and include self-repair of DNA damage, efficient cellular damage clearance mechanisms such as hydrolysis of damaged proteins, and overexpression of repair proteins and effective clearance of reactive oxygen species (ROS). Thus, *Deinococcus* cells are highly resistant to oxidation, radiation, and desiccation. Some of the mechanisms of radiation resistance are briefed by the authors as follows.

Shashidhar et al. (2010) have evaluated role of enzymatic and nonenzymatic antioxidant systems in the radiation resistance of *Deinococcus*. Antioxidant enzymes and antioxidant metabolitics appear to have different roles in the oxidative stress resistance responses of radiation-resistant bacteria belonging to the *Deinococcus-Thermus* group.

8.2 Efficient Repair of DNA Damages

The multiple copies of the genome of *Deinococcus* enable efficient repair of double-strand breaks by homologous recombination. It uses normal and intact homologous DNA, both of which are double-stranded DNA molecules. In extended synthesis-dependent annealing under ionizing radiation, the genome of *D. radiodurans* breaks into a large number of DNA fragments which are partially homologous to each other and can be used to synthesized new DNA strands, after which long linear DNA intermediates are transformed into circular genome replication intermediates to complete the repair.

8.3 Excision Repair

Excision repair includes base excision repair (BER) and nucleotide excision repair (NER). BER corrects small sessions in the DNA double helix structure caused by spontaneous decay, deamination, oxidation, or methylation of DNA. BER is initiated by DNA glycosylases, which cleave the bonds between deoxyribose and modified or mismatched DNA bases. These enzymes initiate base excision and repair of a large number of base lesions, each of which is recognized by one or more overlapping specific DNA glycosylases. The DNA glycosylases recognize and remove damaged bases, leaving a base-free site that is further processed by short or long patch repair which involves different proteins (Krokan and Bjoras 2013).

8.4 Efficient Cellular Mechanisms

Hydrolysis of damaged proteins – protein degradation includes different cellular responses to environmental stimuli and removal of potentially toxic damaged proteins or protein aggregates. ATP-dependent proteases play an important role in various stress reactions, enabling bacteria to survive DNA damage, heat shock, or ROS. Under stress conditions, the proteolytic activity of *Deinococcus* cells is greatly enhanced, which is beneficial to the removal of damaged and misfolded proteins.

8.5 Removal of ROS

Reactive Oxygen Species (ROS) are by-products of water irradiation and in an extreme atmosphere more ROS is produced in cells. Free oxygen radicals can destroy DNA, RNA, and proteins, thus threatening fundamental cellular processes and survival. The genome of *Deinococcus* encodes two peroxidases, three catalases, four superoxide dismutases, and two Dps (DNA protection during starvation) proteins. (Slade and Radman 2011; Zeng et al. 2017). Together, their activity against ROS is very high. DPS is a functional protective protein that blinds to Fe^{2+} and oxidizes it to Fe^{3+} avoiding production of excess radicals. It has also a high binding affinity for DNA thus preventing hydroxyl radicals from destroying DNA.

Kotiranta et al. (1999) studied radiation sensitivity of *Bacillus cereus* with and without crystalline surface protein layer. The radiation sensitivity of four strains of *Bacillus cereus* was investigated with attention to bacterial surface structure. All four strains were sensitive to radiation with gamma rays ($D_{10} = 0.4$ kGy). No crystalline surface protein layer could be detected on the cell surface. When cultured on solid media, an s-layer covered the cell of the two strains, and they were 2.6 times as resistant to radiation as the two reference strains without an s-layer. Electron microscopy, SDS-PAGE, Western blot, and fluorescent antibody staining indicated that the higher resistance to radiation of the critical strains from plate cultures was associated with the presence of the S-layer on the cell surface.

Gabani and Singh (2012) have reviewed the strategies by which microorganisms thrive in extreme radiation environments and have discussed the potential uses in biotechnology and the therapeutic industry. Some microorganisms thrive under different types of radiation due to defensive mechanisms provided by primary and secondary metabolic products such as extremolytes and extremozymes. Extremolytes including scytonemin, mycosporine-like amino acids, shinorine, porphyra-334, playthine, biopterin, and phlorotannin are able to absorb a wide spectrum of radiation while protecting the organism's DNA from being damaged. The possible commercial applications of extremolytes include anticancer drugs, antioxidants, cell-cycle-blocking agents, and sunscreens.

8.6 Radiophiles, Radioresistant and Radiotolerant Microorganisms Reported from India

Radiophilic microorganisms have been isolated from different habitats, ecological niches, fishes, prawns, marine sediment samples, ore samples, microbial mat deposits around hot springs, wetlands, radon-contaminated groundwater, irradiation cell pool water, etc. They have been listed in Table 8.1.

Some of the investigations made by researchers from India are described under diversity, biotechnology, and sustainability.

Table 8.1 Radiophiles, radioresistant and radiotolerant microorganisms reported from India

S. No.	Name of the organism/group	Source	Diversity/ biotechnology	Reference
1	*Deinococcus indicus* sp. nov.	Arsenic-contaminated aquifer in Chakdah district of West Bengal	Arsenic- and radiation-resistant	Suresh et al. (2004)
2	*Deinococcus mumbaiensis* sp. nov.	Contaminated glucose yeast agar plate in laboratory, Mumbai, Maharashtra	Radiation-resistant	Shashidhar and Bandekar (2006)
3	*Deinococcus piscis* sp. nov.	Marine fish, Mumbai, Maharashtra	Radiation-resistant	Shashidhar and Bandekar (2009)
4	*Anabaena* sp. strain L-31 *Anabaena torulosa*	Nitrogen-fixing cultures	Radiation-resistant	Singh et al. (2010)
5	*Anabena* sp. strain PCC7120	Nitrogen-fixing cyanobacterium	Radiation-resistant	Singh et al. (2013)
6	*Deinococcus* sp. strain RL	Sediments of hot water spring of Manikaran, Himachal Pradesh	Draft genome sequence of moderately thermophilic radiation-resistant strain	Singh et al. (2013) Mahato et al. (2014)
7	*Micrococcus caseolyticus*	Radiation-processed semi-dried corns	Radiation-resistant	Karani et al. (2014)
8	*Bacillus cereus*	Thorium ore sample, Manvala, Kuruchi, Tamilnadu	Radiation-resistant	Aravindh et al. (2015)
9	*Deinococcus enclensis* sp. nov.	Marine sediment sample from Chorao Island, Goa	Radiation-resistant	Thorat et al. (2015)
10	*Arthrobacter alpinus* ERGS4:06	Sikkim Himalaya	Radiotolerant, cold-tolerant, complete gene sequence	Kumar et al. (2016)
11	Gram-positive rod, some resembling *Actinobacteria*	Irradiation cell pool water of type IV gamma radiation processing plant at Vashi, Navi Mumbai, Maharashtra	Radiation-resistant	Kumbhare et al. (2018)
12	*Deinococcus* sp. strain SQ	Microbial mat deposits around hot spring at Manikaran, Himachal Pradesh	Draft genome sequence	Talwar et al. (2019)
13	*Deinococcus indicus* DR1	Dadri wetland	Radiation-resistant	Chauhan et al. (2019)

(continued)

Table 8.1 (continued)

S. No.	Name of the organism/group	Source	Diversity/ biotechnology	Reference
14	*Bacillus* sp., *Stenostrophomonas* spp. 1 *Brevibacillus* sp. 1 *Chryseobacterium* sp., *Escherichia* sp. *Microbacterium* sp.	Radon-contaminated groundwater, Tantloi	Radiation-tolerant desiccation-tolerant Metal-tolerant	Nayak et al. (2020)

8.6.1 Diversity

Musilova et al. (2015) have isolated radiation-resistant bacteria from Mars analog Antarctic dry valleys and discussed on correlation between radiation desiccation resistances. Extreme radiation-resistant microorganisms can survive doses of ionizing radiation far greater than that are present in the natural environment. Radiation resistance is believed to be an incidental adaption to desiccation resistance, as both hazards cause similar cellular damage. Desert soils are therefore promising targets to prospect for new radiation resistance strains. The authors claim as their study is the first study to isolate radiation-resistance microbes by using gamma- ray exposure preselection from the extreme cold desert of the Antarctic Dry Valleys (a Martian surface analogue). Halomonads, identified by 16S rRNA gene sequencing, were the most numerous survivors of the highest irradiation exposures. They were studied for the first time for both their desiccation and irradiation survival characteristics. The association between desiccation and radiation resistance has not been investigated quantitively before for a broad diversity of microorganism. A meta-analysis of scientific literature was conducted to gather a larger data set. A strong correlation was found between desiccation and radiation resistance indicating that an increase in the discussion resistance of 5 days corresponds to an increase in the room-temperature irradiation survival of 1 kGy. Irradiation at −79 °C (representative of average Martian surface temperature) increases the microbial radiation resistance ninefold. Thus, the survival of the cold-, desiccation-, and radiation-resistant organisms isolated here has implications for the potential habitability of dormant or cryopreserved life on Mars.

A total 26 isolates were identified by 16S rRNA gene sequencing. The strains belonged to three phyla: Firmicutes (*Staphylococcus* sp.) Proteobacteria (*Rhodabacter* sp., *Herbaspirillum sp. Halomonas* sp.), and Bacteroidetes (*Hymenobacter* sp.). *Halomonas* strains were selected for further characterization of their radiation and desiccation resistance since they are also halophiles. This is the first study of Halomonads for desiccation and radiation resistance.

Dominik (2016) has discussed effect of radiation and radiotolerant microorganisms other than *Deinococcus radiodurans*. Ionizing radiation damages DNA, proteins, and lipids in cells in a direct (10–20% DNA damage) and indirect manner (80–90%) causing water radiolysis and a redox potential increase (oxido-reductive stress). Hydrogen peroxide and ozone are generated. Hydroxyl radical (OH^-) is the most reactive oxygen species (ROS). Radiotolerant microorganisms sustain at high doses of radiation in a negative state. One of the most resistant and extensively studied microbes is *Deinococcus radiodurans*. This bacterium can reconstitute its genome shattered to dozens of fragments (double-strand breaks) as a result of the exposure to radiation or desiccation. Other radioresistant microorganisms include bacteria: *Acinetobacter radioresistens*, *Rubrobacter radiotolerans*, *Kineococcus radiotolerans*, *Ralstonia* sp., and *Burkholderia* sp. (living in biofilm communities from spent fuel pools), archaea: *Thermococcus gammatolerans*. Radiotolerant or radioresistant microbial species can potentially be used for bioremediation of radioactive environment contamination and for nuclear waste management (bioprecipitation, biosorption, bioaccumulation of uranium or other radioisotopes).

Gerber et al. (2015) have reviewed biology, physiology, and biotechnological potential of *Deinococcus*. *Deinococcus* spp. are among the most radiation-resistant microorganisms discovered so far. They exhibit a remarkable resistance to a range of damages caused by ionizing radiation, desiccation, UV radiation, and oxidizing agents. *Deinococcus* originally identified as *Micrococcus* are coccoid or rod-shaped non-sporulating bacteria. *Deinococcus radiodurans* R1 was the first discovered deinobacteria and was isolated in 1956 (Anderson et al. 1956) from X-ray irradiated canned meat. More than 60 species have been isolated from very diverse environments such as air dust, air samples, activated sludge, desert soils, arsenic-polluted water, cold environments in Antarctica, hot springs, biofilms, radioactive sites, and feces. These species grow at temperature ranging from 4 to 55 °C and have been grouped in a distinct eubacterial phylogenetic lineage related to the *Thermus* genus. No members of the genus *Deinococcus* have been implicated as pathogens. The cells of this genus are able to assimilate all of the common sugars and synthesize the biological macromolecules necessary for growth from cheap carbon sources without auxotrophy. They are able to grow in minimal media. *Deinococcus radiodurans* R1 is the most extensively studied species followed by *D. geothermalis*; other species include *D. grandis*, *D. indicus*, *D. deserti*, *D. gobiensis*, *D. hopiensis*, and *D. protealyticus*. *Deinococcus radiodurans* exhibits resistance to oxidative stress, which can be caused by reactive oxygen species produced by respiration, exposure to irradiation, or chemical agents. It can survive high doses of ionizing radiation without considerable irreversible DNA or protein damage and can reassemble its chromosome with a high fidelity. The resistance of this bacterium to these stresses is due to multiple converging mechanisms. Two different pathways for the assimilation of glucose, the pentose phosphate and glycolysis, are active in *Deinococcus*. The pentose phosphate pathway and the generated NADPH are directly involved in repair mechanisms and resistance to oxidative stress.

Deinococcus is extremely resistant to desiccation. It is believed that resistance to ionizing radiation and desiccation are closely linked evolutionary processes because desiccation resistance appears to require extensive DNA repair. *Deinococcus* often forms tetrads. The cell envelope of *Deinococcus* differs depending on the species and exhibits an unusual structure and composition. *D. indicus*, *D*, *grandis*, and *D. deserti* stain Gram-negative, whereas a majority of the other *Deinococcus* species stain Gram-positive.

Chauhan et al. (2017) have reported a draft genome sequence of *Deinococcus indicus* strain DR1, a red-pigmented arsenic- and radiation-resistant bacterium, isolated from a water sample of the Dadri wetland, Uttar Pradesh, India, which may provide useful information regarding the genes and pathways involved in heavy metal bioremediation.

Mahato et al. (2014) have reported draft genome sequence of a moderately thermophilic *Deinococcus* sp. strain RL isolated from sediments of hot water spring in Manikaran, India.

Thorat et al. (2015) have described a new species of *Deinococcus.* A novel pale-pink colored strain designated NIO-1023 (T) was isolated from a marine sediment sample from Chorao Island, Goa, India. Polyphasic approach was followed to investigate taxonomic position of the isolate. The cells were Gram-stain positive, coccus-shaped, and non-spore forming. Phylogenetic analyses using the 16S rRNA gene sequence of the isolate indicated that the organism belongs to the genus *Deinococcus.* The DNA-DNA relatedness and chemotaxonomic data revealed that the strain NIO-1023(T) is a novel species and a new species as *Deinococcus enclensis* sp. nov. is proposed. The type strain is NIO-1023(T) = DSM25127 (T) = NCIM5456 (T).

Shashidhar and Bandekar (2006) have described a new species of *Deinococcus.* A radiation-resistant, Gram-negative, and pleomorphic bacterium (CON-1) was isolated from a contaminated glucose yeast extract agar plate in the laboratory, Bhabha Atomic Research Centre, Mumbai, India. It was red-pigmented, nonmotile, non-sporulating, and aerobic and contained MK-8 as respiratory quinone. The cell wall of this bacterium is of ornithine. Based on chemotaxonomic, phylogenetic analysis based on 16S rRNA gene sequencing, the strain CON-1 was proposed as a new species as *Deinococcus mumbaiensis* sp. nov. The type strain of the new species is CON-1 (MTCC 7297 (T) = DSM 17424 (M).

Shashidhar and Bandekar (2009) isolated a radiation-resistant, Gram-stain-positive, nonmotile, non-sporulating, aerobic, coccoid bacterium strain 3ax (T) from a marine fish (Black pomfret, *Parastromateus niger*) with radiation as a selective pressure. Phylogenetic analysis based on 16S rRNA gene sequence showed that strain 3ax (T) belonged to genus *Deinococcus.* The predominant respiratory quinone was MK-8. The strain could grow at 20–42 °C (optimum 35 °C), pH 6–9 (optimum pH 7). The cell wall contained ornithine. Based on genotypic, phenotypic, and chemotaxonomic characteristics, the strain 3 ax (T) was identified as a novel species for which *Deinococcus piscis* sp. nov. is proposed. The type strain is 3 ax (T) (=MTCC9123(T) = DSM19767 (T)).

Chauhan et al. (2019) have discussed amino acid-dependent alterations in cell wall and cell morphology of *Deinococcus indicus* DR1. The authors isolated a red-pigmented, aerobic *Deinococcus indicus* DR1 from Dadri wetland, India. This strain exhibited cell-morphology transition from rod-shaped cells to multicell chains in a growth-medium-dependent way. In response to addition of 1% casamino acids in the minimal growth medium, rod-shaped cells formed multicell chains. Addition to all 20 amino acids to the minimal medium was able to recapitulate the phenotype. The transition from rod shape to multicell chains is due to delay in daughter cell separation after cell division.

Nayak et al. (2020) have isolated and characterized indigenous bacteria from radon-contaminated groundwater of Tantloi, India. The isolates were also studied for their response to radiation, oxidative stress, as well as heavy metal tolerance and removal. Total 16 bacterial isolates were identified as *Bacillus* spp., *Stenotrophomonas* spp., *Brevibacillus* spp., *Chryseobacterium* spp., *Escherichia* sp., and *Microbacterium* sp., which showed high-salinity tolerance properties. *Bacillus* spp. could tolerate 1 kGy of gamma radiation, 10 mM H_2O_2, 7 days of desiccation, and different heavy metals (Cu, Pb, Cr, Zn, and As).

Shukla et al. (2007) have isolated ionizing radiation (IR)-resistant bacteria from various habitats as common and stressed. IR-resistant isolates exhibited varying degrees of resistance to gamma radiation and were classified as highly and moderately radiation-resistant. Resistance to ultraviolet (UV) radiation correlated well with gamma-radiation resistance. A comparable desiccation resistance for all the highly and moderately radiation-resistant isolates was also observed. However, salt tolerance failed to correlate with IR resistance, including a divergent evolution of the salt tolerance and radiation resistance. Characterization of the amplified rDNA restriction analysis profiling attested to the clustering of these isolates showed that the bacteria with similar-resistance physiologies clustered together and belonged to related genera. Hydrogen peroxide and mitomycin survival patterns of the isolates indicated the roles of oxidative stress tolerance in desiccation survival and recombination repair in higher radiation resistance respectively.

Singh et al. (2010) have reported an unusual radioresistance of nitrogen-fixing cultures from *Anabena* strains. A nitrogen-fixing culture of two species of the filamentous, heterocystous cyanobacterium *Anabena* namely *Anabena* sp. strain L31 and *Anabena torulosa* were found to be highly tolerant to ^{60}Co gamma radiation. No adverse effect on diazotrophic growth and metabolism were observed up to a dose of 5 kGy. In *Anabena* L31, exposure to 6 kGy of gamma radiation resulted in genome disintegration but did not reduce viability; all the affected parameters recovered after a short lag, without any discernable post-irradiation phenotype.

Karani et al. (2014) isolated radiation-resistant *Macrococcus caseolyticus* (A) from radiation-processed semidried prawns. The organism was identified based on morphological, biochemical, and 16SrRNA gene sequencing. DNA-DNA hybridization studies with *M. caseolyticus* DSM20597^T further confirmed the isolate as *M. caseobyticus*. This is the first report on radiation resistance of *M. caseolyticus*. *Macrococcus caseolyticus* (A) is more resistant to gamma and UV radiation stress than *M. caseolyticus* JCSC5402 and *S.aureus* MTCC 96. However, the strain is sensitive to heat as well as desiccation stress.

Suresh et al. (2004) have isolated an arsenic- and radiation-resistant bacterium, strain Wt/la(T), from water from an arsenic-contaminated aquifer located in the Chakdah district of West Bengal, India. The bacterium was Gram-negative, rod-shaped, nonmotile, non-sporulating and red-pigmented. Based on phenotypic and chemotaxonomic characteristics, the strain Wt/la (T) was identified as a member of genus *Deinococcus.* 16S rRNA gene sequencing and DNA-DNA relatedness indicated as the strain Wt/la (T) as a novel sp. for which *Deinococcus indicus* sp. was proposed.

Shrivastava et al. (2010) have studied survival of phosphate-solubilizing bacteria against DNA-damaging agents. Phosphate-solubilizing bacteria (PSBS) were isolated from different rhizosphere soils of various agroecological regions of India. These isolates exhibited synthesis of pyrroloquinoline quinone (PQQ), production of gluconic acid, and release of phosphorous from insoluble tricalcium phosphate. The bacterial isolates producing PQQ also showed higher tolerance to ultraviolet C radiation and mitomycin C as compared to *Escherichia coli* but were less tolerant than *Deinococcus radiodurans.* PSB isolates showed higher tolerance to DNA damage when grown in absence of inorganic phosphate. Higher tolerance to ultraviolet C radiation and oxidative stress in these PSBS grown under PQQ synthesis inducible conditions namely photo-starvation, might suggest the possible additional role of this redox cofactor in the survival of these isolates under extreme abiotic conditions.

Talwar et al. (2019) have reported draft genome sequence of *Deinococcus* sp. strain S9, a red-pigmented and moderately thermophilic bacterium isolated from microbial mat deposits around the hot springs at Manikaran, Himachal Pradesh, India.

Aravindh et al. (2015) have investigated self-assembled surface layer protein on cell wall of extremophilic *Bacillus cereus* MVK04 isolated from thorium ore sample. Extremophilic nature of bacteria present in thorium-rich mine ore sample from Manavalakuruchi, Tamilnadu, India, was studied. Six bacterial strains were isolated from these ore samples and screened for their resistance to varying pH, uranium, and gamma radiation. *Deinococcus radiodurans* ATCC 13939 was used as a positive control and E. coli MTCC1687 as a negative control. Among the six strains, MVK04 strain was found to resist lower pH, uranium concentration, and gamma radiation. Based on biochemical characterization, 16S rRNA gene sequencing and MALDI-TOF fingerprinting studies, the strain was identified as *Bacillus cereus*. The strain was further investigated for the presence of surface layer protein (S-layer protein) on the cell wall, which may be the possible reason for its extremophilic characteristics and its escape from lethal effect of higher concentration of uranium, lower pH, and increased gamma radiation.

Kumbhare et al. (2018) have isolated radiation-resistant bacteria from irradiation cell pool water of Type IV gamma irradiator. Radiation processing plant at Vashi, Navi Mumbai, Maharashtra, India, is one of the Type IV, wet storage-type irradiators, with maximum design capacity of 1 MCi 60 Co source activity. The authors isolated bacteria from the irradiation cell pool water. The pool water was subjected to filtration by passing through 0.45 micron cellulose nitrate membrane. These membranes were exposed on soybean casein digest agar to gamma radiation dose of 3,4, and 5 kGy in gamma chamber GC-4000. The membranes were incubated at 30–35 °C for 2–3 days. It was observed that 3.3% and 0.58% of the organisms were Gram-positive rods, some of them arranged in long chains resembling Actinobacteria. The radiation sensitivity studies showed that two of the isolates were having D10 values of more than 4 kGy and one isolate showing D10 value as high as 4.7 kGy.

8.6.2 Biotechnology

Singh and Gabani (2011) have discussed potential applications of extremolytes the novel compounds produced by extremophiles, particularly here radiation-resistant microorganisms, for human therapeutics including anticancer drugs, antioxidants cell cycle – blocking agents, anticholestatic drugs etc. It is hypothesized that the majority of ultraviolet radiation (UVR)-resistant microorganisms can be used to develop anticancer drugs to prevent skin damage from UVR.

8.6.2.1 Application of *Deinococcus*

Removal of heavy metals: *Deinococcus* has been studied for the remediation of heavy metal pollution because of its high resistance, tolerance to heavy metal ions, and absorption and degradation ability.

Synthesis of gold nanoparticles: The hydroxyl tetraterpenoid deoxy xanthine (DX) from *Deinococcus* can be used to synthesize functionalized gold nanoparticles (DX-AuNPs) through biotransformation, which includes production of ROS in cancer cells by upregulating the expression of certain genes, thus loading to the apoptosis of cancer cells (Tian et al. 2018).

Dhanapal et al. (2021) have investigated DNA repair pathway and stress-responsive genes expression in *Deinococcus indicus*, a novel bacterium isolated from West Bengal, India, and known for its UV radiation and heavy metal tolerance. Based on available whole genome sequence of *D. indicus*, quantitative real time PCR (qPCR) was performed to determine the expression pattern of multiple stress-responsive genes upon various environmental extremities. Among them, UV-responsive genes like UVrD and UVsE showed elevated expression when subjected to UV-C radiation at different time intervals. Similarly, when exposed to arsenic and chromium, ArsR and ArsB exhibited considerably higher level of expression. Overall, this study opened the functional information on stress tolerance genes that aid to understand the DNA-damaged recovery mechanism towards elucidation of DNA repair pathways.

Prasad et al. (2005) have reported detection of N6-methyladenine in the genome of *D. radiodurans* strain R1 using immunochemical techniques. Methylation of DNA is known to be involved in DNA repair mechanism in bacteria. *D. radiodurans* strain R1 on exposure to high radiation undergoes significant DNA damage, which is repaired without mutation. However, the presence of modified nucleotides has not been reported in its genome. The authors have reported that *D. radiodurans* cell extract exhibited a DNA adenine methyltransferase activity which was reduced in the early postirradiation recovery phase. This N6-methyladenine is not a part of GATC restriction-modification system.

Misra et al. (2006) have studied an exonuclease 1-sensitive DNA repair pathway in *Deinococcus radiodurans*, as the major determinant of radiation resistance. The authors have exploited the known roles of exonuclease 1 in DNA recombination to elucidate the nature of recombination involved in DNA double-strand break repair during postirradiation recovery of *D. radiodurans*. Transgenic *Deinococcus* cells expressing exonuclease functions of *Escherichia coli* showed significant reduction in gamma radiation radioresistance while the resistance to far-UV and hydrogen peroxide remained unaffected. The overexpression of *E. coli* exonuclease 1 in *Deinococcus* inhibited DNA double-strand break repair.

Joshi et al. (2004) have reported protein recycling as the major component of postirradiation recovery in *Deinococcus radiodurans* strain R1. Exposure to 6 kGy dose of ^{60}Co gamma rays resulted in immediate arrest of growth followed by complete recovery of *Deinococcus radiodurans* strain R1 cells. Selective degradation and resynthesis of several predicted highly expressed proteins (including major chaperons, key TCA cycle enzymes, and a few stress proteins) and several hypothetical proteins marked the lag period, preceding resumption of growth. A major exercise in protein recycling appears to be an integral component of postirradiation recovery in *D. radiodurans* and complements the extensive DNA repair, characteristic of this extremely radioresistant bacterium.

Misra et al. (2012) have investigated bioremediation of heavy metals, from acidic/neutral aqueous wastes using recombinant *Deinococcus radiodurans*. Genetically engineered *D. radiodurans* (DR) cells expressing a nonspecific acid phosphatase PhoN in high radiation environment were found to be stable and have metal bioremediation ability. The lyophilized recombinant DRPhoN cells retained PhoN activity and uranium precipitation ability. Such cells also displayed an extended shelf life of 6 months during storage at room temperature and showed surface-associated precipitation of uranium as well as other metals like cadmium. Lyophilized cells immobilized in polyacrylamide gels could be used for uranium bioprecipitation in a flow through system resulting in 70% removal of 1 mm input uranium solution and loading of 1 g uranium/g dry weight cell.

Basu and Apte (2012) have investigated extraordinary radioresistance of *Deinococcus radiodurans* which primarily originates from its efficient DNA repair ability. The kinetics of proteomic changes included by a 6 kGy dose of gamma irradiation was mapped during the postirradiation growth arrest phase by two-dimensional protein electrophoresis coupled with mass spectrometry. The results revealed that at least 37 proteins displayed either enhanced or de novo expression in the first 1 h of postirradiation recovery. All of the radiation-responsive proteins were identified and they belong to major functional categories of DNA repair, oxidative stress alleviation, and protein translation or folding.

Singh et al. (2013) have described high radiation and desiccation tolerance of nitrogen-fixing cultures of the cyanobacterium *Anabaena* sp. strain PCC7120. The organism was found to tolerate very high doses of ^{60}Co-gamma radiation or prolong desiccation. Post-stressed cells remained intact and reviewed all the vital functions. Remarkable capacity to repair highly disintegrated genome and recycle the damage proteome appeared to underlie such high radioresistance and desiccation tolerance.

The close similarity observed between the cellular response to irradiation or desiccation stress lends strong support to the notion that tolerance to these stresses may involve similar mechanism.

Desai et al. (2011) have characterized role of the radS (a putative sensor histidine kinase)/radR (cognate response regulator) two-component system in radiation resistance of *Deinococcus radiodurans* which shows extraordinary tolerance to DNA damage and exhibits differential gene expression and protein recycling. The author identified a putative response regulator, the DRB0091 (radR) ORF, from a pool of DNA-binding proteins induced in response to gamma radiation in this bacterium. *radR* is located upstream of DRB0090 which encodes a putative sensor histidine kinase (radS) on the mega plasmid. Deletion of these genes both individually and together resulted in hypersensitivity to DNA-damaging agents and a delayed or altered double-strand break repair. The results indicated the functional interaction of radS and radR and suggested role for the radS/radR two-component system in the radiation resistance of this bacterium.

Narasimha et al. (2016) have investigated mechanisms involved in induction of radiation-responsive genes of *Deinococcus radiodurans* (DEIRA). In response to DNA damage, the organism selectively expresses a number of genes, which are involved in DNA repair.

Rajpurohit et al. (2008) have discussed involvement of protein kinase activity inducer in DNA double-strand break repair in radioresistance of *Deinococcus radiodurans*. Transgenic bacteria producing pyrroloquinoline quinone, a known cofactor for dehydrogenases and an inducer of a periplasmic protein kinase activity, show resistance to both oxidative stress and protection from nonoxidative effects of radiation and DNA-damaging agents. *Deinococcus radiodurans* R1 encodes an active pyrroloquinoline quinone synthesis and constitutive synthesis of pyrroloquinoline quinone occurring in wild-type bacteria. Disruption of genomic copy of PQQE resulted in a cell that lacked this cofactor. The mutant showed decrease in a gamma radiation resistance and mitomycin C tolerance compared to wild-type cells. Expression of pyrroloquinoline quinone (PQQ) synthesis in *trans* showed that there was a functional complementation of gamma resistance and mitomycin C tolerance in the *PQQE* mutant. The results suggests that pyrroloquinoline quinone has regulatory role as a cofactor for dehydrogenases and an inducer of selected protein kinase activity in radiation resistance and DNA strand break repair in radioresistant bacteria.

Appukuttan et al. (2006, 2007) have carried out genetic engineering of *Deinococcus radiodurans* R1 for bioprecipitation of uranium from dilute nuclear waste. The authors have constructed a *Deinococcus radiodurans* strain harboring *PhoN,* a gene encoding a nonspecific acid phosphate obtained from a local isolate of *Salmonella enterica* serovar typhi. The recombinant strain expressed an ~27 KDa active PhoN protein and efficiently precipitated over 90% of the uranium from 0.8 mm uranyl nitrate solution in 6 h. The engineered strain retained uranium bioprecipitation ability even after exposure to 6 KGy ^{60} Co gamma rays. The PhoN expressing *D.radiodurans* offers an effective and eco-friendly in situ approach to biorecovery of uranium from dilute nuclear waste.

8.6.3 Sustainability

Daly (2000) has suggested engineering radiation-resistant bacteria for bioremediation of radioactive waste. 70 million cubic meters of ground and three trillion liters of groundwater have been contaminated by leaking radioactive waste generated in the United States during the cold war. A cleanup technology was being developed based on the radiation-resistant bacterium *Deinococcus radiodurans* being engineered to express bioremediating functions.

Gogada et al. (2015) aimed to engineer bacteria for removal of Co in contaminated effluents. Radioactive cobalt (^{60}Co) is known as a major contributor for person-sievert budgetary because of its long half-life and high gamma energy values. Some bacterial Ni-Co transporters (NiCoT) were described to have preferential uptake for cobalt. These genes were expressed in *Deinococcus radiodurans* in reason of its high resistance to its radiation. Using qualitative real-time PCR, the authors showed that the expression of these genes is induced by cobalt and nickel. The engineered *Deinococcus radiodurans* are one with NiCoT genes could be used for bioremoval of trace cobalt from spent decontamination solutions of nuclear power reactors.

Rastogi et al. (2015) has reviewed multiple defense systems in cyanobacteria in response to solar UV radiation. Increase in solar ultraviolet (UV) radiations on Earth's surface is of concern due to its negative impact on all sun-exposed photosynthetic light farms including cyanobacteria. UV (280–400 nm) can affect a number of physiological and biochemical processes in cyanobacteria, either directly or indirectly by the generation of reactive oxygen species. However, many cyanobacteria are able to develop several defense mechanisms against the damaging effect imposed by increased UV radiation. Their strategies include the restoration of genetic material by means of excision repair and photoreactivation, repair and resynthesis of D1 and D2 proteins of PSII reaction center, activation of antioxidant systems, and biosynthesis of several photoprotectants such as mycosporine-like amino acids, scytonemin, carotenoids, and polyamines. Cyanobacteria also produce high molecular weight heteropolysaccharides which serve as a boundary between the cells and surrounding environment and play an important role in mitigation strategy against desiccation and harmful effects of UV radiation.

8.6.4 Contribution to the Knowledge of Radiophilic Microbial Species, New to Science, from India

Researchers from India have isolated some novel species of radiophilic microorganisms from different radiation contaminated/exposed sites from India. These are listed in Table 8.2.

It is likely that much more investigations would add to the knowledge of radiophilic microorganisms from India.

Table 8.2 Radiophilic microbial species, new to science, from India

S. No.	Name of the organism/group	Source	Reference
1	*Deinococcus indicus* sp. nov.	Arsenic-contaminated aquifer in Chakdah district of West Bengal	Suresh et al. (2004)
2	*Deinococcus mumbaiensis* sp. nov.	Contaminated glucose yeast agar plate in laboratory, Mumbai, Maharashtra	Shashidhar and Bandekar (2006)
3	*Deinococcus piscis* sp. nov.	Marine fish, Mumbai, Maharashtra	Shashidhar and Bandekar (2009)
4	*Deinococcus enclensis* sp. nov.	Marine sediment sample from Chorao Island, Goa	Thorat et al. (2015)

References

Anderson A, Nordan H, Cain R, Parish G, Duggan D (1956) Studies on a radioresistant *Micrococcus*. I. Isolation, morphology, cultural characteristics and resistance to gamma radiation. Food Technol 10:575–578

Appukuttan D, Rao AS, Apte SK (2006) Engineering of *Deinococcus radiodurans* R1 for bioprecipitation of uranium from dilute nuclear waste. Appl Environ Microbiol 72(12): 7873–7878. https://doi.org/10.1128/AEM.01362-06

Appukuttan D, Rao A, Apte S (2007) Engineering of *Deinococcus radiodurans* R1 for Bioprecipitation of Uranium from Dilute Nuclear Waste. Appl Environ Microbiol 73 (4):1393–1393

Aravindh S, Selvakumar R, Ravichandran J, Kamachi MU, Anandbabu C, Raj B (2015) Extremophilic *Bacillus cereus* MVK04 isolated from thorium ore sample possesses self-assembled surface layer protein on cell wall to resist extreme environments. Geomicrobiol J 32(5):445–452. https://doi.org/10.1080/01490451.2014.957425

Basu B, Apte SK (2012) Gamma radiation-induced proteome of *Deinococcus radiodurans* primarily targets DNA repair and oxidative stress alleviation. Mol Cell Proteomics 11(1): M111.011734. https://doi.org/10.1074/mcp.M111.011734

Battista JR (1997) Against all odds: the survival strategies of *Deinococcus radiodurans*. Annu Rev Microbiol 51:203–224

Chauhan D, Srivastava PA, Yennamalli RM, Priyadarshini R (2017) Draft genome sequence of *Deinococcus indicus* DR1, a novel strain isolated from a freshwater wetland. Genome Announc 5(31):e00754–e00717. https://doi.org/10.1128/genomeA.00754-17

Chauhan D, Srivastava PA, Ritzl B, Yennamalli RM, Cava F, Priyadarshini R (2019) Amino acid-dependent alterations in cell wall and cell morphology of *Deinococcus indicus* DR1. Front Microbiol 10:1449. https://doi.org/10.3389/fmicb.2019.01449

Daly M (2000) Engineering radiation-resistant bacteria for environmental biotechnology. Curr Opin Biotechnol 11:280–285. https://doi.org/10.1016/S0958-1669(00)00096-3

Desai SS, Rajpurohit YS, Misra HS, Deobagkar DN (2011) Characterization of the role of the RadS/RadR two-component system in the radiation resistance of *Deinococcus radiodurans*. Microbiology 157:2974–2982. https://doi.org/10.1099/mic.0.049361-0

Dhanapal AR, Venkidasamy B, Ramatchandirane PS (2021) Molecular characterization of stress tolerance genes associated with *D. indicus* strain under extreme environment conditions. Environ Geochem Health 43(12):4905–4917. https://doi.org/10.1007/s10653-020-00788-9

Dominik MM (2016) Radiotolerant microorganisms - characterization of selected species and their potential usage. Adv Microbiol 55(2):182–194

Gabani P, Singh O (2012) Radiation-resistant extremophiles and their potential in biotechnology and therapeutics. Appl Microbiol Biotechnol 97(3). https://doi.org/10.1007/s00253-012-4642-7

Gerber E, Bernard R, Castang S, Chabot N, Coze F, Dreux-Zigha A, Hauser E, Hivin P, Joseph P, Lazarelli C, Letellier G, Olive J, Leonetti J-P (2015) *Deinococcus* as new chassis for industrial biotechnology: biology, physiology and tools. J Appl Microbiol 119(1):1–10. https://doi.org/10.1111/jam.12808

Gogada R, Singh SS, Lunavat SK, Pamarthi MM, Rodrigue A, Vadivelu B, Phanithi PB, Gopala V, Apte SK (2015) Engineered *Deinococcus radiodurans* R1 with NiCoT genes for bioremoval of trace cobalt from spent decontamination solutions of nuclear power reactors. Appl Microbiol Biotechnol 99(21):9203–9213

https://serc.carleton.edu/microbelife/extreme/radiation/index.html

Jin M, Xiao A, Zhu L, Zhang Z, Huang H, Jiang L (2019) The diversity and commonalities of the radiation-resistance mechanisms of *Deinococcus* and its up-to-date applications. AMB Expr 9: 138. https://doi.org/10.1186/s13568-019-0862-x

Joshi B, Schmid R, Altendorf K, Apte SK (2004) Protein recycling is a major component of post-irradiation recovery in *Deinococcus radiodurans* strain R1. Biochem Biophys Res Commun 320(4):1112–1117

Karani M, Shashidhar R, Kakatkar A, Gautam RK, Sukhi S, Pansare-Godambe L, Bandekar J (2014) Radiation-resistant *Macrococcus caseolyticus* (A) isolated from radiation-processed semidried prawns. Can J Microbiol 61(1). https://doi.org/10.1139/cjm-2014-0636

Kotiranta AK, Ito H, Haapasalo MPP, Lounatmaa K (1999) Radiation sensitivity of *Bacillus cereus* with and without a crystalline surface protein layer. FEMS Microbiol Lett 179(2):275–280. https://doi.org/10.1111/j.1574-6968.1999.tb08738.x

Krokan HE, Bjoras M (2013) Base excision repair. Cold Spring Harb Perspect Biol 5(4):a012583. https://doi.org/10.1101/cshperspect.a012583

Kumar R, Singh D, Swarnkar MK, Singh AK, Kumar S (2016) Complete genome sequence of *Arthrobacter alpinus* ERGS4:06, a yellow pigmented bacterium tolerant to cold and radiations isolated from Sikkim Himalaya. J Biotechnol 220:86–87. https://doi.org/10.1016/j.jbiotec.2016.01.016

Kumbhare M, Singh R, Srivastava P, Shukla P (2018) Investigation of radiation resistant bacteria from irradiation cell pool water of Type IV gamma irradiator. India: National Association for Applications of Radioisotopes and Radiation in Industry, NAARRI international conference - advanced applications of radiation technology; Mumbai (India); 5–7 Mar 2018, 49(48):26

Mahato NK, Tripathi C, Verma H, Singh N, Lal R (2014) Draft genome sequence of *Deinococcus* sp. strain RL isolated from sediments of a hot water spring. Genome Announc 2(4):e00703–e00714. https://doi.org/10.1128/genomeA.00703-14

Misra H, Khairnar N, Kota S, Shrivastava S, Joshi V, Apte S (2006) An exonuclease I-sensitive DNA repair pathway in *Deinococcus radiodurans*: a major determinant of radiation resistance. Mol Microbiol 59:1308–1316. https://doi.org/10.1111/j.1365-2958.2005.05005.x

Misra CS, Appukuttan D, Kantamreddi VSS, Rao AS, Apte SK (2012) Recombinant *D. radiodurans* cells for bioremediation of heavy metals from acidic/neutral aqueous wastes. Bioengineered 3(1):44–48. https://doi.org/10.4161/bbug.3.1.18878

Musilova M, Wright G, Ward JM, Dartnell LR (2015) Isolation of radiation-resistant bacteria from mars analog Antarctic Dry Valleys by preselection, and the correlation between radiation and desiccation resistance. Astrobiology 15(12):1076–1090. https://doi.org/10.1089/ast.2014.1278

Narasimha A, Basu B, Apte SK (2016) Radiation responsive promoters of *Deinococcus radiodurans*. J Radiat Cancer Res 7(1):41

Nayak T, De D, Barman C, Karmakar P, Deb A, Dhal PK (2020) Characterization of indigenous bacteria from radon-rich groundwater and their tolerance to physicochemical stress. Int J Environ Sci Technol 17:1627–1636

Prasad BJ, Sabnis K, Deobagkar DD, Deobagkar DN (2005) *Deinococcus radiodurans* strain R1 contains N6-methyladenine in its genome. Biochem Biophys Res Commun 335(2):412–416. https://doi.org/10.1016/j.bbrc.2005.07.099

Rajpurohit YS, Gopalakrishnan R, Misra HS (2008) Involvement of a protein kinase activity inducer in DNA double strand break repair and radioresistance of *Deinococcus radiodurans*. J Bacteriol 190(11):3948–3954. https://doi.org/10.1128/JB.00026-08

Rastogi RP, Madamwar D, Incharoensakdi A (2015) Multiple defence systems in cyanobacteria in response to solar UV radiation, Chapter 6. In: Douglas Davison D (ed) Cyanobacteria. Nova Science Publisher. ISBN: 978-1-61470-925-1

Romanovskaia VA, Rokitko PV, Malashenko IR (2000) Unique properties of highly radioresistant bacteria. Mikrobiol Z 62(1):40–63

Shashidhar R, Bandekar JR (2006) *Deinococcus mumbaiensis* sp. nov., a radiation-resistant pleomorphic bacterium isolated from Mumbai, India. FEMS Microbiol Lett 254(2):275–280. https://doi.org/10.1111/j.1574-6968.2005.00033.x

Shashidhar R, Bandekar JR (2009) *Deinococcus piscis* sp. nov., a radiation-resistant bacterium isolated from a marine fish. Int J Syst Evol Microbiol 59(Pt 11):2714–2717. https://doi.org/10.1099/ijs.0.003046-0

Shashidhar R, Kumar SA, Misra HS, Bandekar JR (2010) Evaluation of the role of enzymatic and nonenzymatic antioxidant systems in the radiation resistance of *Deinococcus*. Can J Microbiol 56(3):195–201. https://doi.org/10.1139/W09-118

Shrivastava M, Rajpurohit YS, Misra HS, D'Souza SF (2010) Survival of phosphate-solubilizing bacteria against DNA damaging agents. Can J Microbiol 56(10):822–830. https://doi.org/10.1139/W10-067

Shukla M, Chaturvedi R, Tamhane D, Vyas P, Archana G, Apte S, Bandekar J, Desai A (2007) Multiple-stress tolerance of ionizing radiation-resistant bacterial isolates obtained from various habitats: correlation between stresses. Curr Microbiol 54(2):142–148. https://doi.org/10.1007/s00284-006-0311-3

Singh OV, Gabani P (2011) Extremophiles: radiation resistance microbial reserves and therapeutic implications. J Appl Microbiol 122(4):1121–1121. https://doi.org/10.1111/j.1365-2672.2011.04971.x

Singh H, Fernandes T, Apte SK (2010) Unusual radioresistance of nitrogen-fixing cultures of anabaena strains. J Biosci 35(3):427–434

Singh H, Anurag K, Apte SK (2013) High radiation and desiccation tolerance of nitrogen-fixing cultures of the cyanobacterium Anabaena sp. strain PCC 7120 emanates from genome/proteome repair capabilities. Photosynth Res 118(1):71–81

Slade D, Radman M (2011) Oxidative stress resistance in *Deinococcus radiodurans*. Microbiol Mol Biol Rev 75:133–191

Suresh K, Reddy GSN, Sengupta S, Shivaji S (2004) *Deinococcus indicus* sp. nov., an arsenic-resistant bacterium from an aquifer in West Bengal, India. Int J Syst Evol Microbiol 54(Pt 2): 457–461. https://doi.org/10.1099/ijs.0.02758-0

Talwar C, Singh AK, Singh DN, Nagar S, Singh Y, Shakarad M, Negi RK, Lal R (2019) Draft genome sequence of *Deinococcus* sp. strain S9, isolated from microbial mat deposits of hot springs located atop the himalayan ranges at manikaran, India. Microbiol Resour Announc 8(28):e00316–e00319. https://doi.org/10.1128/MRA.00316-19

Thorat MN, Mawlankar R, Sonalkar VV, Venkataramana V, Joseph N, Shouche YS, Dastager SG (2015) *Deinococcus enclensis* sp. nov., isolated from a marine sediment sample. Antonie Van Leeuwenhoek 107(1):141–148. https://doi.org/10.1007/s10482-014-0311-4

Tian B, Li JL, Pang RJ, Dai S, Li T, Weng YL, Jin Y, Hua YJ (2018) Gold nanoparticles biosynthesized and functionalized using a hydroxylated tetraterpenoid trigger gene expression changes and apoptosis in cancer cells. ACS Appl Mater Inter 10:37353–37363

Zeng Y, Ma Y, Xiao FZ, Wang W, He S (2017) Knockout of *pprM*, decreases resistance to desiccation and oxidation in *Deinococcus radiodurans*. Indian J Microbiol 57:1–6

Piezophilic or Barophilic Microorganisms 9

Abstract

Piezophilic microorganisms grow optimally under high hydrostatic pressure as 10 MPa (Mega Pascal unit) when tested over all permissible temperatures. Barophiles was the term originally used. They thrive in extreme high-pressure environments, exceeding up to 150 MPa (1 MPa = 10 bar). They are found in marine trenches, deep-sea sediments, hydrothermal vents, ocean floors, and rocks deep under the Earth. They can also resist extreme temperature, pH, and high salt concentration; thus they are polyextremophiles. Their barostability is attributed to genomic adaptation (robust DNA repair systems), proteomic adaptations (substitution of all other amino acids with arginine), stabilization, and activation of enzymes by high pressures. *Pyrococcus yayanosii* is the most pressure-resistant organism while *Thermococcus barophilus* is both thermophilic and barophilic; *Psychromonas hadalis* is obligately piezophilic and optimally grows at 6 °C and pressure 60 MPa.

In India, piezophilic microorganisms such as nitrifying bacteria are reported from Central Indian Basin; some fungi like *Aspergillus*, *Trichothesium*, and *Sagenomella* from deep-sea sediments of Central Indian Basin; and bacteria like *Acinetobacter*, *Halomonas*, *Sphingomonas*, and *Shewanella* from deep drilling at Koyana, Western India. One novel genus and one novel species of piezophilic microorganism new to Science have been reported from India.

Keywords

Piezophilic · Barophilic · MPa · Deep-sea sediments · Hydrothermal vents · Barostability · Genomic adaptation · Arginine · *Thermococcus piezophilus* · *Shewanella* · *Psychromonas* · *Sagenomella* · *Pyrococcus*

P. P. Kanekar, S. P. Kanekar, *Diversity and Biotechnology of Extremophilic Microorganisms from India*, Microorganisms for Sustainability 41, https://doi.org/10.1007/978-981-19-1573-4_9

Barophilic microorganisms thrive in extremely high-pressure environments, often exceeding up to 150 MPa (Zeng et al. 2009). *Pyrococcus yayanosii* is the most pressure-resistant microorganism which can survive up to 150 MPa (MPa is a MegaPascal unit. One MPa equals to 1,000,000 Pascal units primarily used for higher range pressure measurement due to its larger value, e.g., 1 MPa = 10 bar).

The habitats of such organisms are marine trenches, deep-sea sediments, hydrothermal vents, ocean floors, and rocks deep under the Earth (Kato et al. 1998; Rothschild and Mancinelli 2001). Most barophilic organisms can also face other environmental extremities in addition to high pressure, such as extremes of temperature, pH, high salt concentration, and scarcity of nutrients (Kumar et al. 2018). Thus, barophiles can also be mixotrophic in nature harboring properties of thermophiles, psychrophiles, alkaliphiles, halophiles, and acidophiles. (Chakravorty et al. 2017). Barostability can be defined by proteins to remain kinetically and thermodynamically stable.

9.1 Factors Responsible for Barostability

9.1.1 Genomic Adaptations

Not much work on genomes from barophilic microorganisms has been reported. Di Giulio (2005) initiated his studies by comparison of homologous genomic sequences of barophilic *Pyrococcus abyssi* and non-barophilic *Pyrococcus furiosus*. He reported that barophilic genomes are poor in GC content and also showed that GC-rich codons are insignificant in barophiles (Di Giulio 2013). For survival in high-pressure environment, barophiles require robust DNA repair systems since high pressure can damage DNA or proteins. Thus, survival necessitates avoidance of damage or high repair rates (Rothschild and Mancinelli 2001). Additionally, pressure-regulated operons (OMPH) have evolved in barophilic genomes which are responsible for high-pressure adaptations in barophiles (Kato et al. 1998; Kato et al. 1995).

9.1.2 Proteomic Adaptations

The extreme pressure adaptation of proteome was explored by Di Giulio (2005) that reported the barophilic *Pyrococcus abyssii* tends to substitute arginine (Arg) for all other amino acids in sequences homologous to non-barophilic *P. furiosus*. He considered Arg to be the “barophilic amino acid.” Yafremava et al. (2013) have also shown that Arg is preferred in barophilic proteins when compared to non-barophilic proteins. Another example of adaption in piezophile, *P. abyssi*, having an increase in small amino acids (Gly, Ser, Thr, Pro, Asp and Asn) across its proteome when compared to that of the related non-piezophile, *P. furiosus* has been found (Yafremava et al. 2013; Di Giulio 2005).

The physical factor like pressure is also involved in the stabilization of proteins like hyperthermophiles survive under high-temperature environment and also survive under high-pressure environments. Because of these environments, hyperthermophiles cannot avoid pressure and temperature such as *Thermococcus barophilus*. All their macromolecular cell components have to be adapted to high pressures (Michels and Clarks 1997). In such microorganisms, the enzymes are stabilized and activated by high pressures. The three-dimensional structure of these enzymes is stabilized mainly by three types of biophysical interactions: hydrogen bond, electrostatic interactions, and hydrophobic interactions. Michels and Clark (1997) reported that the pressure favors the smallest volume, and high percentage of hydrophobic interactions in the enzyme structure is the main rationale behind the protein stabilization under high pressure.

9.2 Piezophiles

A piezophile (from Greek 'piezo" for pressure and "phile" for loving) is an organism with optimal growth under high hydrostatic pressure, or more operationally, an organism that has its maximum rate of growth at a hydrostatic pressure equal or above 10 MPa (=99 atm = 1450 PSI) when tested over all permissible temperatures (Yayanos 2008). Originally, the term barophiles was used for these organisms, but since the prefix "baro" stands for weight, the term piezophiles should be given preference (Yayanos 1995). Like all definitions of extremophiles, the definition of piezophiles is anthropocentric and humans consider that moderate values for hydrostatic pressure are those around 1 atm (=0.1 MPa = 14.7 PSI). Hydropiezophile is defined as an organism that has its maximum rate of growth above 50 MPa (=493 atm = 7252 PSI) (Fang et al. 2017). The current record for highest hydrostatic pressure where growth was observed is 130 MPa (=1283 atm = 18,855 PSI) by the archaea *Thermococcus piezophilus* (Dalmasso et al. 2016). Obligate piezophiles refer to organisms that are unable to grow under lower hydrostatic pressure such as 0.1 MPa. In contrast, piezotolerant organisms are those that have their maximum rate of growth at a hydrostatic pressure under 10 MPa but that are able to grow at lower rates under higher hydrostatic pressures.

Most of the Earth's biosphere (in terms of volume) is subject to high hydrostatic pressure and the piezosphere comprises the deep sea (at the depth of 1000 m and greater) plus the deep subsurface (which can extend up to 5000 m beneath the seafloor or the continental surface). (Fang et al. 2017; Kieft 2016). The deep sea has a mean temperature around 1 to 3 °C, and it is dominated by psychropiezophiles in contrast to the deep-subsurface and hydrothermal vents in seafloor which are dominated by thermopiezophiles that prosper on temperatures above 45 °C (113F).

The high pressure experienced by this organism can cause the normal fluid cell membrane to become waxy and relatively impermeable to nutrients. The high pressure decreases the ability of the subunits of multimeric proteins to interact. Thus, large protein complexes must interact to decrease pressure-related effects and regulate processes such as protein and DNA synthesis, which are sensitive to high pressure. Piezophilic bacteria have a high proportion of fatty acids in their cytoplasmic membrane, which allows membranes to remain functional and keep from gelling at high pressure (Madigan et al. 2014).

9.3 Diversity

Nogi et al. (2007) isolated an obligately piezophilic bacterium from sediment collected from the bottom of the Japan Trench at a depth of 7542 m. The isolated strain K41GT was closely affiliated with members of the genus *Psychromonas* on the basis of 16S rRNA gene sequence analysis. The optimal temperature for the growth of the organism was 6 °C and pressure 60 MPa. Based on the taxonomic differences observed, the strain K41GT was considered to represent a novel obligately piezophilic *Psychromanas* species for which the name *Psychromonas hadalis* (type strain K4IGT = JCM 11830 T = ATCC BAA-638 T) was proposed. This is the second species of obligately piezophilic bacteria to be proposed in the genus *Psychromonas*.

Psychrophilic and piezophilic bacteria have been isolated from deep-sea environments. Their physiological adaptations, which enable them to survive at high pressure and low temperature, make them interesting. They contribute to our understanding of the interaction between the deep-sea environments and its microbial inhabitants. All piezophilic bacteria isolated to date fall into the class Gamma proteobacteria, according to phylogenetic classification based on 16S rRNa gene sequence information (Delong et al. 1997; Nogi and Kato 1999; Nogi et al. 2002) Margesin and Nogi (2004) reported that 11 cultivated psychrophilic and piezophilic deep-sea species were affiliated with genera within the Gamma proteobacteria: Shewanella, Photobacterium, Colwellia, Moritella, and Psychromonas.

Kato et al. (1998) have described obligately barophilic bacteria isolated from a sample of the world's deepest sediment, obtained by the unmanned deep-sea submersible Kaiko in the Mariana Trench, Challenger Deep at a depth of 10898 m. From the results of phylogenetic analysis based on 16S rRNA gene sequences, DNA-relatedness study, and analysis of fatty acid composition, the strain DB21MT-2 appeared to be highly similar to *Shewanella benthica* and the strain DB21MT-5 to be closely related to the genus *Moritella*. The optimum pressure conditions for growth of these strains were 70 MPa for strain DBMT-2 and 80 MPa for strain DBMT-5, and no growth was observed at pressure less than 50 MPa with either strains. This is the first evidence of the existence of an extreme barophile bacterium of the genus *Moritella* isolated from deep-sea environment.

Kato and Bartlett (1997) have discussed molecular biological aspects of barophilic bacteria. Many microorganisms from the deep sea display high-pressure adapted (also described as barophilic or piezophilic) growth characteristic. Phylogenetic studies have revealed that a large proportion of the barophilic bacteria currently in culture collections belong to a distinct subgroup of the genus *Shewanella*, referred to as the "barophile branch." Many of the basic properties of barophiles that enable their survival at extremes of pressure remain to be elucidated. However, several genes whose expression is regulated by pressure or which appear to be critical to bar adaption have been uncovered. One such operon whose presence appears to be restricted to the "barophile" branch has been identified in DNA samples obtained from sediments recovered in the deepest ocean trench. In the case of another set of pressure-regulated genes, regulatory elements required for pressure signaling have been uncovered.

9.4 Piezophilic or Barophilic Microorganisms Reported from India

Researchers from India have investigated piezophilic microorganisms from deep-sea sediments in Indian Ocean, Central Indian Basin, deep drilling at Koyana, Western India, etc. These are listed in Table 9.1.

9.4.1 Work on Peizophilic Microbial Diversity from India

Raghukumar and Damare (2011) have reviewed role of Labyrinthulomycetes in marine ecosystems, their diversity, habitats, biomass and productivity. Earlier studies contributed enormously to the cultured diversity of Labyrinthulomycetes. In recent years, their uncultured diversity has been demonstrated in exotic environments like deep sea and anoxic waters. Their findings emphasize the need for novel culture methods to grow these organisms. Many species seem to be substrate-specific in their occurrence. The biomass of Labyrinthulomycetes in the water column may often match or even exceed that of bacteria, although such occurrences seem to be seasonal. The high biomass and production of several degradative enzymes indicate their importance as remineralizers in the ocean. However, the mechanisms by which they overcome bacterial competition are not clear. It is likely that they occupy special niches, such as marine aggregates. One role of the Labyrinthulomycetes is suggested as that of "leftover scavenging" following bacterial growth.

Cao et al. (2016) have described a piezophilic sulfate-reducing bacterium from the Indian Ocean. A novel sulfate-reducing bacterium, strain J2T, was isolated from a serpentinized peridotite sample from the Indian Ocean. Phylogenetic analysis based on 16S rRNA gene sequences showed that the strain J2T clustered with the genus *Desulfovibrio* within the family Desulfovibrionaceae, but it showed low (87.95%) similarity to the type species *Desulfovibrio desulfuricans* DSM642T. Cells were Gram-staining-negative, anaerobic, motile vibrios, able to grow at salinities ranging from 0.2 to 6% (optimum 2.5%), pH 5 to 8 (optimum pH 6.5 to 7), and at temperature between 9 and 40 °C (optimum 30–35 °C). The strain J2T was piezophilic, growing optimally at 10 MPa (range 0–30 MPa). Based on genotypic and phenotypic characteristics, it was observed that the strain J2T represents a novel species of a novel genus in the family Desulfovibrionaceae, for which the name

Table 9.1 Piezophilic or barophilic microorganisms reported from India

Sr. No.	Organism/group	Source	Diversity/ biotechnology	Reference
	Nitrifying bacteria exhibiting both phases of nitrification under high pressure	Central Indian Basin, sediment cores at ~5000 m depth	Nitrification	Ram et al. (2001)
1	*Aspergillus sydowii*	Deep-sea sediments from Chagos Trench, Indian Ocean	Barotolerant	Raghukumar et al. (2004)
2	*Aspergillus* sp. *Exophiala* sp. *Trichothesium* sp. *Tilletiopsis* sp. *Sagenomella* sp. *Capronia* sp.	Deep-sea sediments of Central Indian Basin at ~5000 m depth	Growth at 200 bar/5 °C	Singh et al. (2010)
3	*Labyrinthulomycetes*	Water column of ocean	Role in marine ecosystem Deep sea and anoxic water	Raghukumar and Damare (2011)
4	*Pseudodesulfovibrio indicus* gen. nov., sp. nov.	Indian Ocean	Piezophilic sulfate reducer growing optimally at pressure 10 MPa	Cao et al. (2016)
5	*Acinetobacter* *Staphylococcus* *Enhydrobacter* *Halomonas* *Micromonospora* *Rhizobium* *Agromyces* *Bacillus* *Sphingomonas* *Shewanella* *Deinococcus* Archaea *Methanobacterium* *Methanosaeta* *Bacilli* Actinobacteria *Clostridia* Bacteriodetes *Thermoleophilia*	Deep drilling at Koyna Western India Deep terrestrial subsurface microbiome in Late Cretaceous Deccan Traps and underlying Archaean basement, India Koyna seismogenic zone, boreholes penetrating thickness of 412–1251 m	Diversity	Dutta et al. (2018)

Pseudodesulfovibrio indicus gen. nov. sp. nov. was proposed with the type strain J2T (=MCCC1A01867T = DSM101483T).

Govil et al. (2019) have reviewed taxonomical diversity of extremophiles in the deep biosphere. Conditions in the deep biosphere approach the limits for life but scientific expeditions undertaken for deep subsurface life suggest that these biospheres host a large fraction of microorganisms, living in water, occupying pore space or colonizing mineral and rock surfaces. Microbes in the deep biosphere are phylogenetically different from most cultured taxa and represent novel microbial assemblages. Investigating these subsurface terrains, which are the deepest and potentially the largest accessible ecosystem on Earth, provides unique opportunity for obtaining direct insights into the deep microscopic world. The authors have discussed current understanding of the subsurface depth and environmental limits of our planet's life processes and factors that potentially define diversity and distribution of life in the deep terrestrial crustal system. Deep terrestrial biospheres include hydrothermal vents, oil reservoirs, caves, and subsurface mines.

Nath and Subbiah (2016) have discussed molecular basis of piezophilic adaptation. Piezophiles are the organisms which can successfully survive at extreme pressure conditions. However, the molecular basis of piezophilic adaptation is still poorly understood. Analysis of the protein sequence adjustments that had taken place during evolution can help to reveal the sequence adaptation parameters responsible for protein functional and structural adaptation at such high-pressure conditions. The authors performed experiments on three different temperature ranges of piezophilic groups, namely, psychrophilic-piezophilic, mesophilic-piezophilic, and thermophilic-piezophilic for the detailed comparative study. The best classification results were obtained with moving up the temperature range from psychrophilic-piezophilic to thermophilic-piezophilic. Based on physicochemical classification of amino acids and using feature-ranking algorithms, hydrophilic and polar amino acid groups have higher discriminative ability for psychrophilic-piezophilic and mesophilic-piezophilic groups along with hydrophobic and nonpolar amino acids for the thermophilic-piezophilic groups.

Singh et al. (2010) have investigated phylogenetic diversity of culturable fungi from the deep-sea sediments of the Central Indian Basin, from a depth of ~5000 m. A total of 16 filamentous fungi and 12 yeasts isolated from 20 sediment cores of ~35 cm length were identified by 18S and ITS sequencing of SSU rDNA. Most of the filamentous fungi were Ascomycota while two were Basidiomycota. Phylogenetic analysis of ascomycetes using 18S sequence data grouped them into seven cultures belonging to *Aspergillus* sp., *Sagenomella* sp., *Exophiala* sp., *Capronia* sp., *Cladosporium* sp., *Acremonium* sp. and *Tritirachium* sp. ITS sequence data grouped isolates into six clusters belonging to *Aspergillus* sp., uncultured member of Hypocreaceae, *Exophiala* sp., uncultured soil fungus, Hypocreales and *Trichothecium* sp. The two basidiomycete isolates were *Tilletiopsis* sp. evident from 18S as well as ITS sequence. Most of the yeast isolates belonged to Basidiomycete and only one isolate belonged to the phylum Ascomycota. Sequences of 18S as well as ITS gave matching identification of most of the yeasts. Filamentous fungi as well as the yeasts grew at 200 bar/5 °C indicating their adaptations to deep-sea conditions. The authors claim this as the first report on isolation of *Sagenomella*, *Exophiala*, *Capronia*, and *Tilletiopsis* spp. from deep-sea sediments. This study reports on the presence of terrestrial fungi as a component of culturable fungi in deep-sea sediments.

Dutta et al. (2018) have explored deep terrestrial subsurface microbiome in Late Cretaceous Deccan traps and underlying Archaean Basement, India. Scientific deep drilling at Koyna, Western India, provides a unique opportunity to explore microbial life within deep biosphere hosted by ~65 Myr old Deccan basalt and Archaean granite basement. Characteristic low organic carbon content, mafic/felsic nature but distinct trend in sulfate and nitrate concentrations demarcates the basaltic and granite zones as distinct ecological habitats. The Deccan traps represents a continental flood basalt province covering an area of over 0.5 million km^2 of lava flows (~65 Ma) with a total thickness of over 2000 m near the eruptive center in Western India. This thick pile of Deccan basalts rests on ~2.5 Ga Archaean crystalline basement. The Koyna-Warna region has gained further importance owing to the reservoir-triggered seismicity ever since the impoundment of the Koyna dam in 1962. Exploratory core drilling was carried out at multiple sites in the Koyna seismogenic zone to ascertain the subsurface geology and structure in the region and carry out seismological investigations. The boreholes penetrated the entire thickness of Deccan traps (412–1251 m) and passed through a few 100 m of the underlying granite basement. Rock cores obtained from three exploratory boreholes in the Koyna-Warna region were used to decipher microbial community structure and function.

Samples were collected from three horizons, that is, Deccan basalt (BS) (~65 Ma), the underlying granite basement rock (GR) (~250 Ma), and the short section between the two formations representing the weathered surface of granite basement affected by first lava flows (referred as a transition zone, TZ).

Vertical distribution of microorganisms: The abundance of bacterial and archaeal 16S rRNA gene copies were determined by quantitative PCR-based approach to delineate vertical distribution as microbes. Bacterial 16S rRNA gene copy numbers were significantly higher at all depths compared to their archaeal counterparts. Bacterial 16S rRNA gene copies per gram of rock vary between 3.4×10^5 and 4.01×10^6 with an average of 1.61×10^6.

Sequences assigned to bacteria cover 54 phyla, 127 classes, 212 orders, 377 families, and 1049 genera whereas sequences assigned to archaea cover 12 phyla, 12 classes, 8 orders, 15 families, and 23 genera. On the basis of average relative abundance analysis at phylum level, *Proteobacteria*, *Actinobacteria*, and *Firmicutes* constitute the three major populations of the deep biosphere of Deccan trap region. *Bacteroidetes*, *Chloroflexi*, *Cyanobacteria*, *Planctomycetes*, *Acidobacteria*, and *Deinococcus-Thermus* are the other phyla constituting the communities. *Euryarchaeota* is the most dominant archaeal phylum detected across all the samples followed by *Thaumarchaeota*. Out of 12 assigned archaeal classes, *Thermoplasmata* exhibited the highest average abundance followed by *Methanobacteria*, *Methanomicrobia*, and *Halobacteria*.

Taxonomic affiliation up to genus level indicated presence of *Acinetobacter* followed by *Corynebacterium*, *Pseudomonas,* and *Staphylococcus* as the most abundantly detected genera. Other genera included *Enhydrobacter*, *Halomonas*, *Micromonospora*, *Rhizobium*, *Agromyces*, *Bacillus*, *Sphingomonas*, *Shewanella*, and *Deinococcus*. Archaeal genera were represented by *Thermoplasma*, *Methanobacterium*, and *Methanosaeta* as the most abundant archaeal genera.

Compared to GR microbiome, BS exists at relatively higher abundance of *Actinobacteria* (*Corynebacteriales* and *Propionibacteriales*), Alpha proteobacteria *(Rhizobiales, Sphinobacteriales, Rhodobacteriales),* and Beta-proteobacteria (*Burkholderiales)* but lower abundance of *Bacilli* and *Gamma-proteobacteria (Pseudomonadales, Enterobacteriales, Oceanospirillales,* and others). TZ microbiome core community is more enriched. *Actinobacteria, alphaproteobacteria,* and *Bacilli* were mainly present. Unique presence of *Clostridia, Bacteroidia, Deltaproteobacteria,* and *Thermoleophilia* was noted.

Ram et al. (2001) studied population of nitrifying bacteria from sediment cores collected from the Central Indian Basin (10°1′–10°2′S, 75°59′–76°2′E) harbored a high population of nitrifying bacteria which ranged from 10^3 to 10^7 cfu per gram dry weight. This high density could contribute to the high nitrate concentration observed in these sediments. The authors report for the first time the unusual property of having both the phases of nitrification, that is, $NH_4+ \rightarrow NO_2$—(phase I) and $NO_2 \rightarrow NO_3$—(phase II) under normal and high-pressure conditions in an autotrophic nitrifying bacterium isolated from the deep sediment cores at 5000 m depth (15–20 cm below the sea floor). Thus, cultured bacterial representatives showing both phases of nitrification have been isolated from this region. Nitrifying bacteria were enumerated on modified Winogradsky's medium for ammonia oxidizers and nitrite oxidizers by substituting the medium with $(NH_4)_2SO_4$ at 2 mM (final concentration) or $NaNO_2$ at OAE5 mM (final concentration) as an energy source. Earlier studies on bacteria from these tangreen mottles at 20 to 25 cm bsf (below sea floor) highlighted their capability of showing different phases of nitrification.

Rao (2011) developed a high-pressure Winogradsky bottle which stimulates deep-sea environments with simple laboratory items. It also showed the corrosion of iron that explains the microbial corrosion of steel sunk ships over a period of time.

Raghukumar et al. (2001) have studied microbial biomass and organic nutrients in the deep-sea sediments of the Central Indian Basin. The microbial standing stock and concentration of organic nutrients in the deep-sea sediments of Central Indian Ocean Basin in the Indian pioneer area was measured to assess the impact of deep-sea mining on the in situ benthic life. The authors collected sediments using box core and grab samples. The total bacterial numbers ranged from 10^{10} to 10^{11} cells per gram dry weight sediment. There was a marginal decrease in the number of bacteria from surface to 30 cm depth, though the subsurface section registered a higher number than did the surface. The highest numbers were encountered at depths of 4–8 cm. The retrievable number of bacteria were two orders less in comparison with the direct total counts of bacteria. An almost homogenous distribution of bacteria, total organic carbon, living biomass, and lipids throughout the depth of cores indicates active microbial and benthic processes in the deep-sea sediments. A uniform distribution of total counts of bacteria, carbohydrates, and total organic carbon in all the cores indicates their stable nature and suggests that they can serve as useful parameters for long-term monitoring of the area after the benthic disturbance.

Raghukumar et al. (2004) have described culturable fungi in a deep-sea sediment core from the Chagos Trench, Indian Ocean. The authors report the occurrence of paleobs for the first time from a deep-sea sediment core obtained from a depth of

5904 m from the Chagos Trench in the Indian Ocean. Culturable fungi, direct counts of bacteria, age of the sediments based on the radiolarian assemblage, total organic carbon, Eh and $CaCO_3$ were determined in these sediments. Culturable fungi were obtained from subsections up to 370 cm depth. The age of the sediments from which fungi were isolated was estimated to range from >0.18 to 0.43 million years (Ma), being the oldest recorded age for recovery of culturable fungi. Bacterial number in the core showed oscillations corresponding to cycles of approximately 100 ka (kilo years). The fungi comprised of non-sporulating forms and a sporulating form identified to be *Aspergillus sydowii.* Germination of spores at 100, 300, and 500 bar hydrostatic pressure at 5 °C confirmed barotolerance of *A. sydowii* and its nativity to deep-sea sediments.

Naik et al. (2016) have investigated relationship of sediment biochemistry, bacterial morphology and activity to geotechnical properties in the Central Indian Basin (CIB). Distribution of biochemical properties, morphological and functional characteristics of bacteria in the extreme sedimentary environment of the Central Indian Basin (CIB) were examined in relation to geotechnical properties and sediment texture. Interrelationships between these parameters in siliceous northern sediments were compared to southern pelagic clay to understand relationships and to appreciate mining implications. Clay impacted lipids negatively and silt positively in the south. Geotechnical properties had the opposite influence on the distribution of coccoidal and rod forms of bacteria. Rods were affected negatively by water content/porosity and positively by wet bulk density in the north suggesting their preferences for deeper niches. Sheer strength affected coccoidal distribution negatively in the south. Competition for liable substrates by coccoidal forms in the surface layers could also have restricted the rods to the deeper regions. Principal component analysis (PCA) further strengthened the authors inference that the distribution of rod forms increased with depth. Enzymatic properties of bacterial isolates also showed that the rods were able to degrade substrates like gelatin and DNA. These observations suggest that they are capable of degrading relatively more recalcitrant compounds in the deeper layers. The minor components, like the rod morphotypes and sand content, could yield a large influence on the variability of other parameters.

9.5 Barophilic Microbial Species, New to Science, from India

Piezophilic sulfate reducer growing optimally at pressure 10 MPa *Pseudodesulfovibrio indicus*, gen. nov., sp. nov. has been reported from Indian Ocean (Cao et al. 2016).

References

Cao J, Gayet N, Zeng X, Shao Z, Jebbar M, Alain K (2016) *Pseudodesulfovibrio indicus* gen. nov., sp. nov., a piezophilic sulfate-reducing bacterium from the Indian Ocean and reclassification of four species of the genus *Desulfovibrio*. Int J Syst Evol Microbiol 66(10):3904–3911. https://doi.org/10.1099/ijsem.0.001286

Chakravorty D, Mohd K, Patra S (2017) Multifactorial level of extremostability of proteins: can they be exploited for protein engineering? Extremophiles 21(1). https://doi.org/10.1007/s00792-016-0908-9

Dalmasso C, Oger P, Selva G, Courtine D, L'Haridon S, Garlaschelli A, Roussel E, Miyazaki J, Reveillaud J, Jebbar M, Takai K, Maignien L, Alain K (2016) *Thermococcus piezophilus* sp. nov., a novel hyperthermophilic and piezophilic archaeon with a broad pressure range for growth, isolated from a deepest hydrothermal vent at the Mid-Cayman Rise. Syst Appl Microbiol 39(7):440–444. https://doi.org/10.1016/j.syapm.2016.08.003

Delong EF, Franks D, G, Yayanos A. A. (1997) Evolutionary relationships of cultivated psychrophilic and barophilic deep sea bacteria. Appl Envirom Microbial 63(5):2102–2108

Di Giulio M (2005) A comparison of proteins from *Pyrococcus furiosus* and *Pyrococcus abyssi*: Barophily in the physicochemical properties of amino acids and in the genetic code. Gene 346:1–6. https://doi.org/10.1016/j.gene.2004.10.008

Di Giulio M (2013) The origin of the genetic code in the ocean abysses: new comparisons confirm old observations. J Theor Biol 333:109–116. https://doi.org/10.1016/j.jtbi.2013.05.019

Dutta A, Gupta DS, Gupta A, Sarkar J, Roy S, Mukherjee A, Sar P (2018) Exploration of deep terrestrial subsurface microbiome in Late Cretaceous Deccan traps and underlying Archean basement, India. Sci Rep 8:17459. https://doi.org/10.1038/s41598-018-35940-0

Fang J, Kato C, Runko GM, Nogi Y, Hori T, Li J, Morono Y, Inagaki F (2017) Predominance of viable spore-forming piezophilic bacteria in high-pressure enrichment cultures from ~1.5 to 2.4 km-deep coal-bearing sediments below the ocean floor. Front Microbiol 8:137. https://doi.org/10.3389/fmicb.2017.00137

Govil T, Rathinam NK, Salem DR, Sani RK (2019) Chapter 35 - Taxonomical diversity of extremophiles in the deep biosphere. In: Microbial diversity in the genomic era, pp 631–656. https://doi.org/10.1016/B978-0-12-814849-5.00035-6

http://exprotdb.com/barophiles.php

https://en.wikipedia.org/wiki/Piezophile

https://serc.carleton.edu/microbelife/extreme/pressure/index.html

Kato C, Bartlett DH (1997) The molecular biology of barophilic bacteria. Extremophiles 1:111–116

Kato C, Sato T, Horikoshi K (1995) Isolation and properties of barophilic and barotolerant bacteria from deep-sea mud samples. Biodivers Conserv 4:1–9

Kato C, Li L, Nogi Y, Nakamura Y, Tamaoka J, Horikoshi K (1998) Extremely barophilic bacteria isolated from the Mariana Trench, Challenger Deep, at a depth of 11,000 meters. Appl Environ Microbiol 64(4):1510–1513. https://doi.org/10.1128/AEM.64.4.1510-1513.1998

Kieft TL (2016) Microbiology of the deep continental biosphere. In: Hurst C (ed) Their world: a diversity of microbial environments. Advances in environmental microbiology, vol 1. Springer, pp 225–249. https://doi.org/10.1007/978-3-319-28071-4_6

Kumar R, Acharya V, Singh D, Kumar S (2018) Strategies for high-altitude adaptation revealed from high-quality draft genome of non-violacein producing *Janthinobacterium lividum* ERGS5:01. Stand Genomic Sci 13:11. https://doi.org/10.1186/s40793-018-0313-3

Madigan MT, Kelly B, Daniel B, Sattley W, Stahl D (2014) Broch biology of microorganism, 15th edn. Pearson. 1056 pages. ISBN-13: 978-0134261928

Margesin R, Nogi Y (2004) Psychropiezophilic microorganism. Cell Mol Biol 50:429–436

Michels PC, Clark DS (1997) Pressure-enhanced activity and stability of a hyperthermophilic protease from a deep-sea methanogen. Appl Environ Microbiol 63:3985–3991

Naik SS, Khadge NH, Valsangkar AB, Das A, Fernandes CEG, LokaBharati PA (2016) Relationship of sediment-biochemistry, bacterial morphology, and activity to geotechnical properties in the central Indian Basin. Mar Georesour Geotechnol 34(1):21–32. https://doi.org/10.1080/1064119X.2014.954680

Nath A, Subbiah K (2016) Insights into the molecular basis of piezophilic adaptation: extraction of piezophilic signatures. J Theor Biol 390:117–126. https://doi.org/10.1016/j.jtbi.2015.11.021

Nogi Y, Kato C (1999) Taxonomic studies of extremely barophilic bacteria isolated from the Mariana Trench end description of *Moritella Yatanosii* sp.nov, a new barophilic bacterial isolate. Extremophiles 3:71–77

Nogi Y, Kato C, Horikoshi K (2002) *Psychromonas kaikoae* sp. nov, a novel piezophilic bacterium from the deepest cold – sleep sediments in the Japan Trench. Int J Syst Evol Microbiol 52:1527–1532

Nogi Y, Hosoya S, Kato C, Horikoshi K (2007) *Psychromonas hadalis* sp. nov., a novel piezophilic bacterium isolated from the bottom of the Japan trench free. Int J Syst Evol Microbiol 57(6). https://doi.org/10.1099/ijs.0.64933-0

Raghukumar S, Damare VS (2011) Increasing evidence for the important role of Labyrinthulomycetes in marine ecosystems. Bot Mar 54(1):3–11

Raghukumar S, Ramaiah N, Raghukumar C (2001) Dynamics of thraustochytrid protists in the water column of the Arabian Sea. Aquat Microb Ecol 24:175–186

Raghukumar C, Raghukumar S, Sheelu G, Gupta SM, Bejugam N, Rao BR (2004) Buried in time: culturable fungi in a deep-sea sediment core from the Chagos Trench, Indian Ocean. Deep Sea Res Pt I Oceanograph Res Pap 51(11):1759–1768. https://doi.org/10.1016/j.dsr.2004.08.002

Ram ASP, LokaBharathi PA, Nair S, Chandramohan D (2001) A deep-sea bacterium with unique nitrifying property. Curr Sci 80(9):1222–1224

Rao TSAA (2011) Designing a barophile enrichment apparatus to culture deep sea microbes. GRIN Verlag, Munich. https://www.grin.com/document/184271

Rothschild LJ, Mancinelli RL (2001) Life in extreme environments. Nature 409:1092–1101. https://doi.org/10.1038/35059215

Singh P, Raghukumar C, Verma P, Shouche Y (2010) Phylogenetic diversity of culturable fungi from the deep-sea sediments of the Central Indian Basin and their growth characteristics. Fungal Divers 40:89–102. https://doi.org/10.1007/s13225-009-0009-5

Yafremava LS, Di Giulio M, Caetano-Anollés G (2013) Comparative analysis of barophily-related amino acid content in protein domains of *Pyrococcus abyssi* and *Pyrococcus furiosus*. Archaea. https://doi.org/10.1155/2013/680436

Yayanos AA (1995) Microbiology to 10500 meter in the deep sea. Annu Rev Microbiol 49(1): 777–805

Yayanos AA (2008) Piezophiles. In: Encyclopedia of life sciences. Wiley, p a0000341. https://doi.org/10.1002/19780470015902.a0000341pub2

Zeng X, Birrien J-L, Fouquet Y, Cherkashov G, Jebbar M, Querellou J, Oger P, Cambon-Bonavita M-A, Xiao X, Prieur D (2009) *Pyrococcus* CH1, an obligate piezophilic hyperthermophile: extending the upper pressure-temperature limits for life. ISME J 3:873–876. https://doi.org/10.1038/ismej.2009.21

Xerophilic and Xerotolerant Microorganisms

10

Abstract

Xerophilic microorganisms are the microbes that can grow at a low availability of water or low water activity (Aw below 0.8). Xerotolerant microorganisms tolerate dry conditions. They occur in arid desert soils, foods preserved by drying, and high sugar foods like jams and honey. The organisms growing in high sugar foods are also called as osmotolerant. Halophiles growing at high concentration of salt (NaCl) are also xerophiles. Some of the xerophilic fungi are able to grow at a water activity of 0.61. Halophilic archaea have been found to grow at water activities from 0.681 to 0.748. Xerophilic fungi include *Aspergillus penicillioides*, *Xeromyces bisporus*, *Eurotium halophilicum*, and *Xerochrysium*. Xerophilic fungi cause food spoilage. To combat osmotic stress, xerophyles produce osmolytes or compatible solutes.

In India, xerophilic fungi have been reported from wall paintings of Ajanta caves, mixed ruminant feed and forage, rhizospheric soil of perennial grass from Thar desert, stored wheat grains, etc. and include *Acremonium*, *Alternaria*, *Aspergillus*, *Cladosporium*, *Drechslera*, *Penicillium*, *Eurotium*, and many more.

Keywords

Xerophiles · Osmotolerant · Water activity (Aw) · Desert soils · Dehydrated foods · High sugar foods · Salted foods · Osmolytes · Desert grasses · Sand dunes · Wall paintings · Xerophilic fungi

A xerophile (from Greek xerox "dry" and philes "loving") is an extremophilic organism that can grow and reproduce in conditions with a low availability of water also known as water activity. Water activity (a_w) is measured as the humidity above a substance relative to the humidity above pure water ($A_w = 1.0$). Xerophiles

P. P. Kanekar, S. P. Kanekar, *Diversity and Biotechnology of Extremophilic Microorganisms from India*, Microorganisms for Sustainability 41,
https://doi.org/10.1007/978-981-19-1573-4_10

are xerotolerant, meaning tolerant to dry conditions. They can often survive in environments with water activity below 0.8, above which is typical for most life on Earth. Typically, xerotolerance is used with matric drying, where a substance has a low water concentration. These environments include arid desert soils. The term osmotolerance is typically applied to organisms that can grow in solutions with high solute concentrations (salt, sugars), such as halophiles.

The common food preservation method of reducing water activities may not prevent the growth of xerophilic organisms, often resulting in food spoilage. Some mold and yeast species are xerophilic. Examples of xerophiles include *Trichosporonoides nigrescens* and cacti.

Grant (2004) has reviewed life at low water activity. Two major types of environment provide habitats for the most xerophilic organisms known: food preserved by some form of dehydration or enhanced sugar levels and hypersaline sites where water availability is limited by a high concentration of salt (usually NaCl). These environments are essentially microbial habitats with high-sugar foods being dominated by xerophilic (sometimes called osmophilic) filamentous fungi and yeasts. Some of the xerophilic fungi are able to grow at water activity $a(w)$ of 0.61, the lowest $a(w)$ value for the growth recorded so far. High-salt environments are almost exclusively populated by prokaryotes, notably the haloarchaea, capable of growing in saturated NaCl ($a(w)$ 0.75). Different strategies are employed for combating the osmotic stress imposed by high levels of solutions in the environment. Eukaryotes and most prokaryotes synthesize or accumulate organic so-called compatible solutes (osmolytes) that have counterbalancing osmotic potential. Some bacteria and haloarchaea counterbalance osmotic stress imposed by accumulating equivalent amount of KCl. Haloarchaea become entrapped and survive for long periods inside halite (NaCl) crystals. They are also found in ancient subterranean halite (NaCl) deposits, leading to speculation about survival over geological time periods.

Stevenson et al. (2015) have discussed on the interesting topic on common water-activity limit for the three domains of life. Archaea and bacteria constitute a majority of life systems on Earth. The most halophilic prokaryotes are known for their ability to multiply at saturated NaCl (water activity (a_w) 0.755) whereas some xerophilic fungi can germinate usually at high-sugar concentrations, at values as low as 0.650–0.605 a_w. The authors have presented evidence that halophilic prokaryotes can grow down to water activities of <0.755 for *Halanaerobium lacusrosei* (0.748), *Halobacterium* strain 004.1 (0.728), *Halobacterium* sp. NRC-1 and *Halococcus morrhuae* (0.717), *Haloquadratum walsbyi* (0.709), *Halococcus salifodinae* (0.693), *Halobacterium noricensi* (0.687), and *Natrinema pallidum* (0.681). The findings suggested that there is a common water-activity limit that is determined by physicochemical constraints for the three domains of life.

Stevenson et al. (2017) have investigated enhancement of fungal germination at the water-activity limit for life with glycerol. For the most extreme fungal xerophiles, metabolic activity and cell division typically halts between 0.700 and 0.640 water activity (approximately 70–64% relative humidity). To see effect of glycerol, the authors carried out experiments using spores from a variety of species including *Aspergillus penicillioides*, *Eurotium halophilicum*, *Xerochrysium* (formerly

Chrysosporium xerophilum), and *Xeromyces bisporus* obtained by growing the cultures on media supplemented with glycerol (up to 189 mg glycerol per g dry spores). The ability of these spores to germinate and the kinetics of germination were then determined in media with water activities from 0.765 to 0.575. *A. penicillioides*, *Eurotium amstelodami*, *E. halophilicum*, *X. xerophilum*, and *X. bisporus* occurred at lower water activities than previously recorded (0.640, 0.685, 0.651, 0.664, and 0.667 respectively). In addition, the kinetics of germination at low water activities were substantially faster than those reported previously. Thus, glycerol appears to have a role in expanding the water-activity limit for microbial cell function since glycerol is present at high concentration in many types of microbial habitats.

Pettersson and Leong (2011) have described xerophilic fungi as yeasts and molds that are capable of growth at or below a water activity (a_w) of 0.85. These microorganisms have developed physiological mechanisms that enable their biochemical pathways to function in environments where little water is available. External conditions at low a_w are sensed by membrane osmosensors and xerophiles, then accumulate glycerol as a compatible solute to balance the internal and external osmotic pressure. They also modify their membranes to retain this glycerol within the cell. Extreme xerophiles compete poorly at high a_w, since they require decreased a_w for growth. Moderate xerophiles include *Aspergillus*, *Penicillium*, and *Eurotium*. Some xerophiles have a preference for salt or sugar substrates, whereas other species can be isolated from both jam and salterns. Xerophiles are widely spread on the fungal tree of life.

Irwin (2020) has taken an overview of extremophiles and their food and medical applications, wherein xerophiles are also discussed. Xerophiles are defined as microorganisms that grow at low water activity a_w levels (0.85). The most halophilic prokaryotes can survive in saturated NaCl which has an a_w activity less than 0.755. However, the organisms that show the highest tolerance of low a_w are xerophilic fungi, which can tolerate values of 0.65 to 0.605. Many of these xerophilic fungi can colonize dried foods and survive in high-sugar environments. They can balance their outside environments by accumulating internal solutes, which allow the cells to maintain a high enough osmotic pressure for survival and growth (a similar strategy to halophiles). The most xerophilic of these food-spoilage fungi known is *Xeromyces bisporus*. This is capable of growth at a_w 0.61–0.62 and has been found in various foods including dried fruits, fruit cake, bakery goods, and animal feed.

Hyperarid desert environments are also habitats for desiccation-resistant microorganisms. These environments can also have extremes of temperature (both heat and cold) and radiation from sunlight. These environments can thus serve as ideal analogs for environments found outside Earth, in particular Mars. The coldest hyperarid desert environments on Earth are the McMurdo Dry Valleys in Antarctica and as such are a good model for Martian soils. Cryptoendolythic microorganisms are the main life forms found here, for example, *Cryomyces antarcticus* CCFEE. This fungus can withstand long-term desiccation, temperature extremes, and high levels of radiation. It withstands Martian-like conditions, including 95% CO_2 and UV radiation of wavelength <200 nm. Studies of the Atacama Desert in Chile, which is one of the driest places on Earth, showed that microbial communities were present, and after sporadic rain, biomolecules indicative of active cells and replicating genomes could be detected.

10.1 Xerophiles Reported from India

Researchers from India have described xerophilic microorganisms from habitats like the Thar Desert, Rajasthan; rhizosphere of desert grasses and man-made environments like wall paintings in caves. These organisms, mostly the fungi, are listed in Table 10.1.

Some of the work on xerophiles by Indian researchers is described.

Rao et al. (2016) have described microbial diversity of soil, sand dunes, and rock substrates of the Thar Monsoon Desert, India. A culture-independent diversity assessment of archaea, bacteria, and fungi in semiarid soils, arid soils, arid sand dunes, and arid cryptoendolithic substrates from six locations in Ajmer, Jaisalmer, Jaipur, and Jodhpur was made. A real-time quantitative PCR approach revealed that bacteria dominated soils and cryptoendoliths while fungi dominated sand dunes. The archaea formed a minor component of all communities. Substrate and climate were found to be most parsimonious predictors. Sequence-based identification of 1240 phylotypes revealed that most taxa were common desert microorganisms. Semiarid soils were dominated by actinobacteria and alpha-proteobacteria, arid soils by chloroflexi and alpha-proteobacteria, sand dunes by ascomycete fungi, and cryptoendoliths by cyanobacteria. This distribution is explained by climatic variables such as mean annual rainfall and maximum annual temperature. Substrate variables were conductivity, soluble salts, Ca^{2+} ions, and pH. Since available moisture is less in desert soils, especially sand, the organisms appear to be xerophilic or xerotolerant, particularly xerophilic fungi obtained in sand dunes.

Chowdhary et al. (2007) have identified diazotrophs in the culturable bacterial community associated with roots of a perennial grass of Thar Desert, India. The Thar Desert, situated in northwestern India, extends over 2.34 million km^2 and about 91% of it falls in the state of Rajasthan. The sandy soils of this desert are poor in organic matter, having rapid infiltration rates of water and rapid oxidation with very low productivity. The climate induces frequent droughts, making farming difficult and unpredictable. The climate is a major threat to the sustainability of the fragile desert ecosystem of the Thar Desert. *Lasiurus sindicus* (locally known as "sewan") is one of the most common and endemic perennial grasses growing in areas with annual rainfall <150 mm. It grows naturally in sandy soils of Thar Desert in Barmer and Jaisalmer districts of Western Rajasthan. It has high nutritive value and is preferentially consumed by cattle in desert. The grass is drought-tolerant. Since desert plants survive extreme variations in temperature and moisture and grow in typically poor soils with low organic content with limited amounts of bioavailable inorganic nutrients, microorganisms associated with desert plants should possess adaptive mechanisms to cope with frequent droughts and various other stresses such as starvation, high osmolarity, high temperature, and desiccation. The authors therefore studied rhizospheric microbial community associated with the desert plant. Bacteria

Table 10.1 Xerophilic and xerotolerant microorganisms reported from India

Sr. No.	Name of the organism/group	Source	Diversity/ biotechnology	Reference
1	*Acremonium indicum* *Alternaria alternata* *Aspergillus flavus* *A. nidulans*, *A. niger*, *A. terreus*, *A. versicolor* *Cladosporium cladosporioides* *C. herbarum* *Curvularia lunata* *C. pallescens* *Chaetomium globosum* *Drechslera australiensis* *D. hawaiiensis* *Emericella nidulans* *Epicoccum nigrum* *Fusarium oxysporum*, *F. moniliforme*, *F. solani* *Macrophomina phaseolina* *Memnoniella echinata* *Mycelia sterilia* *Paecilomyces variotii* *Rhizopus nigricans* *Stachybotrys atra* *Trichoderma harzianum*	Wall paintings of Ajanta caves, Aurangabad, Maharashtra	Diversity	Agrawal et al. (1988)
2	*Aspergillus versicolor* *Penicillium citrinum* *Eurotium sp.* *Wallemia sebi* *Aspergillus penicillioides* *Fusarium moniliforme* *Cladosporium* sp.	Mixed ruminant feed and forage samples collected from Eastern India with low moisture content	Diversity	Phillips et al. (1996)
3	*Azospirillum* sp. *Rhizobium* sp.	Rhizospheric soil of a perennial grass, *Lasiurus sindicus* from		Chowdhury et al. (2007)

(continued)

Table 10.1 (continued)

Sr. No.	Name of the organism/group	Source	Diversity/ biotechnology	Reference
	Agrobacterium tumefaciens *Inquilinus limosus* *Ralstonia* sp. *Varivorax paradoxus* *Bordetella petrii* *Pseudomonas pseudoalcaligenes* *Stenotrophomonas* sp. *Staphylococcus warneri* *Bacillus* sp. *Micrococcus luteus* *Microbacterium* sp.	Thar Desert Barmer and Jaisalmer districts of Western Rajasthan	Diazotrophs in culturable diversity	
4	*Aspergillus flavus* *Penicillium* spp. *Paecilomyces* spp. *Mucor* spp.	Marketed tea samples from local markets of Tamil Nadu	Diversity	Al-Sohaibani et al. (2011)
5	1240 phylotypes Actinobacteria, Alpha-proteobacteria Chloroflexi Ascomycete fungi Cyanobacteria	Semiarid soils, arid soils, arid sand dunes, arid cryptoendolithic (rock) substrates from six locations in Ajmer, Jaisalmer, Jaipur and Jodhpur, Thar Monsoon Desert, Rajasthan	Diversity	Rao et al. (2016)
6	*Aspergillus flavus* *Rhizopus oryzae* *Eurotium amstelodami*	Stored wheat grains in three warehouses of the food Corporation of India located in three agroclimatic zones (Paschim Medinipur, Bankura, Purulia) of West Bengal	Diversity	Kumari et al. (2019)

were isolated from the roots of three *L. sindicus* plants that survived three consecutive droughts, using nitrogen-free semisolid medium. Based on analysis of 16S rRNA gene sequences, the organisms were identified as *Azospirillum* sp., *Rhizobium* sp., *Agrobacterium tumefaciens*, and *Inquilinus limosus* (α-proteobacteria); *Ralstonia* sp., *Variovorax paradoxus*, and *Bordetella petrii* (β-proteobacteria); and *Pseudomonas pseudoalcaligenes*, *Stenotrophomonas* sp. (γ-proteobacteria). The occurrence of nifH sequences in *Azospirillum* sp., *Rhizobium* sp., and *P. pseudoalcaligenes* showed the possibility of supplying biologically fixed nitrogen by the root-associated diazotrophs to the host plant. Gram-stain-positive bacteria were identified as *Staphylococcus warneri*, *Bacillus* sp., *Micrococcus luteus*, and *Microbacterium* sp. These bacteria isolated from a drought-tolerant grass *Lasiurus sindicus* rhizosphere soil of Thar Desert could be classified as xerotolerant bacteria.

Al-Sohaibani et al. (2011) have reported xerophilic aflatoxigenic fungi from black tea. The authors have isolated xerophilic and aflatoxigenic fungi associated with marketed tea from 21 different tea samples collected from local market of Tamil Nadu, India. The results indicated 0.38% contaminated by *Aspergillus flavus*, 0.30% by *Penicillium* spp., 0.14% by *Paecilomyces* spp. and 0.19% by *Mucor* spp. The author had used MY50G media for isolation of xerophilic fungi since supplementation with glucose lowers the a_w value less than 0.89 (MY50G = Malt extract yeast extract 50% glucose agar).

Phillips et al. (1996) have reported mycoflora of range of mixed ruminants' feed and forage samples collected in Eastern India and Bangladesh. Most fungi isolated from the samples were those isolated with low moisture contents. The most frequently isolated fungi were *Aspergillus versicolor* (Vuill) Tiraboschi, *Penicillium citrinum* Thorn, *Eurotium* species, *Wallemia sebi* (Fr.) v. Arx, *Aspergillus penicilloides* Speg. and yeasts, *Fusarium moniliforme* Sheldon, and *Cladosporium* species were prevalent in straw but not other feeds. Thermotolerant species were found in rice straw.

Ranjana Kumari et al. (2019) have investigated fungal diversity of stored wheat grains. Nine stored wheat grain samples in three warehouses of the Food Corporation of India located in three agroclimatic zones (Paschim Medinipur, Bankura, and Purulia) of West Bengal in India. Stored grain contains low moisture, hence the fungi isolated from the wheat grains could be xerophilic or xerotolerant (although there is no such mention by the authors). The majority of fungal isolates belonged to *Aspergillus flavus* 35% followed by *Rhizopus oryzae* 13% and *Eurotium amstelodami* 9%.

Garg and Dhawan (1994) have reviewed biodeterioration of wall paintings. The microorganisms develop on the paintings only if the relative humidity of the surroundings exceeds 75% and temperature ranges between 5 and 40 °C. Fungi grow on paintings only when the relative humidity exceeds 70%. Several fungal species have been isolated from paint films. *Aureobasidium pullulans*, *Cladosporium herbarum*, and species of *Alternaria*, *Penicillium*, and *Aspergillus* are common on exterior paints in temperate climates. Of these, *Aureobasidium pullulans* is the most important in causing deterioration of paint. It is tolerant to high temperature, grows over a wide range of pH, is resistant to strong sunlight, and has a low requirement for water.

A systematic study of these microorganisms on Indian wall paintings has been little studied earlier (Tilak et al. 1970; Agrawal et al. 1988). Deterioration of murals by heterotrophic microorganisms is largely the result of accumulation of dust on the paintings which absorb water (Tilak et al. 1970). Tilak et al. (1970) have described fungal species *Aspergillus* spp. *Camposporium* spp., *Cladosporium spp.*, *Curvularia* spp., *Helminthosporium* spp., *Nigrospora* spp., *Pithomyces* spp., *Pseudotorula* spp. as aeroflora of Ajanta wall paintings, India, and *Alternaria* spp., *Aspergillus* spp., *Helminthosporium* spp., *Nigrospora* spp., *Pithomyces* spp. as aeroflora of Ellora wall paintings, India.

Agrawal et al. (1988) have described several fungal species as *Acremonium indicum*, *Alternaria alternata*, *Aspergillus flavus*, *A. nidulans*, *A. niger*, *A. terreus*,

A. versicolor, *Cladosporium cladosporioides*, *C. herbarum*, *Curvularia lunata*, *C. pallescens*, *Chaetomium globosum*, *Drechslera australiensis*, *D. hawaiiensis*, *Emericella nidulans*, *Epicoccum nigrum*, *Fusarium oxysporum*, *F. moniliforme*, *F. solani*, *Macrophomina phaseoniella*, *Memnoniella echinata*, *Mycelia sterilia*, *Paecilomyces variotii*, *Rhizopus nigricans*, *Stachybotrys atra*, *Trichoderma harzianum*, and some bacteria from wall paintings of Ajanta caves, India.

Biotechnological applications or role in environmental sustainability of xerophilic microorganisms has not been much reported so far but may appear in future.

References

Agrawal OP, Dhawan S, Garg KL, Shaheen F, Pathak N, Misra A (1988) Study of biodeterioration of the Ajanta wall paintings. Int Biodeterior 24(2):121–129

Al-Sohaibani S, Murugan K, Lakshimi G, Anandraj K (2011) Xerophilic aflatoxigenic black tea fungi and their inhibition by *Elettaria cardamomum* and *Syzygium aromaticum* extracts. Saudi J Biol Sci 18(4):387–394. https://doi.org/10.1016/j.sjbs.2011.06.005

Chowdhury SP, Schmid M, Hartmann A, Tripathi AK (2007) Identification of diazotrophs in the culturable bacterial community associated with roots of *Lasiurus sindicus*, a perennial grass of Thar Desert, India. Microb Ecol 54:82–90. https://doi.org/10.1007/s00248-006-9174-1

Garg KL, Dhawan S (1994) Biodeterioration of wall paintings: the Indian experience. In: Singh J (ed) Building mycology. Chapman and Hall, London, pp 239–260. ISBN 0419190201

Grant WD (2004) Life at low water activity. Philos Trans R Soc Lond Ser B Biol Sci 359(1448): 1249–1267. https://doi.org/10.1098/rstb.2004.1502

https://en.wikipedia.org/wiki/Xerophile

Irwin JA (2020) Chapter 6 - Overview of extremophiles and their food and medical applications. In: Physiological and biotechnological aspects of extremophiles, pp 65–87

Kumari R, Jayachandran LE, Ghosh AK (2019) Investigation of diversity and dominance of fungal biota in stored wheat grains from governmental warehouses in West Bengal, India. J Sci Food Agric 99(7):3490–3500

Pettersson VO, Leong S-L (2011) Fungal Xerophiles (Osmophiles). https://doi.org/10.1002/9780470015902.a0000376.pub2

Phillips SI, Wareing PW, Dutta A, Panigrahi S, Medlock V (1996) The mycoflora and incidence of aflatoxin, zearalenone and sterigmatocystin in dairy feed and forage samples from Eastern India and Bangladesh. Mycopathologia 133:15–21

Rao S, Chan Y, Bugler-Lacap DC, Bhatnagar A, Bhatnagar M, Pointing SB (2016) Microbial diversity in soil, sand dune and rock substrates of the thar Monsoon Desert, India. Indian J Microbiol 56(1):35–45. https://doi.org/10.1007/s12088-015-0549-1

Stevenson A, Cray JA, Williams JP, Santos R, Sahay R, Neuenkirchen N, McClure CD, Grant IR, Houghton JDR, Quinn JP, Timson DJ, Patil SV, Singhal RS, Anton J, Dijksterhuis J, Hocking AD, Lievens B, Rangel DEN, Voytek MA, Gunde-Cimerman N, Oren A, Timmis KN, McGenity TJ, Hallsworth JE (2015) Is there a common water-activity limit for the three domains of life? ISME J 9:1333–1351

Stevenson A, Hamill PG, Medina A, Kminek G, Rummel GD, Dijksterhuis J, Timson DJ, Magan N, Leong S-LL, Hallsworth JE (2017) Glycerol enhances fungal germination at the water-activity limit for life. Environ Microbiol 19(3):947–967

Tilak ST, Sharma BRN, SenGupta SR, Kulkarni RL (1970) The deterioration and microbiological studies of Ajanta and Ellora paintings. In: Conservation of Cultural Property in India. Proceedings of the 5th Seminar, New Delhi, pp 77–82

The manufacturer's authorised representative in the EU is Springer Nature Customer Service Centre GmbH, Europaplatz 3, 69115 Heidelberg, Germany. If you have any concerns regarding our products, please contact ProductSafety@springernature.com

Printed and bound by CPI Group (UK) Ltd, Croydon, CR0 4YY
30/09/2026
02225317-0008